Springer

Tokyo
Berlin
Heidelberg
New York
Barcelona
Budapest
Hong Kong
London
Milan
Paris
Singapore

D. Shindo · K. Hiraga

High-Resolution Electron Microscopy for Materials Science

With 203 Figures

 Springer

Daisuke Shindo, Dr.
Professor, Institute for Advanced Materials Processing, Tohoku University
2-1-1 Katahira, Aoba-ku, Sendai, Miyagi, 980-8577 Japan

Kenji Hiraga, Dr.
Professor, Institute for Materials Research, Tohoku University
2-1-1 Katahira, Aoba-ku, Sendai, Miyagi, 980-8577 Japan

Front cover:
High-resolution electron microscope (HREM) image of $W_8Ta_2O_{29}$ observed with an imaging plate. An accurately traced intensity profile at a thin region near the crystal edge shows intensity minima at higher potential positions (H) of metal atoms projected along the beam direction. Peak intensity at lower potential positions (L) between the metals is strongly enhanced at a thicker region because of the dynamical diffraction effect, and this effect is well reproduced in simulated intensity profiles for increases in thickness (t) at the upper left.

Back cover:
HREM image of a characteristic lattice defect consisting of stacking faults in Si. The Z-shaped defect appropriately indicates the end of the book.

ISBN-13: 978-4-431-70234-4   e-ISBN-13: 978-4-431-68422-0
DOI: 10.1007/978-4-431-68422-0

Library of Congress Cataloging-in-Publication Data

Shindō, D. (Daisuke), 1953–
    [Zairyo Hyoka no tameno Kobunkaino Denshi Kenbikyou Ho.   English]
    High-resolution electron microscopy for materials science / D.
Shindo, K. Hiraga.
        p.   cm.
    Includes bibliographical references and index.
    ISBN-13: 978-4-431-70234-4
    1. Materials—Microscopy. 2. Electron microscopy. I. Hiraga, K.
(Kenji), 1939– . II. Title.
TA417.23.S53   1998
620.1′1299—DC21   98-20291
                                                        CIP

This translation is based on the Japanese original.
D. Shindo, K. Hiraga; High Resolution Electron Microscopy for Materials Analysis
published by Kyoritsu Shuppan
1996 © Daisuke Shindo and Kenji Hiraga

Printed on acid-free paper

© Springer-Verlag Tokyo 1998

Typesetting: Best-set Typesetter Ltd., Hong Kong

SPIN: 10678114

Preface

With rapid improvement of current transmission electron microscopes, their maximum resolution reaches almost 0.1 nm. The mode of observing crystal structure directly by electron microscopes utilizing such high resolution is called high-resolution electron microscopy and is one of the most powerful methods for investigating internal structures of various materials on an atomic scale. In fact, publication of high-resolution electron microscope images is becoming much more common in scientific papers and also in journals. However, most of these high-resolution images were taken under rather poor imaging conditions, and precise information on crystal structure could not be obtained. Thus, clear understanding of optimum imaging conditions and appropriate interpretation of high-resolution images become more and more important.

Taking into account the importance of high-resolution electron microscopy for materials science as noted above, the authors plainly explain not only the basis but also the practice and application of high-resolution electron microscopy. In addition to the fundamental formulation of the imaging process of high-resolution electron microscopy, details of image simulation, which are indispensable for interpretation of high-resolution images, are presented in Chapter 1. In practice, the authors make detailed statements on appropriate imaging conditions for observing "lattice images" and "structure images," and then they clearly show how to extract structural information from these observed images in Chapter 2. In Chapter 3, typical examples of high-resolution images, which have been taken by the authors and their colleagues, cover not only various structure defects such as dislocations, interfaces, and surfaces but also various materials such as composite ceramics, high-Tc superconductors, ordered alloys, and quasicrystals. Furthermore, the newest instruments and techniques in high-resolution electron microscopy, such as the imaging plate and quantitative high-resolution electron microscopy, are also included in Chapter 4.

The authors are grateful to Emeritus Professors of Tohoku University Drs. M. Hirabayashi and D. Watanabe for useful discussion on high-resolution electron microscopy. For obtaining high-resolution electron microscope images presented in this book, the authors had invaluable collaboration with their colleagues the late Mr. H. Ohta, Dr. O. Terasaki, Mr. E. Aoyagi, Dr. N. Ohnishi, Dr. K. Lee, Dr. T. Oku, Dr. B.T. Lee, Dr. T. Ohsuna, Dr. W. Sun, Dr. G.S. Park, Mr. H. Emori, and Dr. H.G. Jeong. The authors also thank Dr. F.J. Lincoln for his useful comments on the manuscript. Special acknowledgments go to President K. Etoh and Dr. T. Oikawa, JEOL Ltd., for their interest and encouragement.

Sendai, March 1998
D. Shindo and K. Hiraga

Contents

Preface V
List of specimens observed by high-resolution electron microscopy.... IX
List of electron microscopes (EMs) used IX

1. Basis of High-Resolution Electron Microscopy
1.1 Principles of Transmission Electron Microscopy 1
1.2 Electron Scattering and Fourier Transform 3
1.3 Formation of High-Resolution Images 4
    1.3.1 High-Resolution Images of Thin Crystals 4
    1.3.2 Resolution of Electron Microscopes 8
    1.3.3 High-Resolution Images of Thick Crystals 9
1.4 Computer Simulation of High-Resolution Images 11
    1.4.1 Simulation Program and Input Parameters 11
    1.4.2 Generalization of Image Simulation 13
    1.4.3 Checking Programs 14
References 15

2. Practice of High-Resolution Electron Microscopy
2.1 Classification of High-Resolution Images 17
    2.1.1 Lattice Fringes 17
    2.1.2 One-Dimensional Structure Images 20
    2.1.3 Two-Dimensional Lattice Images 21
    2.1.4 Two-Dimensional Structure Images 25
    2.1.5 Special Images 29
2.2 Practice in Observing High-Resolution Images 31
    2.2.1 Points to Note Before Observation 31
    2.2.2 Points to Note During Observation 33
    2.2.3 Selection of Good Images 36
    2.2.4 Points to Note in the Interpretation of Images 38
    2.2.5 Training for the Observation of
         High-Resolution Images 39
References 39

3. Application of High-Resolution Electron Microscopy
3.1 High-Resolution Images of Various Defects 41
    3.1.1 Dislocations 41
    3.1.2 Grain Boundaries and Interfaces Between
         Different Phases 54
    3.1.3 Surfaces 69
    3.1.4 Other Structural Defects 75
3.2 High-Resolution Images of Various Materials 83
    3.2.1 Ceramics 83

3.2.2 Superconducting Oxides 91
3.2.3 Ordered Alloys 99
3.2.4 Quasicrystals 113
References 127

4. Peripheral Instruments and Techniques for High-Resolution Electron Microscopy

4.1 Image Processing 129
    4.1.1 Input and Output of High-Resolution Images 129
    4.1.2 Practice in Processing High-Resolution Images 131
4.2 Quantitative Analysis 138
    4.2.1 Principles of New Recording Systems 138
    4.2.2 Characteristics of New Recording Systems 139
    4.2.3 Quantitative High-Resolution Electron
          Microscopy 142
4.3 Electron Diffraction 147
    4.3.1 Basis of Electron Diffraction 147
    4.3.2 Practice of Electron Diffraction 149
    4.3.3 Electron Diffraction Patterns of Various Structures 152
4.4 Weak-Beam Method 156
    4.4.1 Principles of the Weak-Beam Method 156
    4.4.2 Weak-Beam Method in Practice 158
4.5 Evaluation of the Performance of Electron
    Microscopes 159
    4.5.1 Evaluation of Basic Parameters in Electron
          Microscopes 159
    4.5.2 Evaluation of the Resolution of Electron
          Microscopes 163
4.6 Specimen Preparation Techniques 164
    4.6.1 Crushing 164
    4.6.2 Electropolishing 164
    4.6.3 Chemical Polishing 165
    4.6.4 Ultramicrotomy 165
    4.6.5 Ion Milling 166
    4.6.6 Focused Ion Beam (FIB) 167
    4.6.7 Vacuum Evaporation 167
References 168

Appendixes

Appendix A. Physical Constants, Conversion Factors
    and Electron Wavelength 169
Appendix B. Geometry of Crystal Lattice 170
Appendix C. Typical Structures in Materials and
    Their Electron Diffraction Patterns 172
Appendix D. Properties of Fourier Transform 181
Appendix E. Sign Conventions 185

Subject Index 187

Intermetallic compounds—high-temperature
    structural materials 48
Tilt boundaries and Σ values 58
Reminiscence: High-resolution observation of
    a $YBa_2Cu_3O_7$ superconductor 94
Symbols for crystal structures in ordered alloys 100
Kikuchi patterns 154

List of specimens observed by high-resolution electron microscopy (*page number*)

Indication of compositions of various compounds follows the common use in each field. When the composition is not clear, it is indicated with elements simply connected with hyphens.

1. Ceramics

Si_3N_4 (*26*), $HIP-Si_3N_4-SiC$ (*55*), $CVD-Si_3N_4$ (*55*),
   $CVD-SiC$ (*23, 56, 57*),
$CVD-Si_3N_4-TiN$ (*66*), Si_3N_4-Al (*68*), Si_3N_4-SiC (*84, 85*),
ZrO_2 (*86*), $Al_2O_3-24vol \% ZrO_2$ (*87, 90*),
   $ZrO_2-24vol \% Al_2O_3$ (*89*)

2. Semiconductors

Si (*44, 45, 46, 50, 51*), $Ga_{0.5}In_{0.5}P$ (*132*)

3. Alloys · Intermetallic compounds

$Fe_{73.5}CuNb_3Si_{13.5}B_9$ (*18*),
$Ni_3(Al, Ti)$ (*47, 59*), $CoTi$ (*48*), Fe_3Al (*52*), $Ni-20.1at\% Mo$ (*109*),
   $Sm-Co$ (*62, 63*),
$Au-24.0at\% Cd$ (*101*), $Au-30.5at\% Cd$ (*111*),
   $Au-32.0at\% Cd$ (*111*),
$Au-20.7at\% Mn$ (*106, 107*), $Au-22.6at\% Mn$ (*102*), $Cu-41.0at\% Au$ (*104*),
   $Cu-50.0at\% Au$ (*105*),
$Al-20wt\% Si-1wt\% Ni$ (*65*),
$Al_{80}Mn_{20}$ (*117*), $Al_{74}Mn_{20}Si_6$ (*118, 120*), $Al_{70}Pd_{20}Mn_{10}$ (*120*),
   $Al-Li-Cu$ (*121*),
Al_3Mn (*124*), $Al_{72}Pd_{18}Cr_{10}$ (*124*), $Al_{70}Pd_{13}Mn_{17}$ (*125, 126*)

4. Inorganic compounds

$YBa_2Cu_3O_7$ (*49, 70, 92*),
$Bi-Sr-Ca-Cu-O$ (*20, 60*),
$TlBa_2CaCu_2O_7$ (*70*), $TlBa_2Ca_3Cu_4O_{11}$ (*7, 71*), $Tl_2Ba_2CuO_6$ (*42, 93, 142*),
$Tl_2Ba_2CaCu_2O_8$ (*96*), $Tl_2Ba_2Ca_3Cu_4O_{12}$ (*96*),
$Pb_2Sr_2Y_{0.5}Ca_{0.5}Cu_3O_8$ (*71*),
EMT Zeolite (*74*), LTL Zeolite (*74*),
$\alpha-Fe_2O_3$ (*72*), $Nb-O-F$ (*76*), Nb_2O_5 (*77*), $WO_3-1/8Ta_2O_3$ (*79*),
   $WO_3-1/4Ta_2O_3$ (*80, 81*),
Sm_2CuO_4 (*98*)

List of electron microscopes (EMs) used

200kV EM: JEM-200 CX, $C_s = 0.8$ mm
400kV EM: JEM-4000 EX, $C_s = 1.0$ mm
1000kV EM: JEM-1000, $C_s = 8.0$ mm
1250kV EM: JEM-ARM 1250, $C_s = 1.6$ mm

1. Basis of High-Resolution Electron Microscopy

1.1 Principles of Transmission Electron Microscopy

We first describe the main principles of electron microscopy. The formation of images in a transmission electron microscope can be understood from an optical ray diagram with an optical objective lens, as shown in Fig. 1.1. When a crystal of lattice spacing d is irradiated with electrons of wavelength λ, diffracted waves will be produced at specific angles 2θ, satisfying the *Bragg condition*, i.e.,

$$2d\sin\theta = \lambda. \tag{1.1}$$

The diffracted waves form diffraction spots on the *back focal plane*. In an electron microscope, the use of electron lenses allows the regular arrangement of the diffraction spots to be projected on a screen and the so-called *electron diffraction pattern* can then be observed. If the transmitted and the diffracted beams interfere on the *image plane*, a magnified image (*electron microscope image*) can be observed. The space where the diffraction pattern forms is called the *reciprocal space*, while the space at the image plane or at a specimen is called the *real space*. As shown in the following section, the scattering from the specimen to the back focal plane, in other words, the transformation from the real space to the reciprocal space, is mathematically given by the *Fourier transform*.

In a transmission electron microscope, by adjusting the electron lenses, i.e., by changing their focal lengths, both electron microscope images (information in real space) and diffraction patterns (information in reciprocal space) can be observed. Thus, in analyses of the microstructures of materials, both observation modes can be successfully combined. For example, in an investigation of electron diffraction patterns, we first observe an electron microscope image. By inserting an aperture (*selected area aperture*) in a specific area and adjusting the electron lenses, we get a diffraction pattern of the area. This observation mode is called *selected area diffraction*. Because a selected area diffraction pattern can be obtained from each grain in a polycrystal, crystal structures and mutual crystal orientation relationships between adjacent grains can easily be clarified. The area selected is usually limited to about $0.1\,\mu m$ in diameter, but the recently developed *microdiffraction* method, where incident electrons are converged on a specimen, can now be used to get a diffraction pattern from an area only a few nm in diameter.

In order to investigate an electron microscope image, we first observe the electron diffraction pattern. Then by passing the transmitted beam or one of the diffracted beams through an aperture (*objective aperture*) and changing to the imaging mode, we can observe the image with enhanced contrast, and precipitates and lattice defects can easily be identified. As shown in Fig. 1.2a, the observation mode using the transmitted beam is called the *bright-field method*, and the image observed is a *bright-field image*. When one diffracted beam is selected (Fig. 1.2b), it is called the *dark-field method*, and the observed image is a *dark-field image*. The contrast in these images is attributed to the change of the amplitude of the transmitted beam or the diffracted beam due to absorption and scattering in the specimens. Thus the image contrast is called the *absorption–diffraction contrast* or the *amplitude contrast*.

It is also possible to form electron microscope images by selecting more than two beams on the back focal plane using a large objective aperture, as shown in Fig. 1.2c. This observation mode is called *high-resolution electron microscopy*, and the image observed is a *high-resolution electron microscope image* (hereafter *high-resolution image* for short). This is what we will discuss in detail in this book. Since the contrasts in high-resolution images are formed because of the differences of phase of the transmitted and diffracted beams, this contrast in high-resolution images is called the

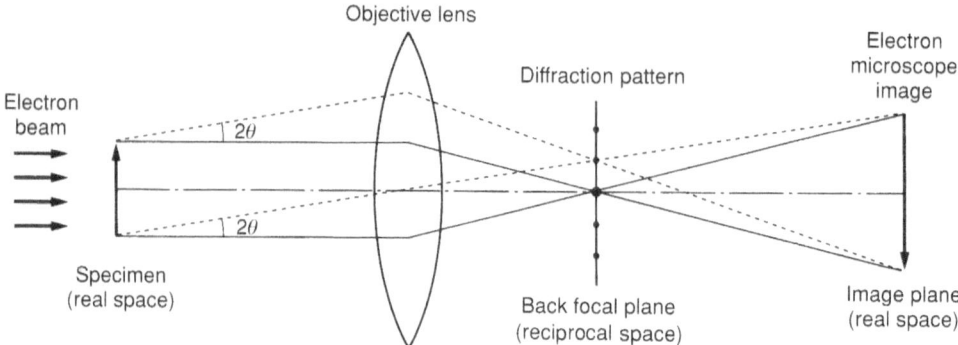

Fig. 1.1. Optical ray diagram with an optical objective lens showing the principle of the imaging process in a transmission electron microscope

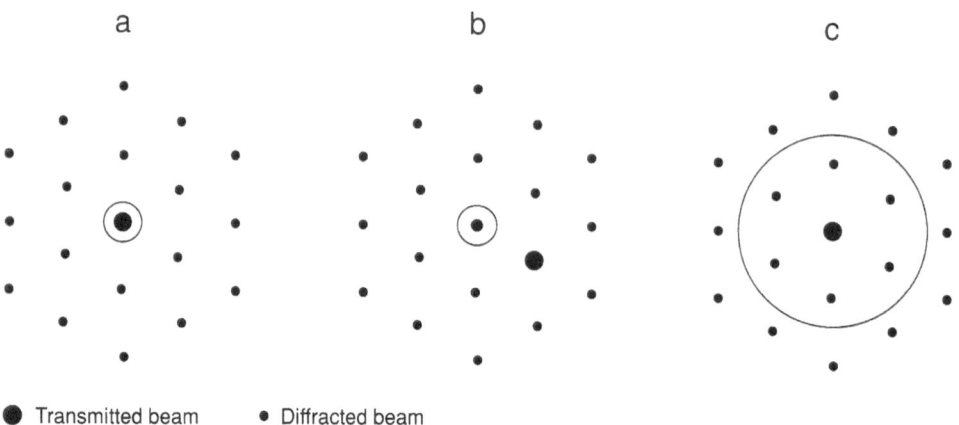

● Transmitted beam   • Diffracted beam

Fig. 1.2. Three observation modes in electron microscopy using an objective aperture. The center of the objective aperture is assumed to be set to the optical axis. **a** Bright-field method; **b** dark-field method; **c** high-resolution electron microscopy (axial illumination)

phase contrast. The observation of high-resolution images under the illumination condition where the transmitted beam is set parallel to the optical axis of the objective lens is called the *axial illumination method*, while it is called the *off-axis illumination method* when the transmitted beam is tilted at an angle to the optical axis. Currently, the axial illumination method is generally used.

High-resolution images result from interference of the electron which is scattered in a specimen and is affected by aberrations in the electron lenses. Thus, in order to find the optimum imaging conditions and interpret high-resolution images appropriately, it is necessary to understand the electron scattering process in a specimen as well as the imaging process through the electron lenses. These processes are discussed in the following.

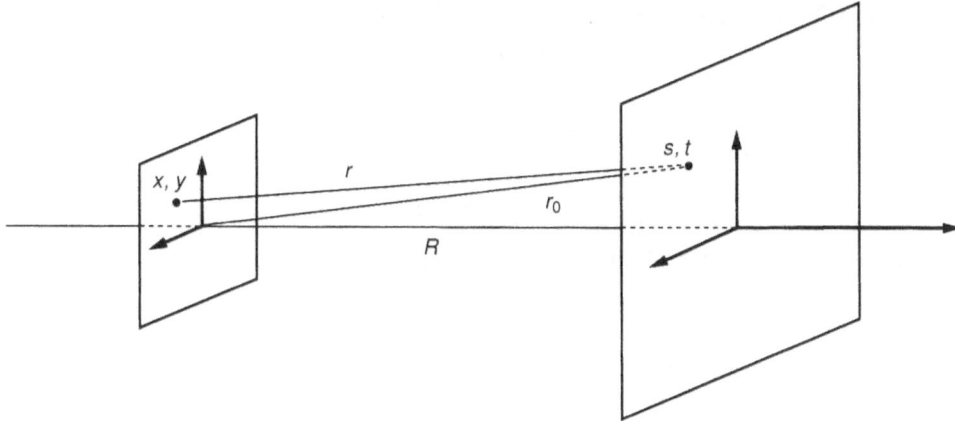

Fig. 1.3. Schematic illustration showing the electron scattering in a specimen

1.2 Electron Scattering and Fourier Transform

The scattering process of incident electrons from a specimen onto the back focal plane forming the diffraction pattern, and the subsequent image formation process through the back focal plane onto the image plane are represented by the Fourier transform. In this section, we will briefly demonstrate that the formulation of these scattering processes is approximately equivalent to the Fourier transform. In order to understand the next section, and also Sect. 1.4 which concerns the computer simulation of high-resolution images, it is important to understand the correspondence between the electron scattering process and the Fourier transform.

Let us consider the scattering of the incident plane wave $\exp(ikr)$ by a specimen with wave number k ($k = 2\pi/\lambda$, where λ is the wavelength).[1] If the effect of the specimen on the plane wave is given by $q(x, y)$, the scattering amplitude at a separate point (s, t) (Fig. 1.3) is given as

$$\Psi(s, t) = c \iint q(x, y) \frac{\exp(ikr)}{r} dx dy \quad (1.2)$$

where c is a constant. Equation 1.2 indicates that the phase and amplitude of the incident plane wave are changed by $q(x, y)$, and that it is scattered as a spherical wave. If we observe the scattering amplitude at a distance which is much

larger than the scale of the specimen (*Fraunhofer diffraction*[2]), i.e., $R >> x, y$, then r can be written as

$$r = \left[R^2 + (x - s)^2 + (y - t)^2 \right]^{1/2}$$
$$\approx \left[R^2 + s^2 + t^2 - 2(sx + ty) \right]^{1/2}$$
$$\approx r_0 - sx/r_0 - ty/r_0. \quad (1.3)$$

Thus, the scattering amplitude of Eq. 1.2 becomes

$$\Psi(u, v) \approx c' \iint q(x, y)$$
$$\exp(-2\pi i(ux + vy)) dx dy \quad (1.4)$$

where c', u, and v are given, respectively, as

$$c' = c \exp(ikr_0)/r_0, \quad u = s/\lambda r_0, \quad v = t/\lambda r_0. \quad (1.5)$$

Since the right-hand side of Eq. 1.4 is a form of the Fourier transform, it can be seen that the scattering amplitude $\Psi(u, v)$ is given by the Fourier transform of $q(x, y)$.

The conditions for Fraunhofer diffraction, as given by Eq. 1.3, are satisfied for electron diffraction with an electron lens in an electron microscope, while in X-ray and neutron diffraction, the distance between the detector and the specimen has to be much larger than the scale of the specimen and so these conditions are approximately satisfied. Appendix D gives some typical examples of $q(x, y)$ and the absolute values of their Fourier transform $|\Psi(u, v)|$ in order to clarify the properties of the Fourier transform.

---

[1] The incident plane wave may also be written as $\exp(-ikr)$, as seen in some literature. The choice of signs in the wave function follows the *sign conventions*, which are given in Appendix E.

[2] When the scattering process in a small angle is observed at a position close to the specimen, it is called *Fresnel diffraction*.

1.3 Formation of High-Resolution Images

In order to understand the correlation between the contrast of a high-resolution image and an atomic arrangement, we first consider the imaging process for very thin crystals. We will then discuss the imaging process for thicker crystals taking into account the *dynamical diffraction effect*.

1.3.1 High-Resolution Images of Thin Crystals

As shown in the ray diagram in Fig. 1.1, the imaging process of a high-resolution image consists of the following three processes:

1. electron scattering in a specimen;
2. formation of diffracted beams at the back focal plane;
3. formation of a high-resolution image at the image plane.

We now discuss the effect of each of these on the incident electron.

1.3.1.1 Electron Scattering in a Specimen

If a specimen is very thin and the absorption is neglected, only the phase of the incident electron wave will be changed, while the amplitude remain unchanged. Under this *phase object approximation*, the specimen is represented as having a *transmission function*

$$q(x, y) = \exp(i\sigma\varphi(x, y)\Delta z). \qquad (1.6)$$

Equation 1.6 indicates that the phase of the incident wave proceeds by $\sigma\varphi(x, y)\Delta z$ over that in a vacuum, where σ is an *interaction constant* given by

$$\sigma = \frac{2\pi}{V\lambda\left(1 + \sqrt{1 - \beta^2}\right)}. \qquad (1.7)$$

Here, V is the accelerating voltage and β is the velocity of the electron relative to the velocity of light, i.e., $\beta = v/c$. λ is the wavelength of the electron, which is given by

$$\lambda = \frac{h}{\sqrt{2m_e eV\left(1 + \dfrac{eV}{2m_e c^2}\right)}}. \qquad (1.8)$$

$\varphi(x, y)\Delta z$ in Eq. 1.6 corresponds to the two-dimensional projected potential in the z-direction

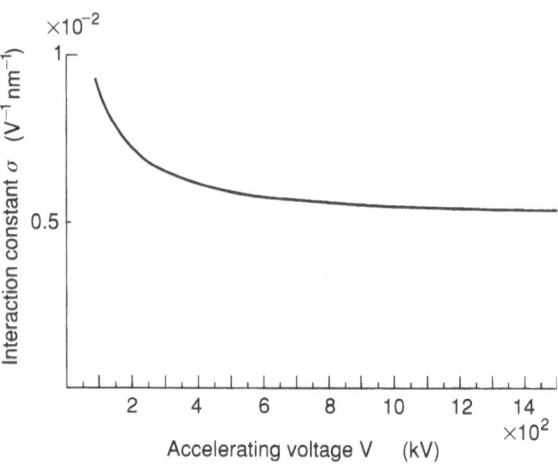

Fig. 1.4. Variation of the interaction constant σ as a function of the accelerating voltage V

with thickness Δz. Here h and m_e are Planck's constant and the mass of the electron, respectively (see Appendix A). The variation of σ as a function of the accelerating voltage V is shown in Fig. 1.4. For example, σ takes 0.00729 ($V^{-1}nm^{-1}$) for 200 kV, while it takes the relatively smaller value 0.00539 ($V^{-1}nm^{-1}$) for 1000 kV (Appendix A). In general, σ takes a smaller value with an increase in accelerating voltage. On the other hand, the inner potential generally becomes larger with an increase in atomic number, although this depends on density. As shown from Table 1.1 [1–4], the mean inner potential takes values in a range from a few volts to about 30 V. Thus, for thin specimens of less than a few nanometers, the phase factor in the exponential term of Eq. 1.6 takes a value much smaller than 1, so the exponential term can be expanded and approximated to (*weak-phase object approximation*)

$$q(x, y) \approx 1 + i\sigma\varphi(x, y)\Delta z. \qquad (1.9)$$

Table 1.1. Mean inner potential of some materials in volts

C ($Z = 6$)	7.8 ± 0.6	[1]
Al (13)	13.0 ± 0.4	[1]
	12.4 ± 1	[2]
	11.9 ± 0.7	[3]
Si (14)	11.5	[4]
Cu (29)	23.5 ± 0.6	[1]
	20.1 ± 1.0	[3]
Ge (32)	15.6 ± 0.8	[3]
Au (79)	21.1 ± 2	[2]

Z = atomic number.

From Eqs. 1.6 and 1.9, the phase change of the incident wave inside a specimen becomes larger with a decrease in the accelerating voltage or with an increase in the crystal potential.

1.3.1.2 Formation of Diffracted Beams at the Back Focal Plane

As shown in Sect. 1.2, the scattering amplitude of the electron wave $\Psi(u, v)$ is given by performing the Fourier transform of the transmission function in Eq. 1.9,

$$\Psi(u, v) = Q(u, v)\exp(i\chi(u, v))$$
$$= \mathscr{F}[q(x, y)]\exp(i\chi(u, v))$$
$$\approx \delta(u, v) + i\mathscr{F}[\sigma\varphi(x, y)\Delta z]\exp(i\chi(u, v))$$
$$(1.10)$$

where $\mathscr{F}$ indicates the Fourier transform. The term $\exp(i\chi(u, v))$ shows the phase change of the scattered wave and is called the *contrast transfer function* (or *phase contrast transfer function*). Here, $\chi(u, v)$ is given by

$$\chi(u, v) = \pi\left\{\Delta f\lambda(u^2 + v^2) - 0.5C_s\lambda^3(u^2 + v^2)^2\right\}.$$
$$(1.11)$$

In this equation, Δf and C_s correspond to a *defocus value* and a *spherical aberration coefficient* of the objective lens, respectively, and the first and second terms of Eq. 1.10 correspond to the transmitted and scattered beams, respectively.

1.3.1.3 Formation of a High-Resolution Image at the Image Plane

The scattering amplitude of electron wave at the image plane can be obtained by a Fourier transform of the scattering amplitude at the back focal plane, i.e.,

$$\psi(x, y) = \mathscr{F}[C(u, v)\Psi(u, v)] \quad (1.12)$$

where $C(u, v)$ indicates the effect of an objective aperture, and is given by

$$C(u, v) = 1, \quad \sqrt{u^2 + v^2} \leq r$$
$$0, \quad \sqrt{u^2 + v^2} > r \quad (1.13)$$

where r indicates the radius of the objective aperture.

If the magnification of the high-resolution image is neglected, the intensity of the high-

resolution image is the square of the scattering amplitude of the electron wave at the image plane.

$$I(x, y) = \psi^*(x, y)\psi(x, y)$$
$$= \left|1 + i\mathscr{F}[C(u, v)\mathscr{F}[\sigma\varphi(x, y)\Delta z]\right.$$
$$\left.\exp(i\chi(u, v))]\right|^2.$$
$$(1.14)$$

We now consider what is shown by the intensity of Eq. 1.14. For simplicity, the effect of an objective aperture is neglected, i.e.,

$$C(u, v) = 1 \quad (1.15)$$

and two typical phase changes with an ideal objective lens,

$$\exp(i\chi(u, v)) = \pm i, \quad u, v \neq 0 \quad (1.16)$$

are assumed. Eventually, we have

$$I(x, y) = \left|1 \mp \sigma\varphi(-x, -y)\Delta z\right|^2$$
$$\approx 1 \mp 2\sigma\varphi(-x, -y)\Delta z. \quad (1.17)$$

Here we used the relation $\varphi(-x, -y) = \mathscr{F}[\mathscr{F}[\varphi(x, y)]]$. According to Eq. 1.17, the high-resolution image directly reflects crystal potential. Here, we note the appearance of a minus sign in the coordinates x, y in φ in Eq. 1.17. Successive operations of the Fourier transform and the inverse Fourier transform on $\varphi(x, y)$ give $\varphi(x, y)$ again, whereas two successive operations of the Fourier transform, corresponding to the scattering process in the z-direction (as shown in Fig. 1.1), give $\varphi(-x, -y)$. This corresponds to a reversed image in the image plane, as is well known in image formation with an optical lens.

Figure 1.5a shows the imaginary part of the contrast transfer function of the objective lens under the *optimum defocus condition* (the so-called *Scherzer focus*) for a 200 kV electron microscope ($C_s = 0.8$ mm, $\Delta f = 53.8$ nm) and a 400 kV electron microscope ($C_s = 1.0$ mm, $\Delta f = 48.7$ nm). It can be seen that the imaginary part takes values around 1 in the range 1.7–4.3 nm^{-1} for 200 kV, and in the range 2.1–5.7 nm^{-1} for 400 kV. Thus, under the Scherzer focus condition we have the image intensity corresponding to one of those given in Eq. 1.17, i.e.,

$$I(x, y) \approx 1 - 2\sigma\varphi(-x, -y)\Delta z. \quad (1.18)$$

Note that under other focus conditions, the contrast transfer function does not take a constant value in a wide range of spatial frequencies, as is

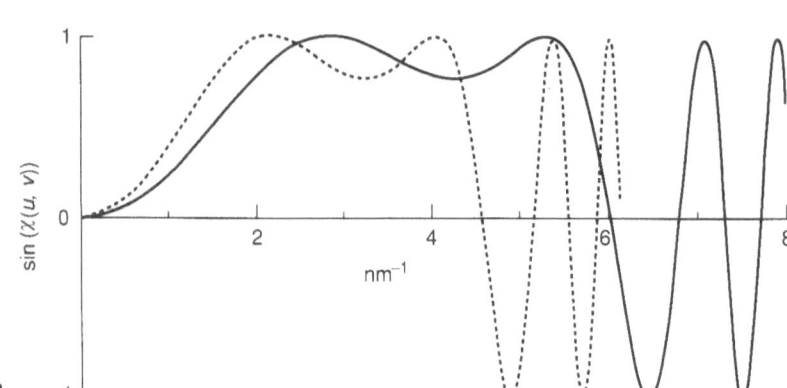

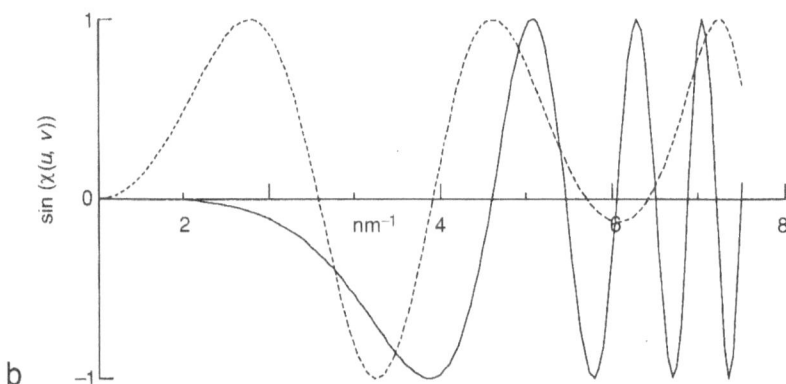

Fig. 1.5. Imaginary part of the contrast transfer function of the objective lens. **a** Under Scherzer focus for a 400 kV electron microscope (*solid line*) and a 200 kV electron microscope (*dotted line*). **b** For a 400 kV electron microscope with $\Delta f = 0$ nm (*solid line*) and $\Delta f = 100$ nm (*dotted line*)

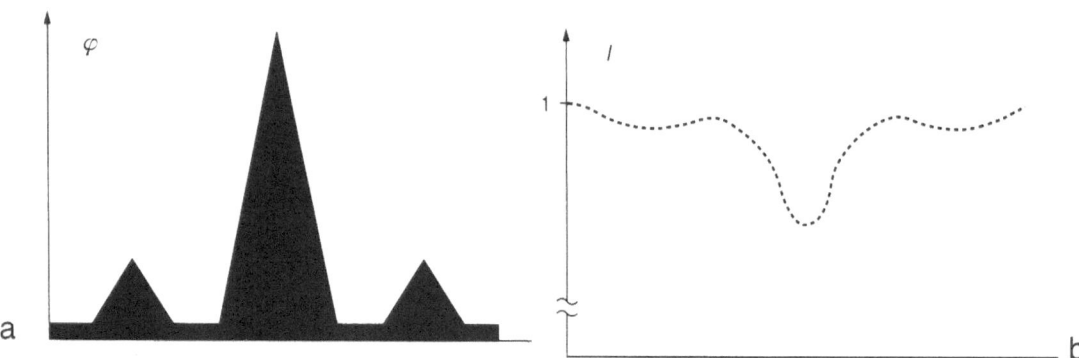

Fig. 1.6. Schematic illustration showing the correspondence of **a** the crystal potential and **b** the intensity distribution of a high-resolution image

shown in Fig. 1.5b. If a heavy and a light atomic column exist, as shown in Fig. 1.6, what does the image contrast look like? Since $\sigma\varphi(x, y)\Delta z$ has a value smaller than 1, and a heavy atomic column has a larger potential than a light atomic column, the image intensity at a heavy atomic column becomes lower. As well as the spherical aberration, several other factors limit the resolution, resulting in a gradation of the image contrast, as shown in

Fig. 1.6b. These factors will be discussed below. Figure 1.7 shows a typical high-resolution image of a superconducting oxide $TlBa_2Ca_3Cu_4O_{11}$. The positions of the heavy atomic columns such as Tl and Ba are seen as large dark spots, while the regions between the metal atomic columns appear bright. The vacant oxygen positions, or the lowest potential regions, appear the brightest, which is consistent with Eq. 1.18 and Fig. 1.6.

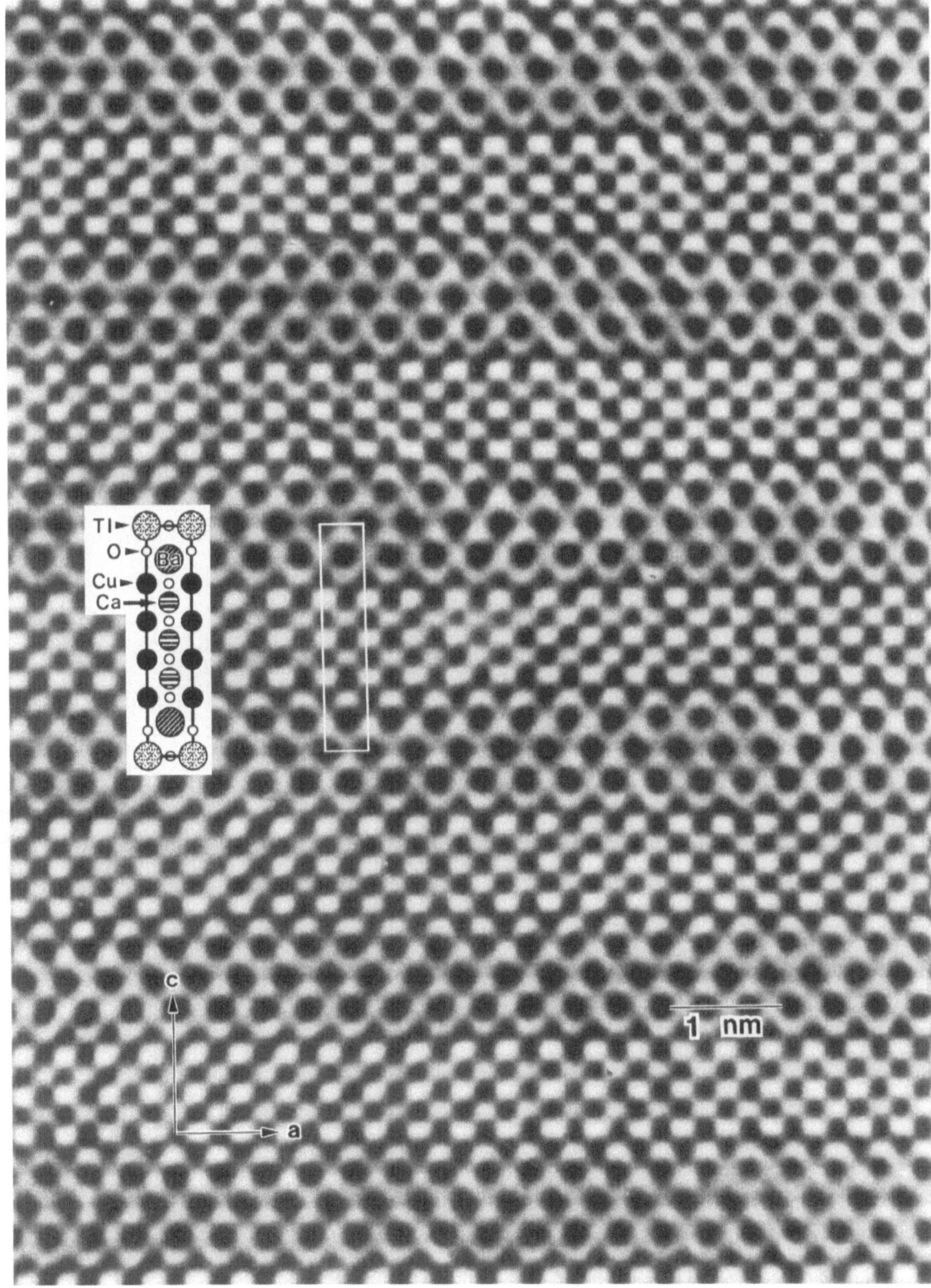

Fig. 1.7. High-resolution image of a superconducting oxide $TlBa_2Ca_3Cu_4O_{11}$.
Specimen: $TlBa_2Ca_3Cu_4O_{11}$; **Preparation**: crushing; **Observation**: 400 kV EM, [010] incidence

1.3.2 Resolution of Electron Microscopes

As assumed in Eq. 1.16 for the ideal case, if the contrast transfer function takes a constant value i for a wide range of spatial frequencies, high-resolution images reflect crystal potential and the resolution is extremely high. For a real case, an optimum defocus condition corresponding to the Scherzer focus [5] can be given as

$$\Delta f = 1.2 (C_s \lambda)^{1/2}. \qquad (1.19)$$

In this equation, a plus sign for Δf is assigned for underfocus (a decrease in the lens current, or weak operation of the objective lens). Under this condition, the limit of the highest spatial frequency where the phases of scattered waves are not significantly disturbed is given by the condition $\chi(u, v) = 0$ in Eq. 1.11, i.e.,

$$\left(u^2 + v^2 \right)^{1/2} = \sqrt{2.4} C_s^{-1/4} \lambda^{-3/4}. \qquad (1.20)$$

Thus, the *resolution limit* of electron microscopes is given by

$$d_s = \left(u^2 + v^2 \right)^{-1/2} = 0.65 C_s^{1/4} \lambda^{3/4}. \qquad (1.21)$$

Generally, the value given by Eq. 1.21 is cited as the *point resolution* of electron microscopes.[3] Resolution depends on both C_s and λ with degrees of 1/4 and 3/4, respectively. Thus it is seen that a decrease in λ is more effective than one in C_s in achieving higher resolution.

In addition to the aberration of the contrast transfer function of the objective lens, the energy fluctuation of incident electrons and their convergence on the specimen result in a deterioration of the resolution of electron microscopes. The aberration due to the fluctuation of incident electron energy is called the *chromatic aberration*, and this causes a fluctuation of the focus in high-resolution images. Fluctuations in both the accelerating voltage and the objective lens current, contribute to the amount of focus fluctuation, i.e.,

$$\Delta = C_c \left[\left(\Delta V_r / V_r \right)^2 + \left(2 \Delta I / I \right)^2 \right]^{1/2}. \qquad (1.22)$$

Here, C_c is called the *chromatic aberration coefficient,* and is another of the important parameters in electron microscopes like the spherical aberration coefficient. Evaluation of this coefficient will be described in Sect. 4.5. In Eq. 1.22, V_r indicates the corrected accelerating voltage taking into account the relativistic effect on the electron mass, and is given by

$$V_r = V \left(1 + \frac{eV}{2 m_e c^2} \right). \qquad (1.23)$$

The effect of damp in scattering amplitudes due to chromatic aberration is given [7] as

$$D = \exp\left(-0.5 \pi^2 \lambda^2 \Delta^2 \left(u^2 + v^2 \right)^2 \right) \qquad (1.24)$$

while the effect due to beam convergence is given [8] by

$$S = \exp\left(-\pi^2 (\alpha/\lambda)^2 \left(C_s \lambda^3 \left(u^2 + v^2 \right)^{3/2} \right. \right.$$
$$\left. \left. - \Delta f \lambda \left(u^2 + v^2 \right)^{1/2} \right)^2 \right) \qquad (1.25)$$

where α indicates beam convergence, defined by the semiangle subtended by the effective source at the specimen. These two functions are considered to be *envelope functions* which limit the contribution of the scattered electrons to high-resolution images. Figure 1.8a shows a contrast transfer function (solid line) and an envelope function (dotted line) for a 400kV electron microscope. Δ is assumed to be 11 nm. Thus an *effective transfer function*, as shown in Fig. 1.8b, clearly indicates the decrease in resolution due to chromatic aberration. The resolution limit, taking into account the chromatic aberration and the beam convergence, can be evaluated from an optical diffractogram or from a diffractogram obtained by the Fourier transform (*digital diffractogram*) of a high-resolution image of an amorphous material (see Sect. 4.5).

The above discussion on the contrast of high-resolution images and the resolution of electron microscopes only applies to ideally thin specimens. For most specimens, such thin crystals can rarely be obtained, and it is necessary to take into account the dynamical diffraction effect.

---

[3] In addition to the point resolution as defined above, there is another resolution specification for electron microscopes which is called the *information-resolution limit* [6]. This expresses the limit of the highest spatial frequency contained in a high-resolution image, independent of the disturbance of the phase due to spherical aberration. Thus, the information-resolution limit is mainly constrained by the electric instabilities of an electron microscope, such as chromatic aberration.

Fig. 1.8. a Contrast transfer function (*solid line*) and an envelope function (*dotted line*) for a 400 kV electron microscope. **b** Effective contrast transfer function

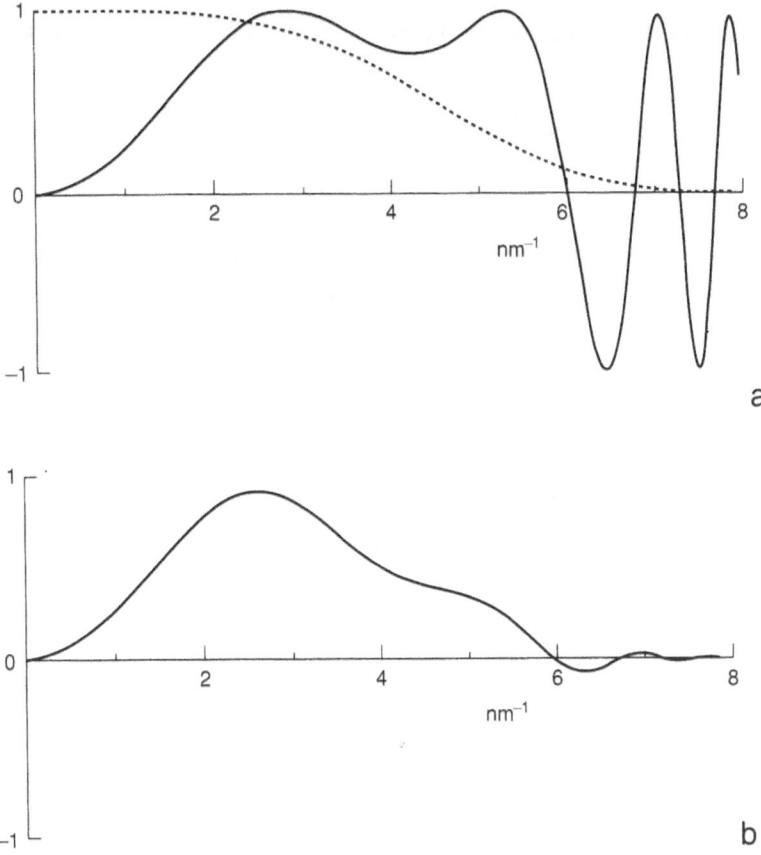

1.3.3 High-Resolution Images of Thick Crystals

Equation 1.18 shows the intensity distribution of a high-resolution image for a very thin crystal. For crystals thicker than 5 nm, the weak-phase object approximation, and even the phase object approximation, cannot be applied, and the phase change in a crystal due to plural scattering should be taken into account. The change in the scattering amplitudes of the transmitted and diffracted beams due to their interaction is called the *dynamical diffraction effect*. There are various theoretical approaches in the study of this effect, such as the column approximation based on differential equations [9, 10], and the Bloch wave approach using matrix equations [11, 12]. The physical optical approach using convolution iteration [13, 14] and the higher-order Born approximation [15] have also been reported. Here we discuss the physical optical approach, the so-called *multi-slice* method proposed by Cowley and Moodie [13], which is widely used for computer simulation of high-resolution electron microscope images.

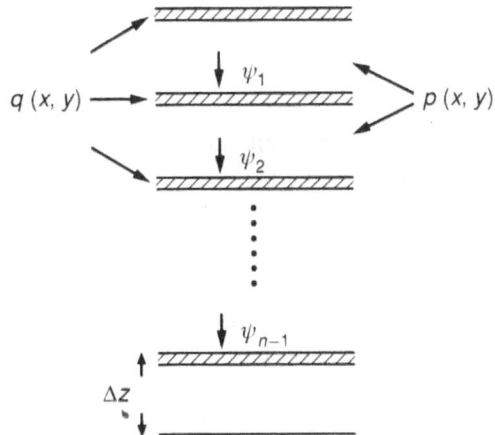

Fig. 1.9. Transmission function $q(x, y)$ and a propagation function $p(x, y)$ in the multi-slice method

As indicated by the name, the specimen is sliced and the effect of each slice on the incident electron wave is considered successively. Usually, the slice thickness is set to be the length of a unit cell, i.e., 0.2–0.5 nm. As shown in Fig. 1.9, the effect on the incident wave can be separated into two processes: the phase change due to the crystal potential of the specimen, and the phase change due to propagation in the slice.

In practice, the effect of the material in the first slice on the incident wave can be explained in the following way. At the top of the first slice, the two-dimensional projected potential causes a phase change in the incident wave as given by Eq. 1.6. Then the propagation of the wave from the top to the bottom of the first slice shows as a small angle scattering in the vacuum. This small angle scattering process is given by the *propagation function* $p(x, y)$, where

$$p(x, y) = \frac{1}{i \Delta z \lambda} \exp\left(\frac{ik(x^2 + y^2)}{2\Delta z}\right). \quad (1.26)$$

Eventually, the scattering amplitudes of the electron wave at the bottom of the first slice can be given by the transmission function and the propagation function as

$$\psi_1(x, y) = q(x, y) * p(x, y) \quad (1.27)$$

where $*$ indicates the convolution operation.[4]

The effect of the second slice on the scattered wave is treated in the same way as the first slice by considering $\Psi_1(x, y)$ to be the incident wave on the second slice. That is, the phase of $\Psi_1(x, y)$ changes due to the crystal potential at the top of the second slice, and then it propagates to the bottom of the second slice with a small scattering angle

$$\psi_2(x, y) = \left(q(x, y)\psi_1(x, y)\right) * p(x, y)$$
$$= q(x, y)\left[q(x, y) * p(x, y)\right] * p(x, y). \quad (1.28)$$

In general, the scattering amplitude at the bottom of a specimen consisting of n slices is given by

$$\psi_n(x, y) = q(x, y)\underbrace{\left[\ldots\underbrace{\left[q(x, y)\underbrace{\left[q(x, y)\right.}_{1}\right.}_{2}\right.}_{n-1}$$
$$\underbrace{\left.* p(x, y)\right]}_{1} \underbrace{* p(x, y)\right]}_{2} \ldots\underbrace{\right] * p(x, y).}_{n-1} \quad (1.29)$$

For thick crystals, it is not possible to discuss the imaging process using the effective transfer function (Sect. 1.3.2) which is applied to thin crystals [16]. In this case, the intensity of high-resolution images is given by

$$I(\mathbf{r}) = \mathscr{F}\left(J(\mathbf{u})\right). \quad (1.30)$$

In the above equation, $\mathbf{u} = (u, v)$ and $J(\mathbf{u})$ is given by

$$J(\mathbf{u}) = \sum T(\mathbf{u}' + \mathbf{u}, \mathbf{u}')\Psi(\mathbf{u}' + \mathbf{u})\Psi^*(\mathbf{u}') \quad (1.31)$$

where $T(\mathbf{u}' + \mathbf{u}, \mathbf{u}')$ is the *transmission cross coefficient* and is given by

$$T(\mathbf{u}' + \mathbf{u}, \mathbf{u}') = C(\mathbf{u}' + \mathbf{u})C(\mathbf{u}')$$
$$\exp\left(i\left[\chi(\mathbf{u}' + \mathbf{u}) - \chi(\mathbf{u}')\right]\right)$$
$$\times D(\mathbf{u}' + \mathbf{u}, \mathbf{u}')S(\mathbf{u}' + \mathbf{u}, \mathbf{u}'). \quad (1.32)$$

Here, $C(\mathbf{u})$ shows the effect of the objective aperture given by Eq. 1.13. D and S correspond to the effects of the chromatic aberration and the convergence of the incident beam, respectively, and are given by

$$D(\mathbf{u}' + \mathbf{u}, \mathbf{u}')$$
$$= \exp\left[-\left(0.5\pi \Delta\lambda\right)^2\left(\left|\mathbf{u}' + \mathbf{u}\right|^2 - \left|\mathbf{u}'\right|^2\right)^2\right] \quad (1.33)$$

$$S(\mathbf{u}' + \mathbf{u}, \mathbf{u}')$$
$$= \exp\left[-\left(\pi q_0\right)^2\left(\left(C_s\lambda^2\left|\mathbf{u}' + \mathbf{u}\right|^2 - \Delta f\right)\lambda(\mathbf{u}' + \mathbf{u})\right.\right.$$
$$\left.\left. - \left(C_s\lambda^2\left|\mathbf{u}'\right|^2 - \Delta f\right)\lambda\mathbf{u}'\right)^2\right] \quad (1.34)$$

where $q_0 = \alpha/\lambda$ (α: beam convergence).

---

[4] The convolution of two functions $f(x)$ and $g(x)$ is defined as

$$h(x) = \int_{-\infty}^{\infty} f(t)g(x - t)\mathrm{d}t = \int_{-\infty}^{\infty} f(x - t)g(t)\mathrm{d}t.$$

1.4 Computer Simulation of High-Resolution Images

As discussed above, high-resolution images are affected not only by various aberrations, but also by the dynamical diffraction effect. Thus, in order to derive structural information from a high-resolution image, it is necessary to simulate the high-resolution image based on a structural model, taking into account these aberrations and the dynamical diffraction effect. Nowadays, various programs are available, and by inputting parameters such as atomic coordinates and experimental conditions, it is easy to obtain simulated images. However, in order to interpret simulated images appropriately, and furthermore to carry out quantitative analyses with great accuracy (as will be discussed in Sect. 4.2), it is important to understand details of image simulation and to be able to make a program yourself. Here, we outline the construction of a program for simulated images and discuss several check points within it.

1.4.1 Simulation Program and Input Parameters

According to the imaging process discussed in Sect. 1.3, the program can be divided into two parts, each of which has several processes.

1. Electron scattering in a specimen:
   a. calculation of a structure factor;
   b. calculation of a transmission function and a propagation function;
   c. evaluation of dynamical diffraction effect based on the multi-slice formulation.
2. Imaging process considering various aberrations:
   a. effect of spherical aberration of an objective lens;
   b. effect of chromatic aberration and beam convergence.

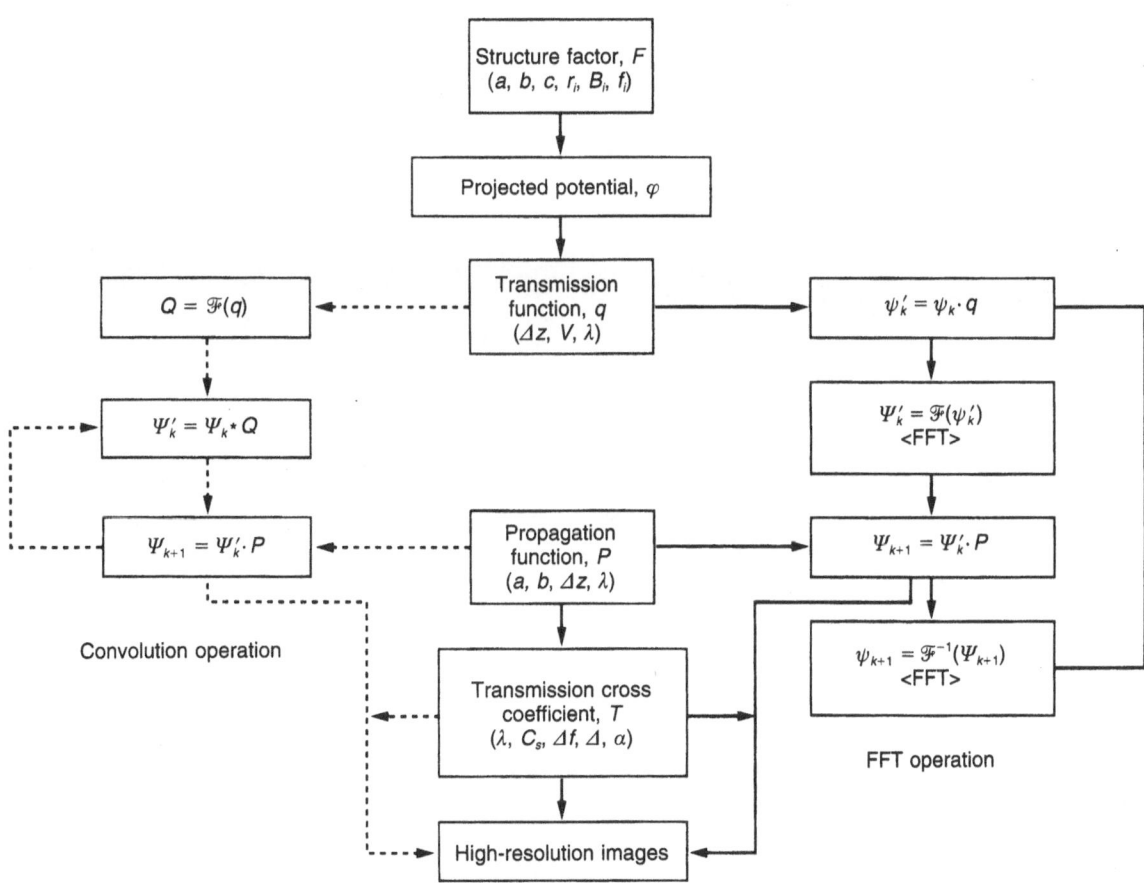

Fig. 1.10. Flow chart of a program for the multi-slice method

The input parameters consist of crystallographic parameters such as lattice constants (a, b, c), atomic coordinates r_i, a temperature factor B_i, and a scattering factor f_i for the ith atom. The parameters regarding the characteristics of an electron microscope and the imaging condition are accelerating voltage V (or wave length λ), a spherical aberration coefficient C_s, a defocus value due to chromatic aberration Δ, and beam convergence α. In the flow chart of the program shown in Fig. 1.10, input parameters are indicated at each step.

As shown in Fig. 1.10, there are two approaches to the iteration of Eq. 1.29 corresponding to the dynamical diffraction effect. One is the convolution method in reciprocal space, as indicated on the left of the flow chart, and the other is the multiplication method using the *fast Fourier transform* (FFT) [17], as indicated on the right. In the

former approach, the convolution method in reciprocal space is given by the Fourier transform of Eq. 1.29:

$$\Psi(u, v) = \left[Q(u, v) * \ldots \left[Q(u, v) * \left[Q(u, v) \right. \right. \right._{n-1} \quad _2 \quad _1$$
$$\left. \left. \left. *Q(u, v)P(u, v) \right] P(u, v) \right] \ldots \right] P(u, v).$$
$$_1 \qquad _2 \qquad _{n-1}$$
$$(1.35)$$

Here we used the following relations of the Fourier transform:

$$\mathcal{F}[q * p] = \mathcal{F}[q] \cdot \mathcal{F}[p] (= Q \cdot P) \qquad (1.36)$$

$$\mathcal{F}[q \cdot p] = \mathcal{F}[q] * \mathcal{F}[p] (= Q * P). \qquad (1.37)$$

The propagation function in reciprocal space can be given as

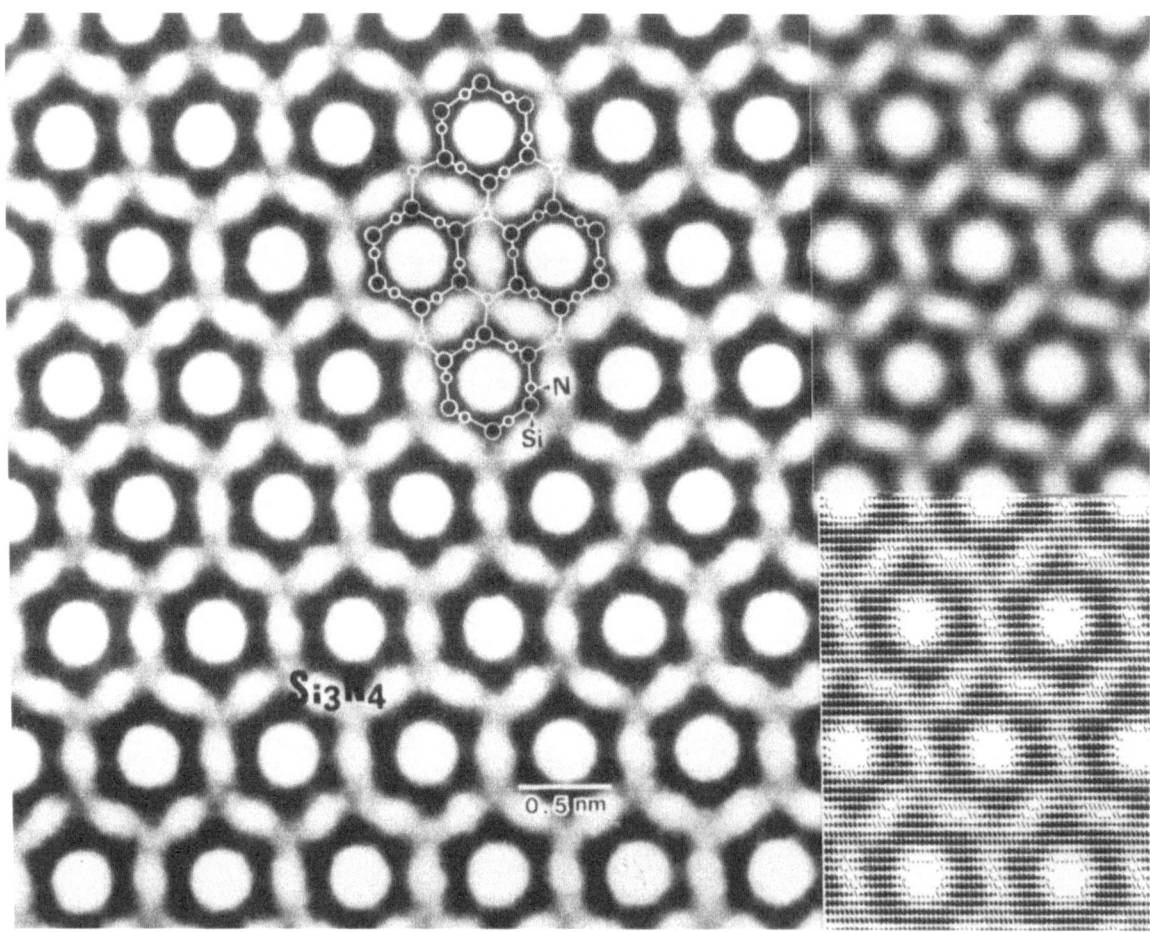

Fig. 1.11. Comparison of a high-resolution image (*left*) of β-Si$_3$N$_4$ with simulated images (*right*). *Upper right*, a half-tone image; *lower right*, an output with keyboard characters

$$P(u, v) = \exp\left(-i\pi\lambda\Delta z(u^2 + v^2)\right) \quad (1.38)$$

and is used for the convolution method.

In the calculation using the FFT, the phase change of the incident electrons due to the operation of the transmission function is carried out in real space, while propagation of the scattered electrons in the vacuum given with the propagation function is carried out in reciprocal space. Eventually, iteration with the multiplication operation through the FFT instead of the convolution operation results in a reduction of computing time. With an increase in sampling points, the advantage of the FFT method over the convolution method becomes greater.

To display the dark contrast of simulated images, keyboard characters (or superimposed characters) used to be used (Fig. 1.11), but now halftone images with gray levels of 256 can be obtained with high-precision printers. However, in order to compare observed images and simulated images quantitatively, it is necessary to evaluate the differences between them using digital data, as will be discussed in Sect. 4.2.

1.4.2 Generalization of Image Simulation

So far we have considered image simulation for a simple perfect crystal. We now address image simulation more generally, taking into account:

1. structural defects;
2. absorption;
3. the inclination of the incident beam;
4. the ionicity of atoms.

1.4.2.1 Structural Defects

If a crystal contains a localized structural defect, a continuous weak diffuse scattering in an electron diffraction pattern is produced, in addition to the strong Bragg reflections. In principle, in order to evaluate this diffuse scattering exactly, it is necessary to calculate the scattering amplitudes at infinitely small sampling points in reciprocal space. In real space, this means an infinitely large unit cell containing a localized defect. However, if the unit cell is large, the number of scattered beams also becomes large, resulting in extremely long computing times. Thus, in the image simulation of structural defects, we usually assume a hypothetical crystal with a unit cell of a finite size. In such a case, the defects are assumed to be arranged periodically in the crystal, and the unit cell should be large enough to neglect the interference of the contrast of the defects. As an example, simulated images of the crystal, including substitutional atoms, are presented in Sect. 3.2.3 (see Fig. 3.53). In general, for a crystal containing lattice defects involving lattice strain, such as interstitial atoms, vacancies, and dislocations, a larger unit cell should be assumed than is necessary for a crystal containing substitutional atoms with little lattice strain. Also, if the lattice defects exist inhomogeneously along the incident beam direction, a different transmission function for each crystal thickness should be assumed, and a large amount of memory will be necessary for the calculation.

1.4.2.2 Absorption

In the transmission function of Eq. 1.6, a specimen is considered to be a phase object and no absorption is included. However, for thick crystals, it is necessary to include the effect of absorption in the simulation. The effect of absorption can be represented as an *absorption function*,

$$\exp\left(-\mu(x, y)\Delta z\right) \quad (1.39)$$

and thus the transmission function can generally be represented as

$$q(x, y) = \exp\left(i\sigma\varphi(x, y)\Delta z - \mu(x, y)\Delta z\right). \quad (1.40)$$

Usually, the introduction of the absorption function does not affect largely the contrast of stimulated images.

1.4.2.3 Inclination of Incident Electron Beam

If the incident electron beam is slightly tilted against the crystal axis by (α_x, α_y), the effect of the inclination of the incident electron beam can be represented by the following term on the propagation function:

$$\exp\left(2\pi i\Delta z(u\cdot\tan\alpha_x + v\cdot\tan\alpha_y)\right). \quad (1.41)$$

1.4.2.4 Ionicity of Atoms

Usually, scattering factors calculated for neutral atoms and available in published tables [18, 19] are used for computer simulations of high-resolution images, but for specimens with strong ionicity, the scattering factors for ionized atoms should be used. However, even if the ionicity is

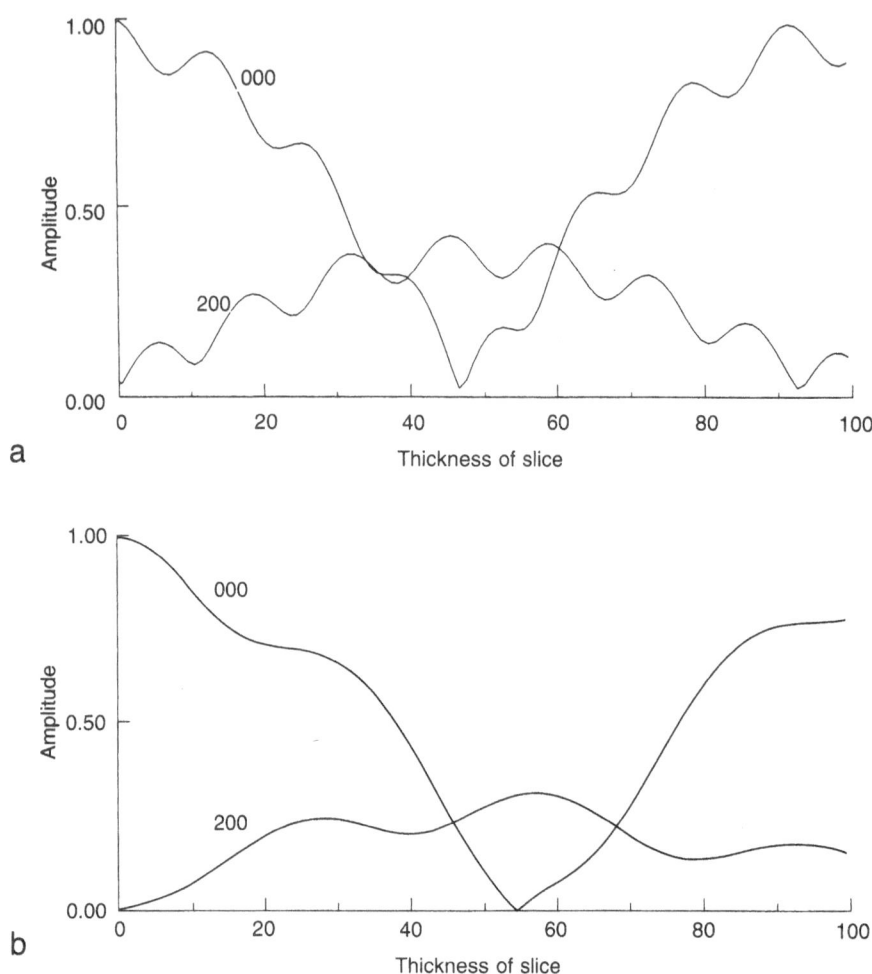

Fig. 1.12. Calculated amplitudes in **a** Au and **b** TiO$_2$ as a function of crystal thickness

Table 1.2. Parameters used for the calculations for Fig. 1.12

	Lattice constant	Space group	One slice thickness (nm)	Accelerating voltage	Temperature factor	Beam direction	Number of beams
Au	$a = b = c = 0.408\,nm^a$	Fm3m	0.408	1 MV	$B = 0.6$	[001]	16×16
TiO$_2$	$a = b = 0.45937\,nm$ $c = 0.29581\,nm$	P4$_2$/mnm	0.29581	200 kV	$B = 0.0$	[001]	64×64

^{a} Unit cell lengths a', b' of Au in projection were considered to be: $a' = a/2 = 0.204\,nm$; $b' = b/2 = 0.204\,nm$.

taken into account, the deviation from scattering factors calculated for neutral atoms is only pronounced at very small scattering angles, and thus no appreciable effect due to the ionicity is observed except for a crystal with a large unit cell.

1.4.3 Checking Programs

Image simulation should be carried out accurately, especially for crystals containing structural defects. Programs of the image simulation can be checked by paying attention to the points listed below.

1.4.3.1 Number of Sampling Points

If the number of diffracted beams is too small, the dynamical diffraction effect cannot be evaluated accurately, and any error in calculation becomes large with an increase in crystal thickness. Thus, the amplitudes of weak, high-order reflections which cannot be neglected, should be included in the calculation. A sufficient number of sampling points means that the calculated amplitudes of the transmitted and diffracted beams do not change even if the number of sampling points is increased further.

1.4.3.2 Thickness of One Slice

If each slice is thicker, the number of slices becomes smaller, resulting in shorter computing time. This can be effective, especially for a thick crystal. However, if the slices are too thick, the phase object approximation of Eq. 1.6 is not applicable. As was discussed above for sampling points, an appropriate choice of thickness for each slice means that the calculated values of the scattering amplitudes do not change even if the thickness of each slice is decreased. Assuming a 200 kV electron microscope, about 0.4 nm is appropriate for the thickness of one slice, except for materials consisting of heavy elements with a high density. According to the discussion in Sect. 1.3.1, the interaction constant becomes smaller with an increase in the accelerating voltage (see Fig. 1.4), and thus each slice may be thicker for a higher accelerating voltage. On the other hand, in a material consisting of light elements with low density, the inner potential becomes smaller and again each slice may be thicker.

1.4.3.3 Mean Inner Potential

As indicated in the flow chart in Fig. 1.10, a structure factor and the projected potential are calculated at the beginning of the computer simulation. At this stage, confirmation of the calculated value of the mean inner potential is useful. For example, the mean inner potential observed for the standard materials listed in Table 1.1 can be compared with a calculated value.

1.4.3.4 Comparison of Scattering Amplitudes

Published data for the amplitudes of diffracted beams calculated for standard materials can be compared with the results of calculations performed with the same input parameters. Figure 1.12 shows the amplitudes of the transmitted (000) and diffracted (200) beams for Au and TiO_2 as a function of crystal thickness. The parameters used in the calculation are given in Table 1.2.

References

1. Keller M (1961) Z Phys 164:274
2. Buhl R (1959) Z Phys 155:395
3. Hoffmann H, Jönsson C (1965) Z Phys 182:360
4. Gaukler KH, Schwarzer R (1971) Optik 33:215
5. Scherzer O (1949) J Appl Phys 20:20
6. Spence JCH (1981) Experimental high-resolution electron microscopy. Clarendon Press, Oxford
7. Fejes PL (1977) Acta Crystallogr A33:109
8. Frank J (1973) Optik 38:519
9. Darwin CG (1914) Philos Mag 27:315
10. Howie A, Whelan MJ (1961) Proc R Soc A263:217
11. Bethe H (1928) Ann Phys 87:55
12. Fujimoto F (1959) J Phys Soc Jpn 14:1558
13. Cowley JM, Moodie AF (1957) Acta Crystallogr 10:609
14. Cowley JM (1981) Diffraction physics, 2nd rev edn. North—Holland, Amsterdam
15. Fujiwara K (1959) J Phys Soc Jpn 14:1513
16. Ishizuka K (1980) Ultramicroscopy 5:315
17. Ishizuka K, Uyeda N (1977) Acta Crystallogr A33:740
18. Doyle PA, Turner PS (1968) Acta Crystallogr A24:390
19. International tables for X-ray crystallography (1974) vol IV. Kynoch Press, Birmingham, UK

2. Practice of High-Resolution Electron Microscopy

2.1 Classification of High-Resolution Images

As mentioned in Chap. 1, high-resolution images are the phase contrast formed by the interference of the transmitted and diffracted beams passing through the back focal plane, so various high-resolution images containing different information may be obtained depending on their scattering amplitudes, which are affected by the diffraction conditions and the crystal thicknesses. Thus, before observing high-resolution images, we should clarify what structural information we need and what experimental conditions we should set up to get such information.

High-resolution images are classified into five groups:

1. lattice fringes;
2. one-dimensional structure images;
3. two-dimensional lattice images (showing the structural information at unit cell scale);
4. two-dimensional structure images (or crystal structure images, showing the structural information at atomic scale);
5. special images.

These images result from different experimental conditions, such as the diffraction conditions and specimen thicknesses. Although the names of lattice fringes, lattice images, and structure images have not yet been defined exactly, the meaning of these names should be clear from the following explanations.

2.1.1 Lattice Fringes

Lattice fringes can be observed through the interference of two waves passing through the backfocal plane, which are sometimes limited by an objective aperture or by the resolution of the electron microscope. To observe lattice fringes there is no need to keep the incident beam exactly parallel to definite lattice planes. In fact, in observa-

tions of small crystal grains and precipitates, it is almost impossible to make the incident beam parallel to a definite crystallographic orientation for every crystal. For example, an electron diffraction pattern of a specimen including small crystals consists of *Debye–Scherrer rings*, as shown in Fig. 2.1c. The Debye–Scherrer rings are formed by many diffraction spots from small crystals with various orientations, and so it is impossible to observe high-resolution images of those small crystals by fixing the appropriate diffraction conditions. Thus, for this type of specimen, high-resolution images are generally observed without carefully controlled diffraction conditions. In high-resolution images, lattice fringes produced by the interference of the transmitted and diffracted beams are observed only in crystals where the lattice spacing is larger than the resolution of the electron microscope and the reflection corresponding to the lattice spacing is excited.

Since this type of image can be observed in a wide range of specimen thicknesses and defocus values, they can be obtained very easily. Lattice fringes are sharp or faint depending on the diffraction conditions of the crystal. Since the diffraction conditions can not be specified, it is hard to obtain structural information from the observed images even with the aid of computer simulation. However, this type of image is useful for investigating the morphology and shapes of micro-crystals, and some information about crystal structure can be obtained from the Debye–Scherrer rings in the diffraction patterns and distances of lattice fringes.

For example, Fig. 2.1 shows the lattice fringes [1] of a soft magnetic material called "FINEMET" [2]. Figure 2.1a is an image of the amorphous state in a melt-quenched specimen, and a "mazy pattern" or "mazy contrast" can be seen which is a characteristic of amorphous materials. Figure 2.1b and c are respectively a high-resolution electron micrograph and an electron diffraction pattern of the FINEMET prepared by annealing the amorphous specimen at 500°C for 1 h. The diffraction pattern shows that the FINEMET consists of

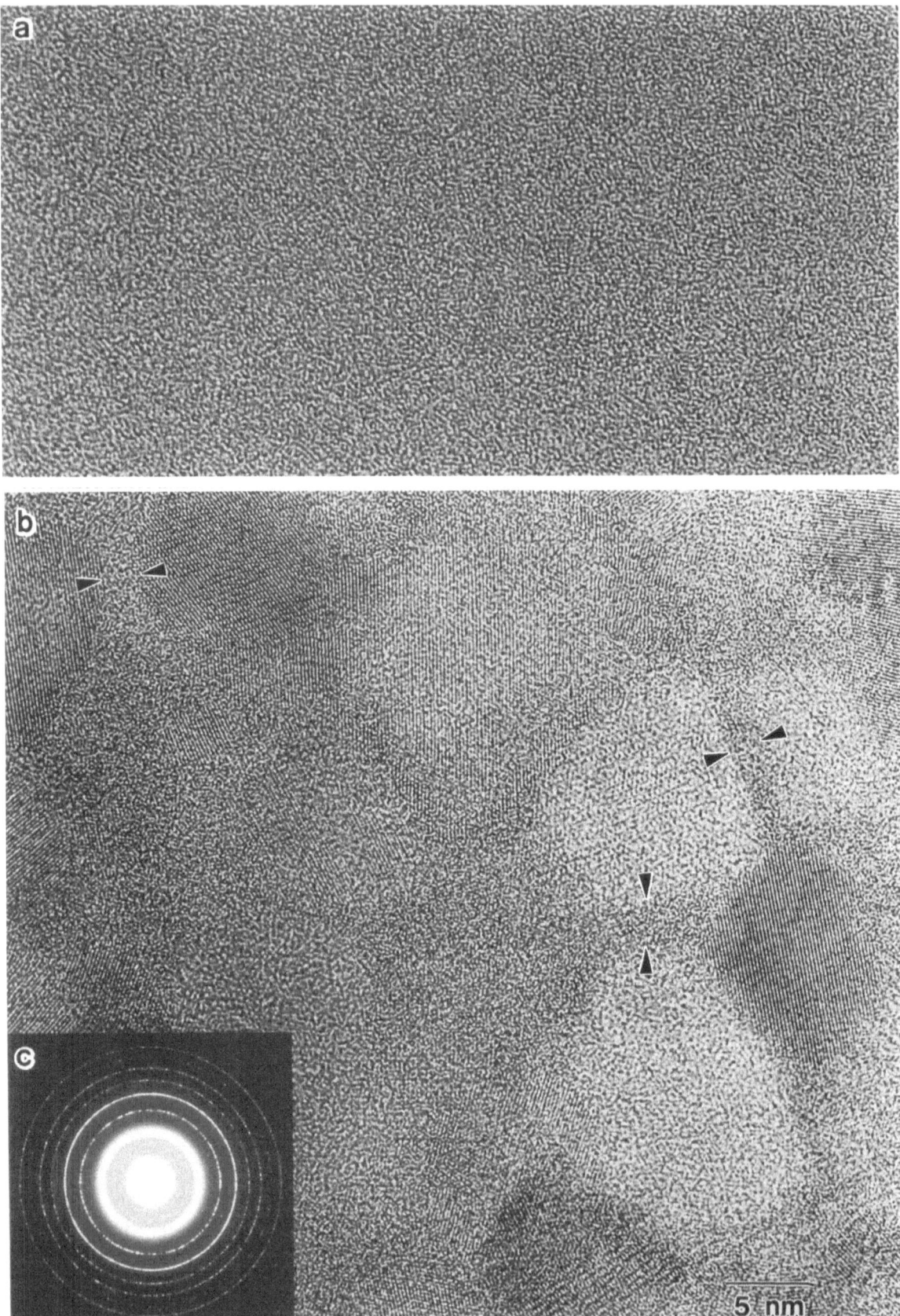

Fig. 2.1. Lattice fringes of the soft magnetic material FINEMET [1]. **a** High-resolution image of the amorphous state in a melt-quenched specimen. **b,c** Lattice fringes and an electron diffraction pattern of FINEMET prepared by annealing the amorphous specimen at 500°C for 1 h.

Specimen: $Fe_{73.5}CuNb_3Si_{13.5}B_9$; **Preparation**: ion milling; **Observation**: 400 kV EM

small crystals of the bcc structure. In the high-resolution micrograph, it is possible to see lattice fringes produced by the interference of the transmitted beam and 110-type reflections in some micro-crystals. The micro-crystals have a spherical shape and are surrounded by amorphous layers with the mazy contrast, as indicated by the arrowheads. The existence of the amorphous layers can also be confirmed by a broad *halo ring* (contributed from the amorphous region) superimposed on the sharp 110 diffraction ring (contributed from the micro-crystals). In the high-resolution image, the regions showing clear lattice fringes are crystals with strongly excited 110-type reflections, and the faint lattice fringes show crystals with weakly excited 110-type reflections. The crystals without lattice fringes have no suitable crystallographic orientations to produce the 110-type reflections, but they are distinguished by bright smooth contrast from the amorphous layers with the mazy contrast. The bright contrast results from the transmitted beam, which is not weakened due to weak scattering into diffracted beams.

2.1.2 One-Dimensional Structure Images

When a crystal is tilted to make the incident beam parallel to definite lattice planes in the crystal, a diffraction pattern with a *systematic condition* (a one-dimensional diffraction pattern with symmetry about the origin) can be obtained. High-resolution images taken with an optimum defocus value using this diffraction condition are different from lattice fringes, and include some information about crystal structures despite the one-dimensional image contrast. That is, you can obtain information about crystal structures by comparing observed images with calculated ones.

Figure 2.2 is a *one-dimensional structure image* of a high-Tc superconducting oxide in a Bi–Sr–Ca–Cu system [3]. In the image, the bright thin lines correspond to Cu–O layers perpendicular to the c-axis, and from their number and distribution, the sequence of Cu–O planes stacked along the c-axis can be seen. That is, a feature of the disordered stacking sequence of the Cu–O planes can be seen, as shown in Fig. 2.2c. Hence, this type of image is useful to visualize stacking sequences of layers in complex layer structures. It is also easier to adjust a systematic diffraction condition to obtain one-dimensional images than to obtain two-dimensional images.

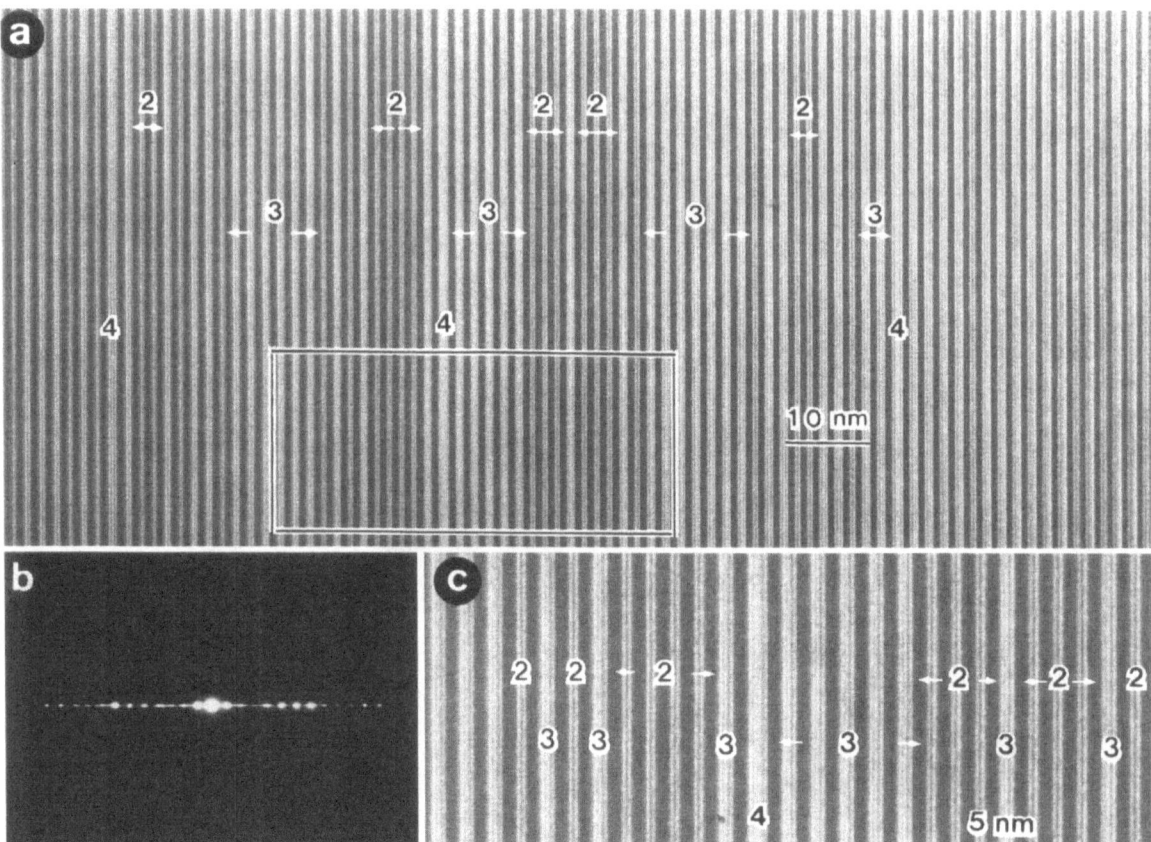

Fig. 2.2. a One-dimensional structure image of a Bi-based superconductor [3]. **b** Electron diffraction pattern. **c** Enlarged micrograph of the framed region in **a**.
Specimen: Bi–Sr–Ca–Cu–O; **Preparation**: ion milling; **Observation**: 400 kV EM.
Remark: The *numbers* in **a** and **c** indicate the number of Cu–O layers

2.1.3 Two-Dimensional Lattice Images

With an incident beam parallel to a crystallo-graphic axis, a diffraction pattern satisfying a two-dimensional diffraction condition can be obtained, as shown in Fig. 2.3. In this type of diffraction pattern, the diffraction spots around the origin (transmitted beam) reflect the size of a unit cell, and so the images produced by the interference between the transmitted beam and the diffraction spots are two-dimensional lattice fringes showing the unit cell size. This type of image includes information about crystal structures on a unit cell scale but not on an atomic scale (atomic arrangement in a unit cell), so they are called *lattice images*.

Figure 2.4 shows changes in amplitudes of 111 and 002 reflections plotted against specimen thickness in the electron diffraction of a Si crystal with [1$\bar{1}$0] incidence. The amplitude of reflection 002, which is caused by the double diffraction due to the dynamical effect (see Sect. 4.3.2), gradually increases with increasing specimen thickness, but the amplitudes of the transmitted beam 000 and the diffracted wave 111 show almost periodic changes with increasing specimen thickness. Figure 2.5 shows the thickness dependency of lattice images of a Si crystal with incidence [0$\bar{1}$1], calculated by computer simulation. Reversed contrasts of bright and dark spots can be seen, but similar images appear periodically with increasing speci-

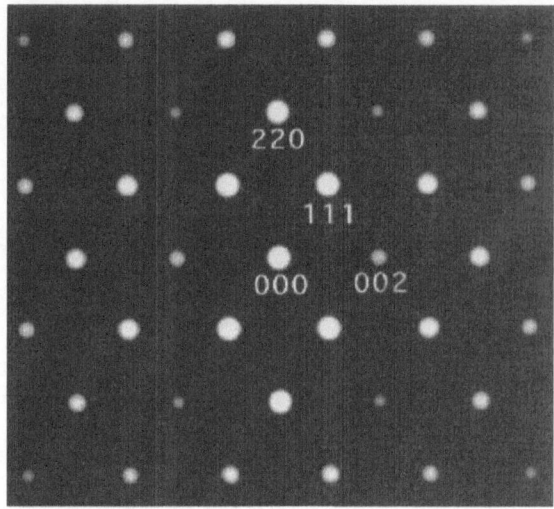

Fig. 2.3. Simulated electron diffraction pattern of a Si crystal with [1$\bar{1}$0] incidence

men thickness. For example, atom positions (dumbbells consisting of two Si atoms) appear as dark spots in Fig. 2.5a and b, while they appear as bright spots in Fig. 2.5e and f. In Fig. 2.5i–k, o and p the atom positions also become bright, and in Fig. 2.5g and m they are dark. Thus, in high-resolution images taken with the transmitted beam and diffracted beams situated around the origin, similar lattice images up to relatively thick regions (a few tens of nanometers) can be seen,

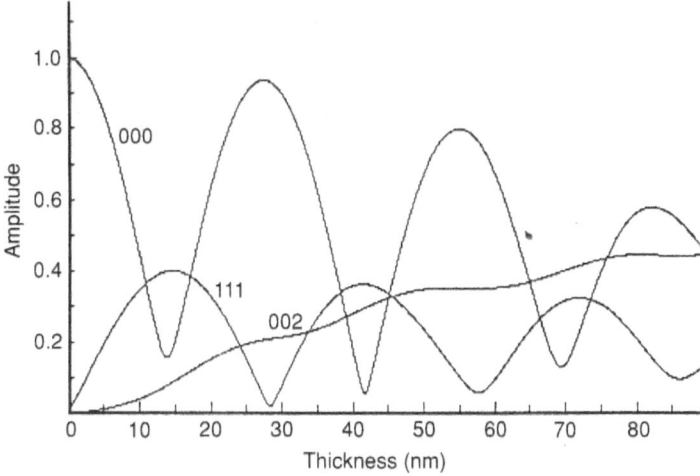

Fig. 2.4. Calculated amplitudes of the transmitted beam, and 111 and 002 reflections against the specimen thickness in the electron diffraction of a Si crystal with [1$\bar{1}$0] incidence. The 002 reflection is a forbidden reflection and appears because of the double diffraction. A 200 kV electron microscope is assumed

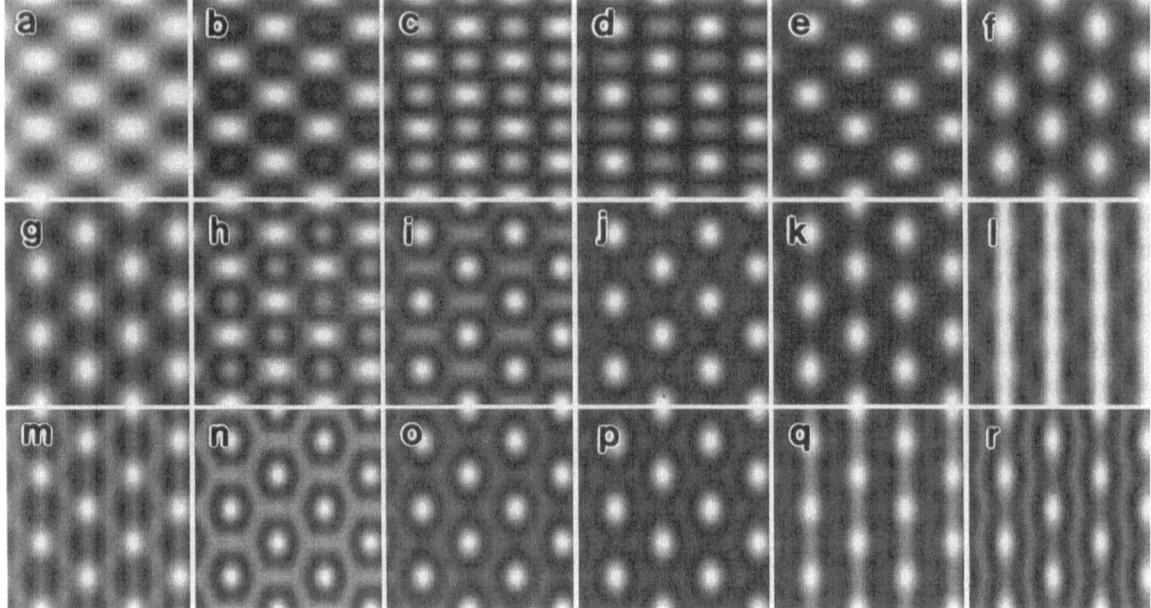

Fig. 2.5. Thickness dependency of lattice images of a Si crystal with incidence [0$\bar{1}$1]. Use of a 200 kV EM and a defocus value of 65 nm are assumed. The thickness changes from **a** 1 nm to **r** 86 nm in steps of 5 nm

although the images sometimes have reversed contrast. Therefore, this type of image is useful for investigating structure defects. Most of the high-resolution images published are this type of lattice image. From two-dimensional lattice images, it is not easy to distinguish whether the atom positions correspond to bright or dark regions. However, if well-known structures exist, e.g., stacking faults in

the observed image, then atom positions can be specified from the contrast of the image [4].

Two-dimensional lattice images can be obtained with a wide defocus region as well as the Scherzer focus, since the observations are performed using only limited diffracted beams. For example, Fig. 2.6 shows a change in the lattice images of a Si crystal; these were calculated with a specimen

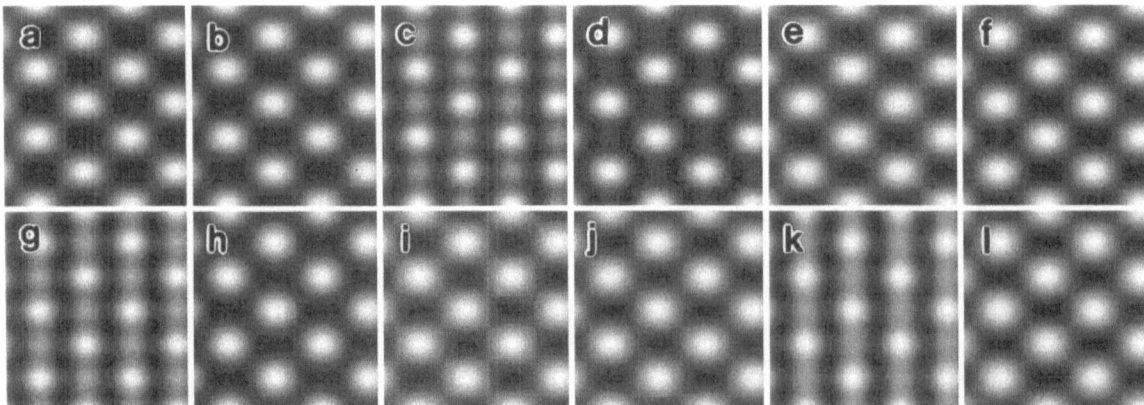

Fig. 2.6. Defocus dependency of lattice images of a Si crystal with incidence [0$\bar{1}$1]. Use of a 200 kV EM and a thickness of 6 nm are assumed. The defocus changes from −20 nm (overfocus) to 90 nm (underfocus) in steps of 10 nm

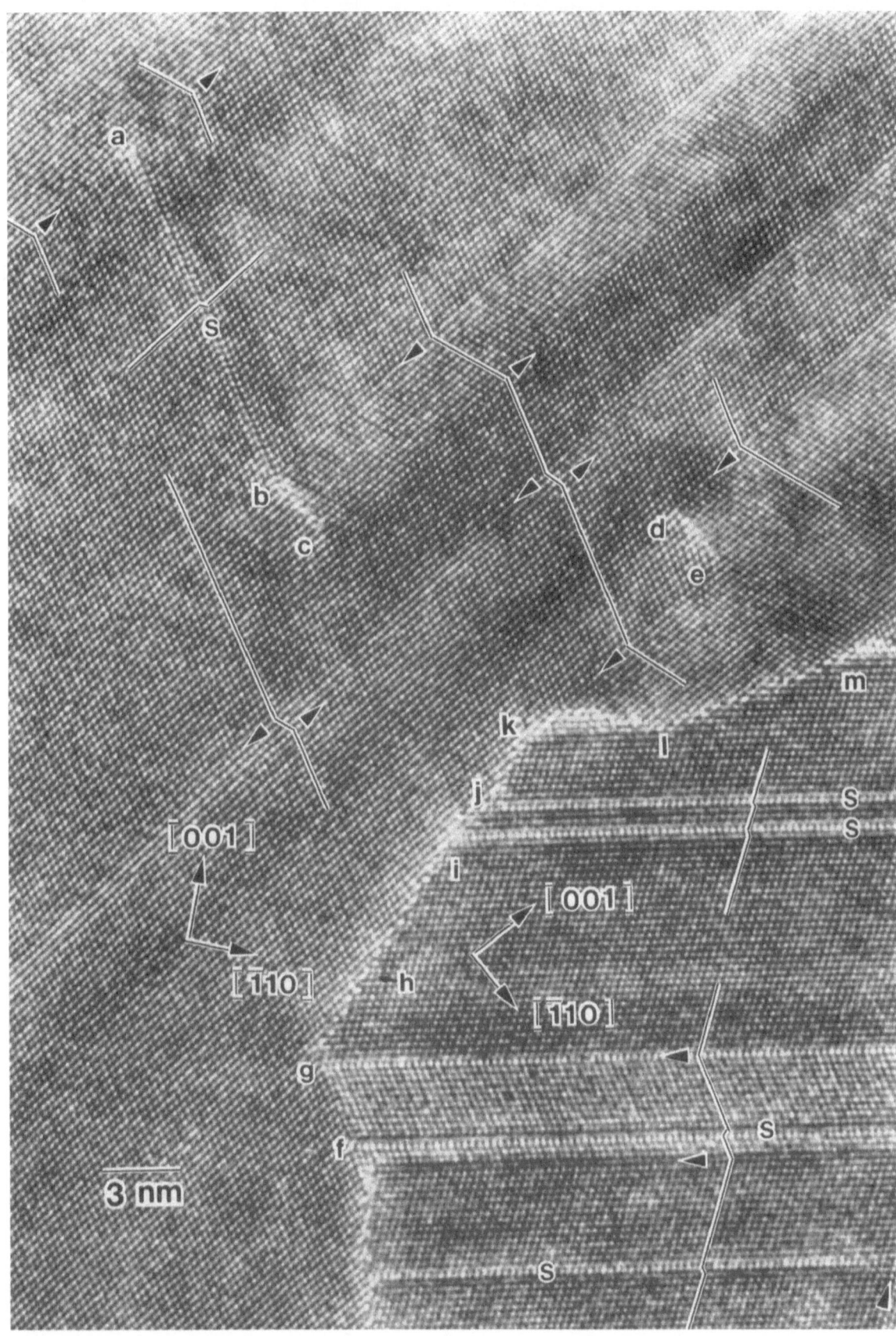

Fig. 2.7. Two-dimensional lattice image of β-SiC.
Specimen: β-SiC prepared by CVD; **Preparation**: ion milling; **Observation**: 200 kV EM, [110] incidence

thickness of 6 nm at various defocus values from −20 nm (overfocus) to 90 nm (underfocus), in steps of 10 nm. Although there are reversals in contrast, similar lattice images can clearly be seen at (a) and (b), from (d) to (f), from (h) to (j), and at (l).

As mentioned above, two-dimensional lattice images can be observed under various conditions, such as different defocus values and crystal thicknesses, and so their observation is much easier than that of structure images, as discussed below. However, if it is the structure of defects in the lattice images which is being studied, the images should be taken under optimum defocus conditions (around the Scherzer focus) and also using thin specimens. It should be noted that in images taken under inadequate conditions, the image contrast of defects is deformed, and it is difficult to get accurate information concerning defect structures from such images.

Figure 2.7 shows a lattice image of β-SiC taken with the incident beam parallel to the [110] axis [5]. The image is formed with 002 and $1\bar{1}1$-type reflections in addition to the transmitted beam. In the image, it is possible to recognize many types of defects, such as tilt grain boundaries from f to m (see Sect. 3.1.2), twin boundaries (arrowheads), stacking faults (s), and dislocations (b–c, d–e).

2.1.4 Two-Dimensional Structure Images

Figure 2.8 is an electron diffraction pattern obtained through computer simulation on β-Si_3N_4 with the incident beam parallel to the c-axis. An objective aperture corresponding to a resolution of 0.17 nm for a 400 kV electron microscope is indicated by the white circle. When high-resolution images are observed with this type of electron diffraction pattern, the greater the number of reflections used for the image formation, the greater the amount of information which can be obtained from the observed image. However, reflections at larger angles above the resolution limit of an electron microscope do not make an accurate contribution to image formation, and mainly result in the background contrast in the image. Thus, by imaging with as many reflections as possible in the range of the resolution, the finest image, containing a great deal of correct information about atomic arrangements inside a unit cell, can be observed.

Figure 2.9a and b are structure images taken with the incident beam parallel to the c-axes of β-Si_3N_4 and α-Si_3N_4. Calculated images with and without atomic arrangements have been inserted into these images. As can be seen from the calculated images, atom positions and channels without atoms are represented in the images as dark and bright regions, respectively. This type of image, which represents atom positions and channels as dark and bright regions, is called a *two-dimensional structure image* (or *structure image*) to distinguish it from two-dimensional lattice images.

Structure images can be obtained only from thin specimens, where the amplitudes of the diffracted beams forming the images are in proportion to the crystal thickness. They cannot be observed in thick specimens, where the amplitudes of the diffracted beams change randomly. We now consider the conditions for observing structure images through computer simulation.

Figure 2.10 shows the thickness dependency of the amplitudes of diffracted beams which contribute to the formation of the structure image of β-Si_3N_4. The amplitudes of all the reflections are approximately proportional to the thickness up to about 8 nm, but at thicknesses of over 10 nm, which is where the amplitudes of some reflections have maxima, the proportional relations in the amplitudes of these reflections break down and do not appear again. It should also be noted that the

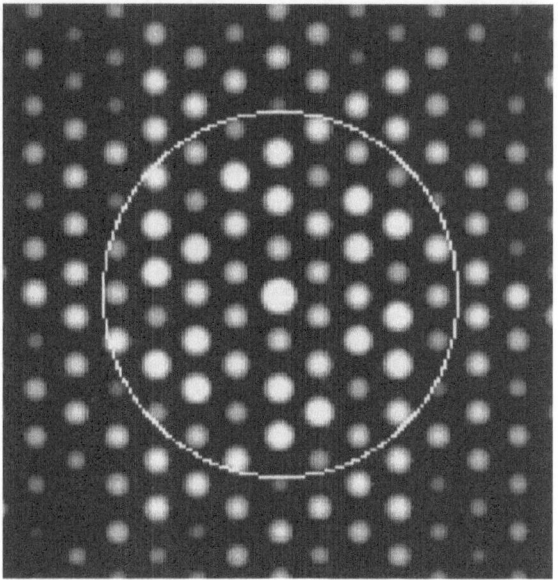

Fig. 2.8. Simulated electron diffraction pattern of β-Si_3N_4 with [001] incidence. An objective aperture corresponding to a resolution of 0.17 nm for a 400 kV EM is indicated by the white circle

phases of the reflections change drastically at a thickness corresponding to the maximum amplitudes, and no images showing a crystal structure are formed in the region over this thickness. This change can clearly be seen in the calculated images in Fig. 2.11, where the thickness dependency of high-resolution images is shown. The simulation shows that the structure images are formed up to a thickness of 7 nm, but that images at thicknesses over 9 nm are considerably deformed. This means that the structure images can be observed up to the thickness where the amplitude of the strongest reflection has a maximum value.

Since structure images are formed by many reflections, the defocus values for observing structure images are limited in the range around the Scherzer focus. Figure 2.12 shows calculated high-resolution images of β-Si_3N_4 for various defocus values. It can be seen that the structure image in Fig. 2.9a appears only in the narrow range from 30 to 50 nm around the Scherzer focus (48.7 nm). It should be noted that the images with reversed contrast are formed in the overfocus region from -40 to -20 nm.

Structure images are observed in relatively thick regions for low-density materials consisting of light atoms. However, in materials with similar densities, structure images are seen in thicker re-

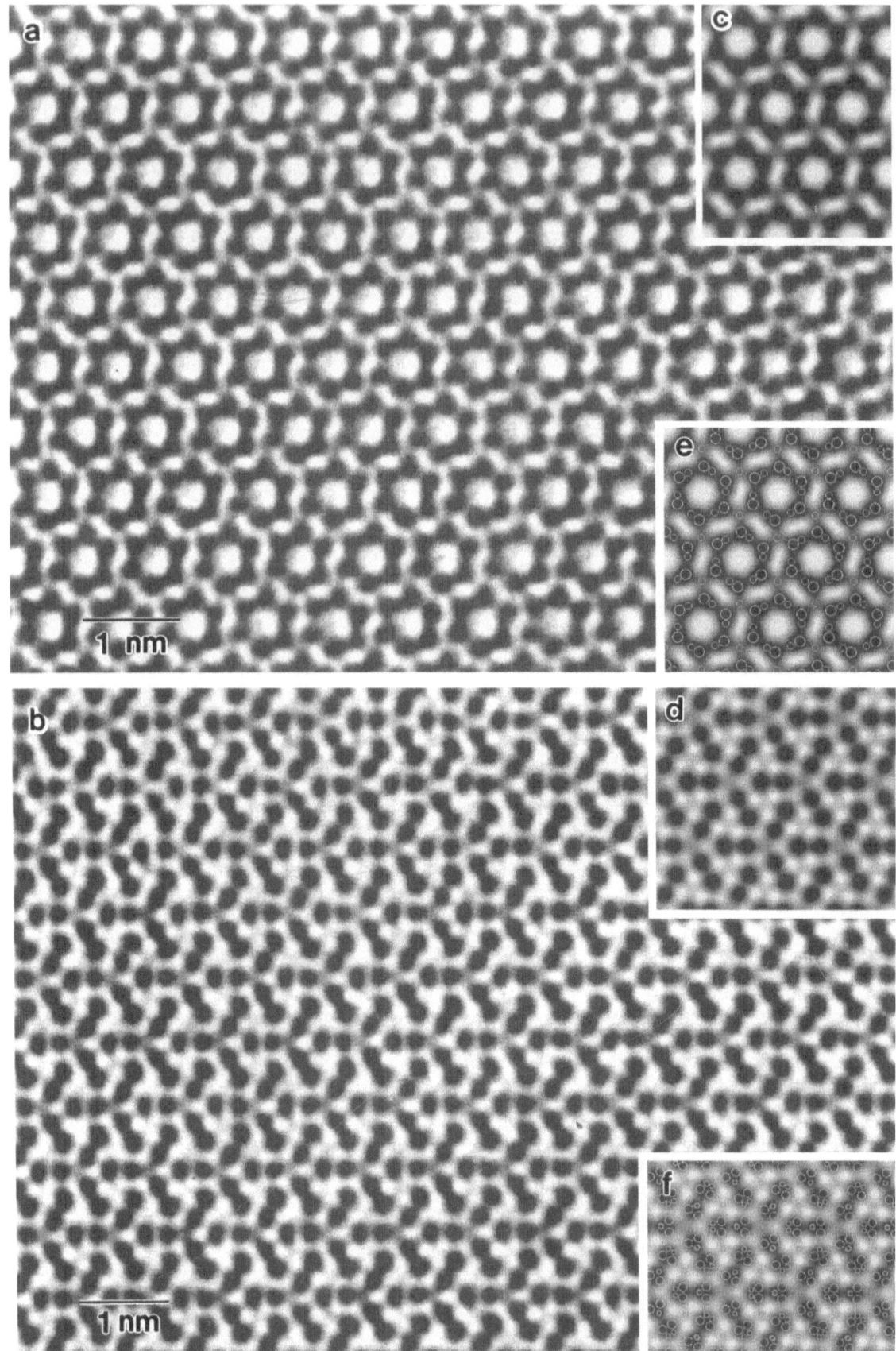

Fig. 2.9. Structure images of **a** β-Si$_3$N$_4$ and **b** α-Si$_3$N$_4$. **c** and **e** are simulated images of β-Si$_3$N$_4$ without and with its atomic arrangement, respectively. **d** and **f** are simulated images of α-Si$_3$N$_4$. In the simulation, a 400 kV EM, a defocus value of 45 nm, and a thickness 3 nm are assumed.

Specimen: β-Si$_3$N$_4$ and α-Si$_3$N$_4$; **Preparation**: ion milling; **Observation**: 400 kV EM, [001] incidence

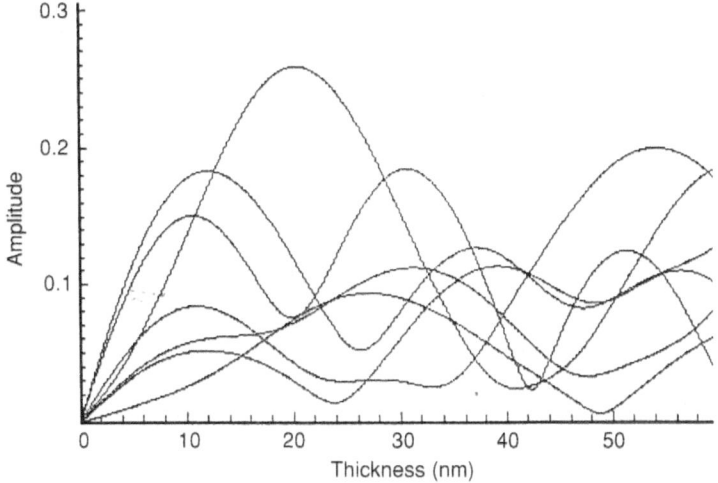

Fig. 2.10. Thickness dependency of amplitudes of diffracted beams in β-Si$_3$N$_4$. A 400 kV EM is assumed

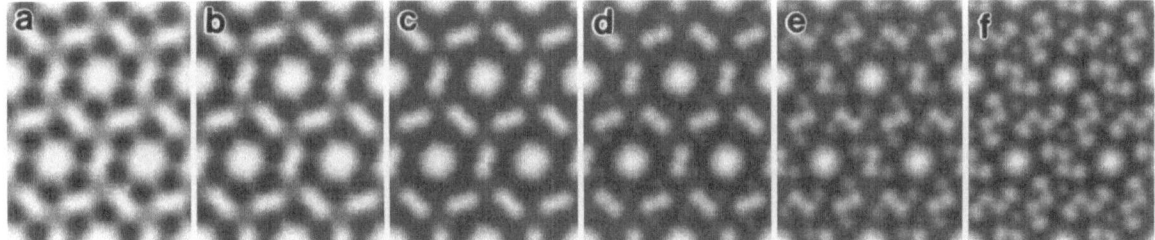

Fig. 2.11. Thickness dependency of high-resolution images of β-Si$_3$N$_4$ with incidence [001]. Use of a 400 kV EM and a defocus value of 45 nm are assumed. The thickness changes from **a** 1 nm to **f** 11 nm in steps of 2 nm

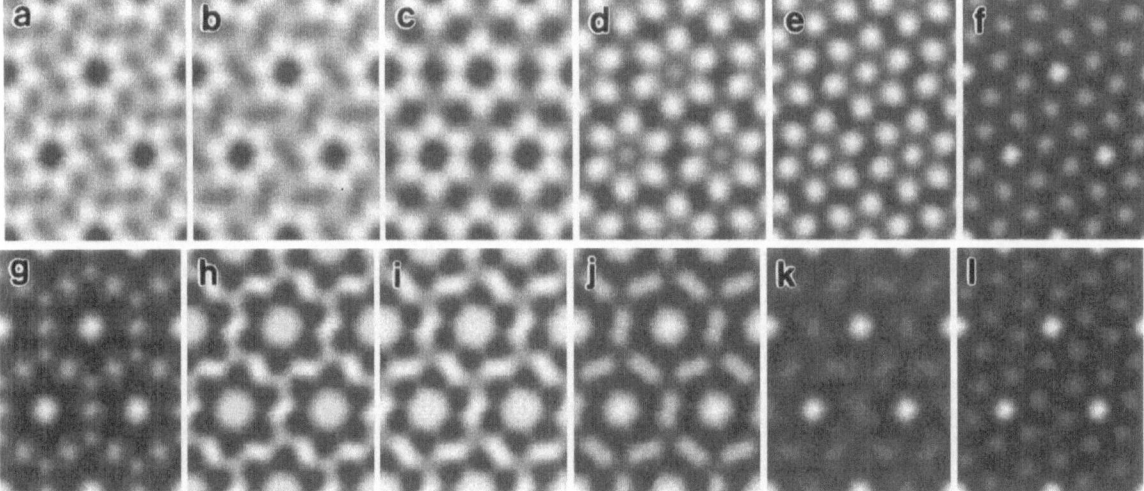

Fig. 2.12. Defocus dependency of high-resolution images of β-Si$_3$N$_4$ with incidence [001]. Use of a 400 kV EM and a thickness of 3 nm are assumed. The defocus changes from −40 nm (overfocus) to 70 nm (underfocus) in steps of 10 nm

gions for structures with a large unit cell having many reflections around the origin but with no particularly strong reflections. For example, in Si_3N_4, the structure image of α-Si_3N_4 can be observed up to a thickness which is twice that for β-Si_3N_4 [6]. In metallic alloys of high density, for which observing structure images is considered to be very difficult, structure images were observed in quasicrystals which produce many reflections around the origin (e.g., see Fig. 3.65a).

2.1.5 Special Images

In high-resolution images taken with specific reflections selected by an objective aperture, there are peculiar images which enhance the specific image contrast corresponding to certain structural information. A typical example of these images is the so-called *superstructure image* [7] (see Sect. 3.2.3). For example, if you take a high-resolution image with a transmission beam and *superlattice reflections* resulting from the ordering of the constituent elements, it is possible to obtain a special image reflecting the ordered arrangement of atoms. This characteristic feature of superstructure images will now be examined using computer simulations.

Figure 2.13a shows an atomic arrangement of an Au_3Cd ordered structure, projected along the [100] axis. The Au_3Cd structure is based on the fcc structure, and Cd and Au atoms form an ordered arrangement in a unit cell composed of four unit cells of the fcc structure. Consequently, in electron diffraction patterns of the Au_3Cd structure (Fig. 2.13b), weak superlattice reflections, which can be indexed by 011, 004, and so on, appear together with strong fundamental reflections which result from the fundamental fcc structure. Figure 2.14 shows a change in image contrast of calculated

high-resolution images as a function of specimen thickness. These images were formed with a transmitted beam and superlattice reflections using the aperture indicated in Fig. 2.13b. In most of the images, Cd atom positions appear as dark or bright spots. For example, the Cd positions are bright spots in images 2.14b and 2.14d–j, while they are dark spots in images 2.14q–u. If this type of image is observed, it is possible to determine an arrangement of Cd atoms directly from the positions of bright or dark spots. Although there is no information about the arrangement of Au atoms in the observed images, the arrangement of Cd and Au atoms can be determined unequivocally because of prior knowledge of the fundamental structure of the alloy, i.e., the fcc structure.

In addition to observations made by selecting the reflections with an objective aperture, superstructure images can also be observed in relatively thick regions (a few tens of nanometers) with strongly excited superlattice reflections whose intensities are not weaker than those of *fundamental reflections* resulting from the dynamical diffraction effect. Since this type of image can be observed from relatively thick regions and at relatively low resolution, they have been widely used for structure analysis of complex ordered structures [8–11] (see Sect. 3.2.3).

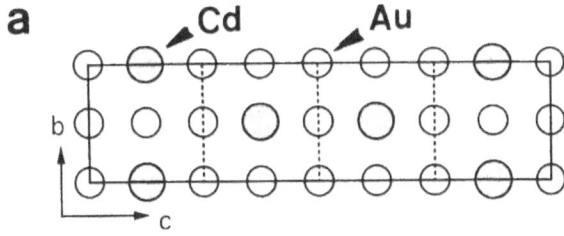

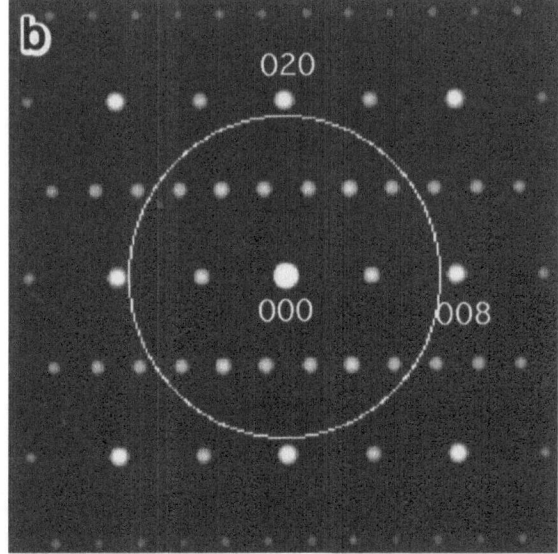

Fig. 2.13. a Atomic arrangement of an Au_3Cd ordered structure. **b** Simulated diffraction pattern of Au_3Cd. The white circle indicates an objective aperture for observing a superstructure image of Au_3Cd

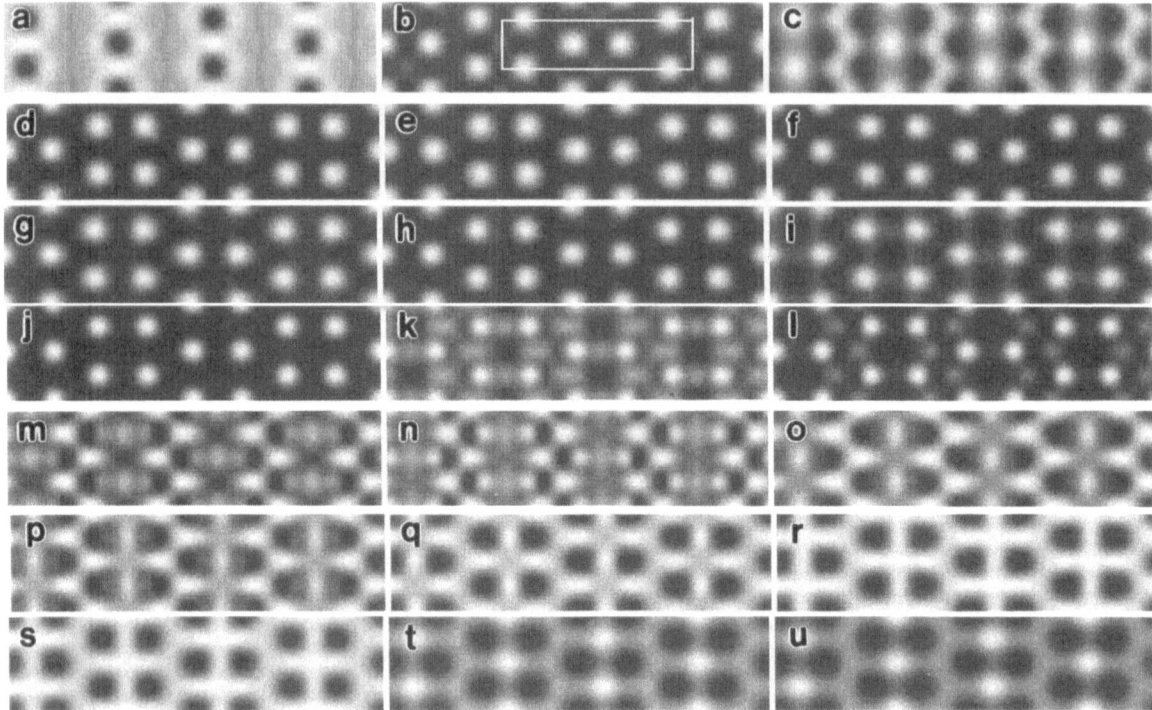

Fig. 2.14. Thickness dependency of high-resolution images of Au_3Cd with [100] incidence. Use of a 400 kV EM and a defocus value of 45 nm are assumed. The thickness changes from **a** 2 nm to **u** 107 nm in steps of 5 nm

2.2 Practice in Observing High-Resolution Images

2.2.1 Points to Note Before Observation

2.2.1.1 Selection of Imaging Conditions for Each Purpose

As mentioned above, a variety of high-resolution images containing different information can be observed, depending on the diffraction conditions, the crystal thickness, the resolution of the electron microscope, and so on. Therefore, before observing high-resolution images, it is necessary to confirm which kind of images are needed to get the information required. The high-resolution images classified above become difficult to observe in order except special images. It should also be noted that images at a higher resolution than is necessary are not only difficult to observe, but sometimes are also difficult to interpret.

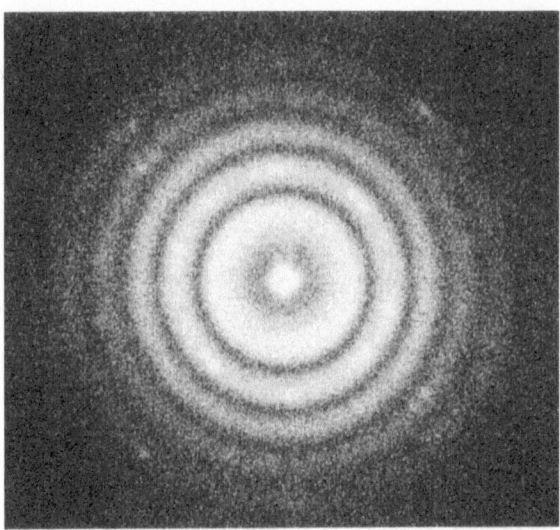

Fig. 2.15. Optical diffractogram of an observed high-resolution image of an amorphous film. The white dots in the diffraction pattern are diffraction spots from small Au particles which can be used for calibration

2.2.1.2 Checks of the Characteristics and Problems of Electron Microscopes

The resolution of an electron microscope depends on the spherical aberration of the objective lens, the chromatic aberration resulting from fluctuations of both an accelerating voltage and an objective lens current, and the mechanical and thermal instabilities of the specimen holder. All these points, except for the spherical aberration of the objective lens, should always be checked by the user. If any faults are found, it is important to contact the manufacturer immediately. The way to check these points is explained below.

To Check the Quality of High-Resolution Images. The condition of an electron microscope can be confirmed by examining the quality of a high-resolution image of an amorphous film. The stability (particularly the alternating current component) of the accelerating voltage and the objective lens current can be assessed from the fineness and sharpness of the image contrast. This stability can also be measured with an electrical instrument, or quantitatively evaluated from a digital or optical diffractogram of an observed high-resolution image of an amorphous film (Fig. 2.15).

To Check Focus Drift. Focus drift (the phenomenon of the focus changing gradually with time) can be checked by observing the mazy contrast in a high-resolution image of an amorphous film, or

the Fresnel fringes at the edge of a specimen; the direct current component of the stability of the accelerating voltage and the objective lens current can then be seen. It is very important to know how long it takes to obtain a stable state and reach the optimum condition after turning on the accelerating voltage and lens current. In particular, it should be noted that it takes a very long time to get to the stable state after turning on the lens current.[1] Fluctuations of temperature due to an air conditioner or the water which is cooling the lenses cause drifting of the focus and the specimen. Therefore, it is very important to keep the room temperature constant, and to protect an electron microscope from the wind from an air conditioner. A high-powered air conditioner produces quick changes in room temperature, especially just after it switches on or off, and this often causes focus drift. Also, the water which is cooling the lenses must be kept at a constant temperature with little fluctuation.

To Check Specimen Drift. It is very important to know how long it takes to reach a stable state after inserting a specimen folder and replenishing the anti-contaminant liquid nitrogen in the cold trap. It should also be noted that it takes some time to

---

[1] It takes a few hours to stabilize the objective lens, and so this current is usually left on.

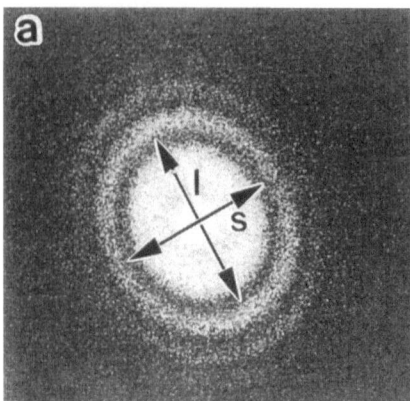 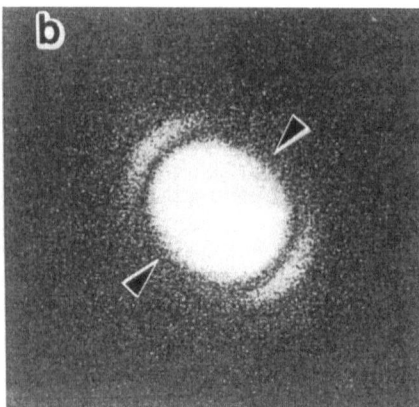

Fig. 2.16. Optical diffractograms of observed high-resolution images of an amorphous film with **a** astigmatism and **b** specimen drift (observed just after moving a specimen)

reach a stable state after moving a specimen and tilting the goniometer. Make frequent checks of how long is necessary to wait before taking images after moving a specimen by observing an image at high magnification. Figure 2.16b is an optical diffractogram of an observed high-resolution image of an amorphous film, taken just after moving a specimen. The disappearance of parts of the bright ring on the lower left and upper right is caused by specimen drift. If the image is taken after waiting for a few minutes, a diffractogram like Fig. 2.15, without specimen drift, can be obtained.

Identification of Optimum Defocus. It is necessary to know the optimum defocus of an objective lens (Scherzer focus) as well as the resolution of the electron microscope. This is known through the spherical aberration coefficient C_s, which is available from the manufacturer (see Sect. 1.3.2).

2.2.2 Points to Note During Observation

After choosing a suitable type of high-resolution image, it is necessary to observe and also to understand the characteristics of the electron microscope being used. Therefore some practical advice on observing high-resolution images is given below.

2.2.2.1 Alignment of an Electron Microscope Before Use

After switching on the lens current and accelerating voltage, first align the illumination system and the voltage center. Then spend a few hours finding good regions for high-resolution observations, and letting the microscope reach the stable state for taking high-resolution images.

2.2.2.2 Finding Thin Crystals

It is most important to find thin crystals for high-resolution observations. In specimens prepared by electropolishing or ion-milling, there is a limited choice of specimen thicknesses, while for crushed specimens there is usually a wide choice of thicknesses and it is necessary to spend time finding thin crystals. If thin crystals with smooth or straight edges like a "sandy beach" are found, then half the job of observing high-resolution images is already done. Good crystals withstand electron irradiation and do not bend, and thus setting the diffraction conditions is easy.

2.2.2.3 Setting Diffraction Conditions

It is important to set appropriate diffraction conditions by observing the electron diffraction pattern (which sometimes shows a Kikuchi pattern; see Sect. 4.3.3) using a selected area aperture as small as possible. Appropriate training will enable an operator to use a specimen goniometer precisely and set the diffraction conditions very quickly. However, it is not easy to set perfect diffraction conditions by observing an electron diffraction pattern, so images should be taken from several regions, and the good images which satisfy the appropriate diffraction conditions should be selected. It is therefore important to take images of a wide region in the film with as low a magnification as possible. Whether the diffraction conditions set were good or bad can be seen from the symmetry of the observed image contrast, or can be judged from the symmetry of a digital diffractogram of the observed image.

2.2.2.4 Correction of Astigmatism

An electric lens always shows some distortion from a true circle due to manufacturing error, so a focal point is elongated in some direction. This means that the focal length differs depending on the direction. This aberration is called *astigmatism*, and can be canceled out electrically. Correction of the astigmatism is the most important job in the final alignment before taking high-resolution images. In practice, since the position of the specimen relative to the objective lens changes when the specimen is tilted, the *astigmatism correction* should be done after setting the diffraction conditions. Also, the correction should be carried out with the objective aperture which will be used for taking the high-resolution images, because the value of the astigmatism changes with the size and position of the aperture.

In general, the astigmatism correction for high-resolution observations is carried out by observing the contrast in a high-resolution image of an amorphous film in a *microgrid* (a thin plastic film reinforced with carbon) or a contamination film at the edge of the specimen. It is usually performed by observing images on a florescent screen or a TV screen, but if a digital or optical diffractogram of the observed image of an amorphous film is used, it can be carried out quantitatively, as follows. Figure 2.16a shows an optical diffractogram of an image of an amorphous film. In the image, the elliptic shape of the bright rings is caused by the astigmatism. That is, they show that defocus values along the directions of the major and minor axes of the ellipse are different. Astigmatism correction makes the ellipse into a circle by using a stigmator.

In cases where the x- and y-axes of a stigmator form an angle of $\pi/4$ in an electron microscope, the electric currents ΔI_x and ΔI_y in the stigmator are given in Eqs. 2.1 and 2.2, where l and s are the major and minor axes, respectively, of an ellipse in an optical or digital diffractogram.

$$\Delta I_x = C_x \left(\frac{1}{s^2} - \frac{1}{l^2} \right) \sin\left(\theta - \pi/4 \right) \quad (2.1)$$

$$\Delta I_y = C_y \left(\frac{1}{s^2} - \frac{1}{l^2} \right) \sin\left(\theta \right) \quad (2.2)$$

where θ (>0) is an angle between the x-axis of the stigmator and the major axis. Coefficients C_x and C_y must be determined experimentally. The relation between coefficients C_x and C_y, magnification M, and the order of the dark ring n is

$$C_x, \; C_y \propto \frac{n}{M^2}. \qquad (2.3)$$

If an electron microscope has a device (the so-called Z-controller) to adjust the specimen position along the electron beam (the z-direction), the specimen can be set at the required position by the Z-controller so that the image can be focused with a defined objective lens current. In this case, it is not necessary to make the astigmatism correction every time.

However, for observations of magnetic materials, the astigmatism correction should be carried out every time the specimen is moved or tilted, even if a Z-controller is used, because magnetic materials disturb the magnetic field of an objective lens.

2.2.2.5 Checking Specimen Drift

While the astigmatism correction is being carried out at high magnification, the specimen drift can also be checked. The effect of specimen drift is the disappearance of part of the rings along the drift direction in a digital diffractogram of an amorphous film, and it can be distinguished from the effect of the astigmatism as shown in Fig. 2.16.

2.2.2.6 Final Setting of a Diffraction Condition

After the astigmatism correction, a diffraction condition should be adjusted again before photographs are taken. Although the diffraction condition has already been set, the specimen is often tilted during astigmatism correction and other mechanical operations, so it is important to check the diffraction condition again at the very final stage.

2.2.2.7 Setting the Magnification

It is very important to take photographs at the appropriate magnification, which should be high enough to clarify the microstructure, but also as low as possible. If the magnification is set unnecessarily high, the specimen area included on the film becomes narrow, and specimen drift caused by the increase in exposure time occurs. For example, if an image is taken with a resolution of about 0.2 nm using conventional EM film, e.g., FG film, the photograph can be taken at a direct magnification

of 400 000, and then magnification by an enlarger gives a fine printed image. All photographs should be taken with an exposure time of only a few seconds.

2.2.2.8 Setting the Optimum Defocus

From the spherical aberration coefficient C_s of an objective lens, an optimum defocus value for thin specimens (Scherzer focus) can be determined (Eq. 1.19). The optimum defocus for actual specimens deviates from the Scherzer focus with increasing specimen thickness. However, even in such cases, the optimum defocus is still around

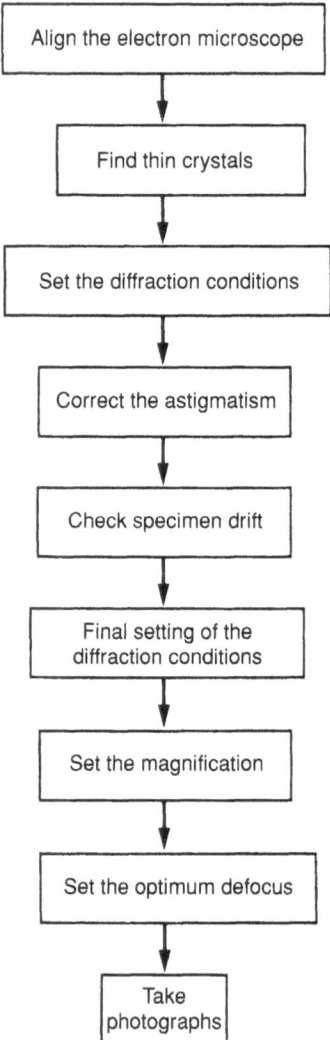

Fig. 2.17. Flow chart showing the process of high-resolution observations

the Scherzer focus value. Therefore, the "just-focused" point is first determined by observing the *Fresnel fringes* at the edge of the specimen, and then the Scherzer focus is set with the focusing knob if the value of a focus change corresponding to a step of the focusing knob is known (a definite value is available from the manufacturer). Around this defocus value, take a few images with changing focus values in steps of 5 or 10nm. If the optimum defocus is determined by observing the high-resolution image contrast of a crystal, it is easy to mistake the defocus setting and to observe images with the wrong defocus values unless the correct image contrast observed with the optimum

defocus is known. It should be noted that high-resolution images can be taken with various defocus values, but images which reflect the real crystal structure are obtained with the optimum defocus. Figure 2.17 is a flow chart showing the process of high-resolution observations, as described above.

When taking photographs, it is also important to include a specimen edge in the film, because the sources which disturb image contrast, such as an inappropriate defocus setting, astigmatism, specimen drift, crystal tilt, etc., can be perceived from the image contrast around the edge (see Sect. 2.2.3).

2.2.3 Selection of Good Images

It is difficult, but very important, to select good images from many observed images to get accurate structural information. In order to make the appropriate selection, the sources which produce deformations in observed image contrast must be known. Again, by taking photographs that include the edge of the specimen on the film, these sources can be identified from the contrast around the edge. Ways of recognizing the sources of deformed images are given below.

2.2.3.1 Astigmatism

In general, an amorphous film of contamination is put at the edge of the specimen. From the image of a contamination film, the existence of astigmatism can be seen. The effect of astigmatism appears as mazy contrast with a unidirectional blur, and this unidirectional contrast is enhanced in fine images of thin regions near the just-focused point (Fig. 2.18). From high-resolution images of crystals, it is hard to distinguish the effect of astigmatism from other effects.

2.2.3.2 Defocus

It is very important to remember the Fresnel fringes at the edge of a specimen, and the image contrast of an amorphous film at the Scherzer focus. For instance, Fig. 2.19 shows high-resolution images of an amorphous film taken with an underfocus value of about 50 nm, near the just-focused point, and with an overfocus value of about −50 nm (the Scherzer focus of the microscope was 48.7 nm). Bright fringes can be seen at the edge of the amorphous film for the underfocused image, and dark fringes for the overfocused image. It is also possible to recognize a very weak contrast of thin regions at the just-focused point. By taking images which include the edges of specimens and checking the image contrast, one can become familiar with the contrast of Fresnel fringes and the mazy contrast of an amorphous film at the Scherzer focus. It should be noted that high-resolution images of crystals with reversed contrast can be observed near the just-focused point and at the slightly overfocused point. If the edge is not included on the film, it is easy to mistake the images with reversed contrast for images observed under optimum defocus.

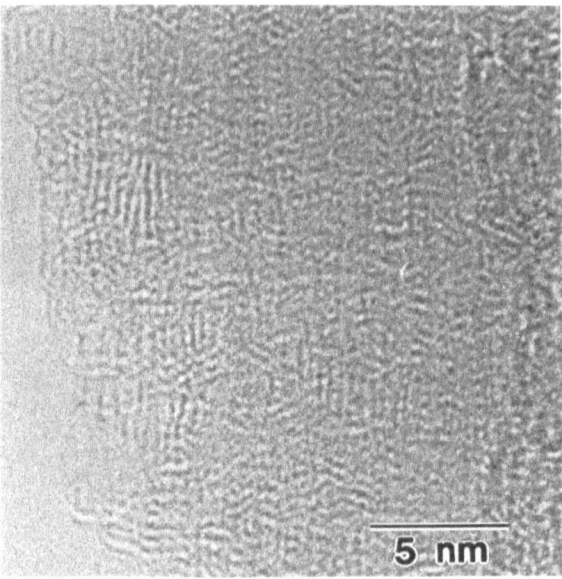

Fig. 2.18. High-resolution image of an amorphous film with astigmatism observed near the just-focused position. Mazy contrast with a unidirectional blur in the vertical direction can be seen

2.2.3.3 Specimen Thickness

In general, specimens are wedge-shaped and thus their thickness increases from the edge to the inside, and there is a change in the image contrast with increasing thickness. The deformation of images due to the misalignment of the incident electron beam against the crystal axis is also enhanced with increasing specimen thickness, so by examining the image contrast changes from the edge to the thicker region, the accuracy of the diffraction condition can be checked.

2.2.3.4 Diffraction Conditions

Even in images which are considered to have been taken with the incident beam exactly parallel to the crystal axis, in most cases the incident beam was slightly tilted. This type of misalignment of the incident beam is not recognizable in images of thin specimens, but can be clearly recognized in images of thicker regions. From structure images taken under strict conditions (for example, Figs. 1.7 and 3.44), the positions of heavy atoms can be determined with an accuracy of about 0.01 nm (see Sect. 3.2.2), but high-resolution images taken under poor conditions can give no information about the positions of atoms. The deformation of images by the misalignment of the incident beam

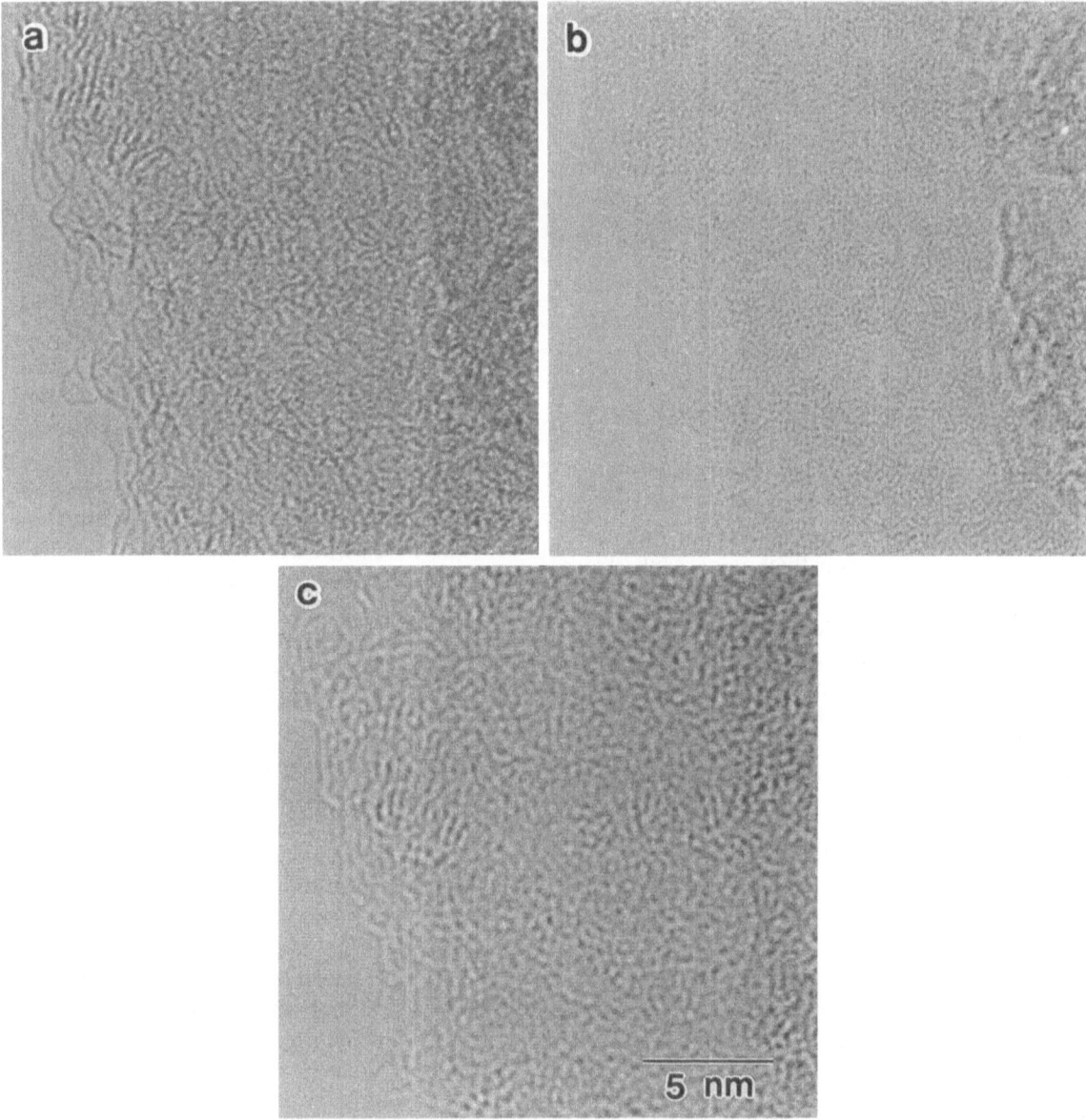

Fig. 2.19. Focus dependency of high-resolution images of an amorphous film. **a** Underfocus of about 50 nm; **b** near the just-focused position; **c** overfocus of about −50 nm

is enhanced in the image contrast of thick regions, and from the asymmetry of the image contrast, this effect can easily be recognized.

2.2.3.5 Specimen Drift

In a specimen holder of the side-entry type, and also if there are mechanical or thermal instabilities in the electron microscope, specimen drift often causes a serious problem. Before taking photographs, this should be checked. The effect of specimen drift can be seen in the movement of the image of a specimen edge.

2.2.4 Points to Note in the Interpretation of Images

Even observed images which seem to be fine have a slight contrast deformation due to imperfect experimental conditions, and thus it is always difficult to obtain accurate information about crystal structures. Some points to note when interpreting observed images are listed below.

2.2.4.1 Abnormal Contrast Around Defects

In the high-resolution image of Fig. 2.7, stacking faults (s), grain boundaries (f–m), dislocations (b–c, d–e), and twin boundaries (arrowheads) can be seen. The image shows bright contrast at the defects compared with the contrast of the perfect regions. This bright contrast, which is enhanced with increasing specimen thickness, results from the differences in the dynamical diffraction effect between the perfect crystal regions and the defect regions. This type of abnormal contrast at defects is weakened in thin regions. Multiple scattering of the electrons in the specimens results in weak reflections being enhanced and strong reflections being relatively weaker. Hence, by taking into account the fact that weak contrast in thin regions is easily enhanced by multiple scattering in thick regions, mistakes in the interpretation of observed images can be avoided.

2.2.4.2 Reversal in the Contrast of High-Resolution Images

In structure images taken from thin specimens under optimum defocus conditions, regions of high potential (atom positions) appear as dark regions, and regions of low potential (channels between atoms) appear as bright regions (see Sect. 1.3.1). However, images with reversed contrast are sometimes observed near the just-focused point and with slight overfocus values (see Fig. 2.12). Confusion in distinguishing the high-resolution images observed at the optimum defocus from other images can easily be avoided by recording images which include a specimen edge. A reversal in contrast in two-dimensional lattice images occurs periodically with increasing specimen thickness (see Fig. 2.5), and thus care should be taken when interpreting bright or dark regions of lattice images as atom positions. However, if the lattice image contains defects, and if the atomic arrangements of the defects, such as stacking faults, are known, it is sometimes possible to determine whether atom positions correspond to dark or bright regions [4].

If it is only necessary to obtain rough information just about the existence of defects and their species, the detailed interpretation of observed images may not be necessary. However, if you need accurate information about atomic arrangements and atom positions at the defects, great care should be taken for high-resolution observations and the interpretation of the observed images. By taking into account any available information about crystal structures which was obtained by other methods, such as electron diffraction or X-ray diffraction, accurate information can be derived from observed high-resolution images.

2.2.5 Training for the Observation of High-Resolution Images

2.2.5.1 Setting Optimum Defocus

A technique that beginners should learn first is how to set the just-focused position ($\Delta f = 0$). If the just-focused point can be determined, a defocus value around the optimum defocus can be found by turning the focus knob in finite steps (see Sect. 2.2.2.8). If a few images are taken while changing the focus in steps of 5 or 10 nm around this defocus value, images taken with the optimum defocus will be found in the observed images.

The just-focused position can be determined by observing Fresnel fringes at the edges of a specimen, or the mazy contrast of a thin amorphous film. At the just-focused position, the Fresnel fringes and the mazy contrast disappear or become very weak. After suitable training, it is possible to determine the just-focused point within 10 nm.

2.2.5.2 Correction of the Astigmatism of an Objective Lens

Astigmatism exists even in lenses made to the highest level of precision, and so the quality of observed images depends strongly on the correction of this astigmatism. The astigmatism of condenser lenses is easily corrected by altering the probe shape of the incident beam, but that of an objective lens must be corrected by taking account of the deformation it causes in the image contrast. Astigmatism produces different focal lengths depending on the direction, and consequently focusing becomes unidirectional. Therefore, the existence of astigmatism results in an image with different defocus values depending on the direction; for instance, there could be a defocus value of 10 nm in one direction and one of 30 nm perpendicular to it. Hence, in a high-resolution image of an amorphous film, high and low spatial frequencies are enhanced in directions at right angles to each other, and eventually the image exhibits a unidirectional blur (see Fig. 2.18). It should be noted that the deformation of the image by astigmatism is visible near the just-focused point and in thin regions.

Methods of correcting astigmatism are given below. These use thin carbon films in a microgrid, or a contamination film at the edge of the specimen. (See also correction with optical or digital diffractograms, as described in Sect. 2.2.2.)

Cases of Small Astigmatism.
1. Find the just-focused position so as to get the image contrast as weak as possible at a magnification larger than 500 000.
2. Turn the x and y "fine knobs" of the stigmator individually, and get a further smooth or weak contrast.
3. Repeat processes 1 and 2.

Cases of Large Astigmatism.
1. Turn the x and y "coarse knobs" of the stigmator so as to see a clear granular image of an amorphous film at a magnification of 200 000–300 000.
2. Go to the correction of small astigmatism.

References

1. Hiraga K, Kohmoto O (1991) Mater Trans JIM 33:868
2. Yoshizawa Y, Yamaguchi K (1990) Mater Trans JIM 31:307
3. Shindo D, Hiraga K, Hirabayashi M, Kobayashi N, Kikuchi M, Kusaba K, Syono Y, Muto Y (1988) Jpn J Appl Phys 27:L2048
4. Sato M, Hiraga K, Sumino K (1980) Jpn J Appl Phys 19:L155
5. Hiraga K (1984) Sci Rep RITU A32:1
6. Hiraga K, Tsuno K, Shindo D, Hirabayashi M, Hayashi S, Hirai T (1983) Philos Mag A47:483
7. Hiraga K, Shindo D, Hirabayashi M (1981) J Appl Crystallogr 14:185
8. Hiraga K, Shindo D, Hirabayashi M, Terasaki O, Watanabe D (1980) Acta Crystallogr B36:2550
9. Terasaki O, Watanabe D, Hiraga K, Shindo D, Hirabayashi M (1981) J Appl Crystallogr 14:392
10. Terasaki O, Watanabe D, Hiraga K, Shindo D, Hirabayashi M (1982) J Appl Crystallogr 15:65
11. Hiraga K, Hirabayashi M, Terasaki O, Watanabe D (1982) Acta Crystallogr A38:269

3. Application of High-Resolution Electron Microscopy

On the basis of the discussions in Chaps. 1 and 2, we now consider how to observe and interpret high-resolution images, with plenty of typical examples. In Sect. 3.1, high-resolution images of various structural defects are presented and explained. X-ray diffraction and neutron diffraction provide accurate structural information averaged throughout a crystal, whereas the advantage of high-resolution electron microscopy over the diffraction methods is that it provides direct information in real space of structural defects which are localized in crystals. High-resolution electron microscopy of lattice defects such as dislocations, interfaces, and surfaces, and also structural defects in non-stoichiometric compounds are presented in this section. Section 3.2 considers the application of high-resolution electron microscopy to various advanced materials such as ceramics and high-Tc superconductors. Structural changes in alloys due to heat treatment and slight compositional changes, and the characteristic structural features of quasicrystals are also explained in detail.

3.1 High-Resolution Images of Various Defects

3.1.1 Dislocations

Dislocations are typical lattice defects which affect the mechanical properties of materials [1]. Since dislocations were first observed by electron microscopy in 1956 [2, 3], dislocations in various materials have been investigated by *dark-field electron microscopy* and by a *weak-beam method*. In these methods, the atomic displacement and lattice distortion around dislocations are imaged through *diffraction contrast*, and it is especially useful to investigate the morphology of dislocations, and also to determine their Burgers vectors. However, the resolution is limited to about 1.5 nm

even with the weak-beam method, and thus the dissociation width and the type of dissociation, i.e., an anti-phase boundary (APB) or a stacking fault, cannot be clarified if the dissociation width is less than 1.5 nm.

However, high-resolution electron microscopy can reveal the dislocation core on the atomic scale. In high-resolution electron microscopy, the crystal potential projected along the incident electron beam is frequently observed, as discussed in Chap. 1, and thus the electron beam should be incident parallel to the atomic columns. In fact there are three choices of incident beam direction for observing dislocations, as indicated in Fig. 3.1, and different information about the dislocations can be obtained in each case, as shown in Table 3.1 [4].

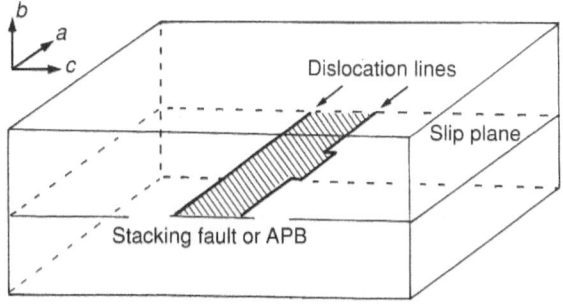

Fig. 3.1. Incident electron beam directions for the observation of high-resolution images of dislocations

Table 3.1. Information obtained from high-resolution images of dislocations

Incident beam direction	Information
$e^-//a$	Dissociation mode
	Dissociation width
$e^-//b$	Dissociation width
	Kink
$e^-//c$	Jog

3.1.1.1 Electron Beam Parallel to the Dislocation Lines (incident along the a-axis in Fig. 3.1)

When the incident beam is parallel to the dislocation lines, the dissociation mode can be clarified directly, and the dissociation width can also be determined accurately on the atomic scale. This observation mode is called an *end-on view*, and is the one most frequently used for investigating dislocations. In order to observe the dislocation core clearly on the atomic scale, the dislocation lines should be straight, and should also be parallel to the incident electron beam. If the dislocation lines are curved, or if there is some relaxation of the lattice around the dislocation core near the crystal surface, sharp high-resolution images cannot be obtained. Figure 3.2a shows a high-resolution image of a high-Tc superconductor $Tl_2Ba_2CuO_6$. As indicated by an arrow, a dislocation exists in the central part of the image and a small lattice strain around the dislocation core can be seen. If the lattice image of the dislocation is seen at an ob-

lique angle, an *extra half-plane* can be found at the arrow. Figure 3.2b shows a processed image obtained by the filtering method (see Sect. 4.1.2), where after making the Fourier transform, the image was obtained through the inverse Fourier transform with $110, \overline{1}10$ reflections, and the transmitted beam. From the processed image, we can clearly see the extra half-plane parallel to the (110) plane. This dislocation is localized, and no dissociations into *partial dislocations* are observed.

A dislocation having a Burgers vector corresponding to the smallest translational vector in the crystal is called a *perfect dislocation*. Since a perfect dislocation has large strain energy in the core, it tends to dissociate into partial dislocations with smaller strain energies (the sum of the Burgers vectors of the partial dislocations equals the Burgers vector of the perfect dislocation). A dislocation splitting into partial dislocations with a stacking fault is called a *dissociated dislocation*. There is a balance between two forces, i.e., one is the repulsive force between the partial disloca-

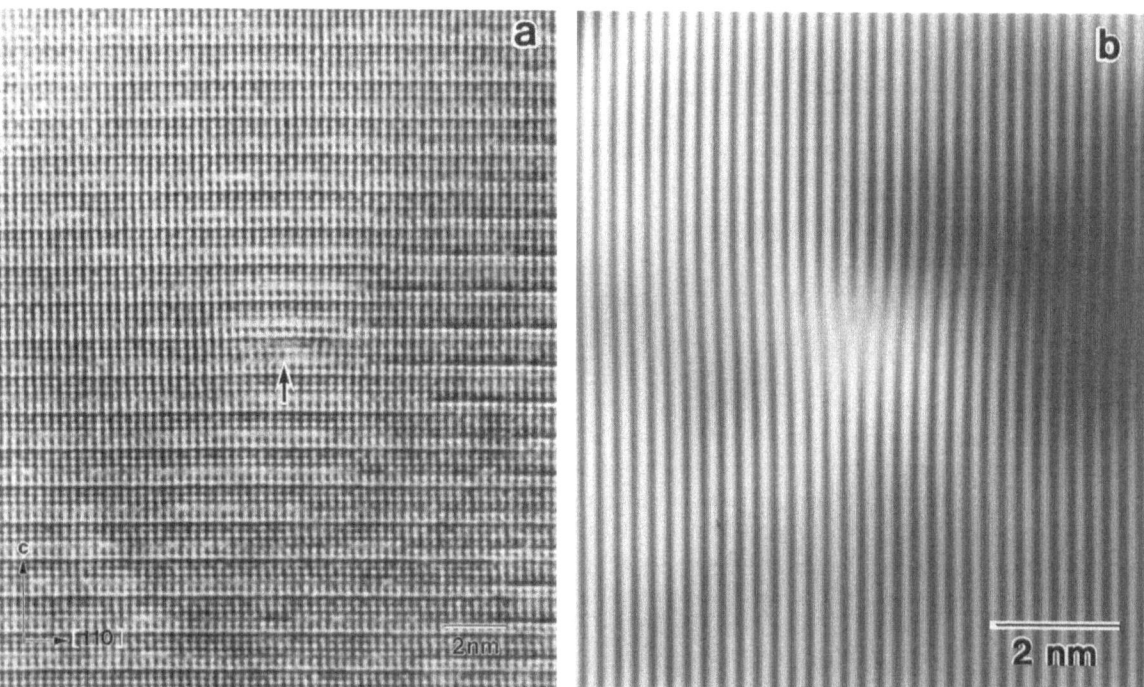

Fig. 3.2. a End-on view of a high-resolution image of a dislocation in an oxide superconductor, and **b** its filtered image.
Specimen: $Tl_2Ba_2CuO_6$; **Preparation**: crushing; **Observation**: 400 kV EM.
Remarks: The dislocation core is localized and no dissociation can be observed; **b** was obtained by making a filtering process on **a**

tions, and the other is the force minimizing the width of the stacking fault. Thus, from measurements of the distances between partial dislocations or the width of the stacking fault, one can determine the stacking fault energy. Figure 3.3 shows a high-resolution image of a 60° dislocation[1] observed in deformed silicon. The image indicates that a 60° perfect dislocation dissociates into 30° (A) and 90° (B) partial dislocations. The length of the stacking faults is 6.87 ± 0.33 nm, and thus the stacking fault energy is evaluated as 48.7 ± 2.4 mJ m^{-2} [5, 6].

The high-resolution image of silicon in Fig. 3.4 shows a characteristic lattice defect, a so-called *Z-type faulted dipole*. As shown in the inset, this defect forms when two dissociated moving dislocations interact each other and connect through the stacking fault AB. Extra half-planes exist above and below the stacking faults being at the top and bottom of the faulted dipole, respectively.

It is well known that when silicon is irradiated with high-energy electrons, dislocation loops form in the crystal. The high-resolution image in Fig. 3.5 shows the formation of dislocation loops by electron irradiation near a Z-type faulted dipole. As indicated in the structure model, which is projected along the electron beam, dislocation loops exist in the areas where the lattice is expanded due to the existence of partial dislocations at the ends of the stacking faults. This is quite consistent with the idea that the dislocation loops consist of interstitial atoms [7].

The Burgers vector of a perfect dislocation in an ordered alloy (see Sect. 3.2.3) is equal to the translational vector of the ordered structure, and the dislocation is generally called a *superlattice dislocation*. In general, a superlattice dislocation dissociates into two *superlattice partial dislocations* which correspond to two perfect dislocations in the basic lattice. Between the superlattice partial dislocations there exists an *anti-phase boundary* (APB, see Sect. 3.2.3). Now there is a balance between the repulsive force of the superlattice

partial dislocations and the force minimizing the width of the anti-phase boundary. A superlattice dislocation may also dissociate into two superlattice partial dislocations with a stacking fault. Various dissociation modes have been reported in many ordered alloys, depending on the basic lattice and also on their ordered structures. We now consider high-resolution images of superlattice dislocations in intermetallic compounds having ordered structures. Dissociation modes and core structures of superlattice dislocations affect the mechanical properties of intermetallic compounds which can be used for structural materials. Figure 3.6 shows a high-resolution image of a screw dislocation in $Ni_3(Al,Ti)$ having an $L1_2$-type structure (see Fig. 3.49) [8]. In Fig. 3.6a, the superlattice dislocation of a Burgers vector a[110] is dissociated into superlattice partial dislocations with a stacking fault 4.8 nm in width. The stacking fault is on the $(1\bar{1}1)$ plane, which corresponds to the slip plane. The superlattice dislocation in Fig. 3.6b shows a dissociation into two superlattice partial dislocations with an anti-phase boundary (see Sect. 3.2.3.2). In this case, since the displacement vector at the anti-phase boundary is parallel to the incident electron beam, no particular phase contrast of the boundary is observed, and only the lattice strain around the dislocation cores of the two superlattice partial dislocations is seen, with diffraction contrast indicated by the arrows. The dissociation width is estimated to be 2.5 nm. It should be noted that the anti-phase boundary is not on the slip plane but on the (001) plane, where the anti-phase boundary is energetically more stable. Thus, in this intermetallic compound, a part of the anti-phase boundary between the superlattice partials, while moving, tends to change from the slip plane to the energetically stable {100} plane by thermal activation at high temperature. This motion of the superlattice partial dislocations and the anti-phase boundary is the so-called *cross-slip*. It is understood that the cross-slip obstructing the movement of the superlattice dislocations in this material results in the abnormal temperature-dependence of the mechanical strength.

Figure 3.7a shows a high-resolution image of an edge dislocation in a CoTi intermetallic compound with a B2-type structure (see Fig. 3.49). This material also shows abnormal temperature-dependence in its mechanical strength, as in $Ni_3(Al,Ti)$. There are two dark regions at the dislocation core, indicated by the arrows, which cor-

---

[1] In semiconductors with a diamond structure like silicon, dislocation lines tend to be parallel to the Burgers vector or at 60° to it. The former is a screw dislocation, while the latter is a mixed dislocation and is usually called a 60° dislocation. Furthermore, dislocation lines of partial dislocations in the semiconductors are at 30° or 90° to their Burgers vectors, and are thus called 30° and 90° dislocations, respectively.

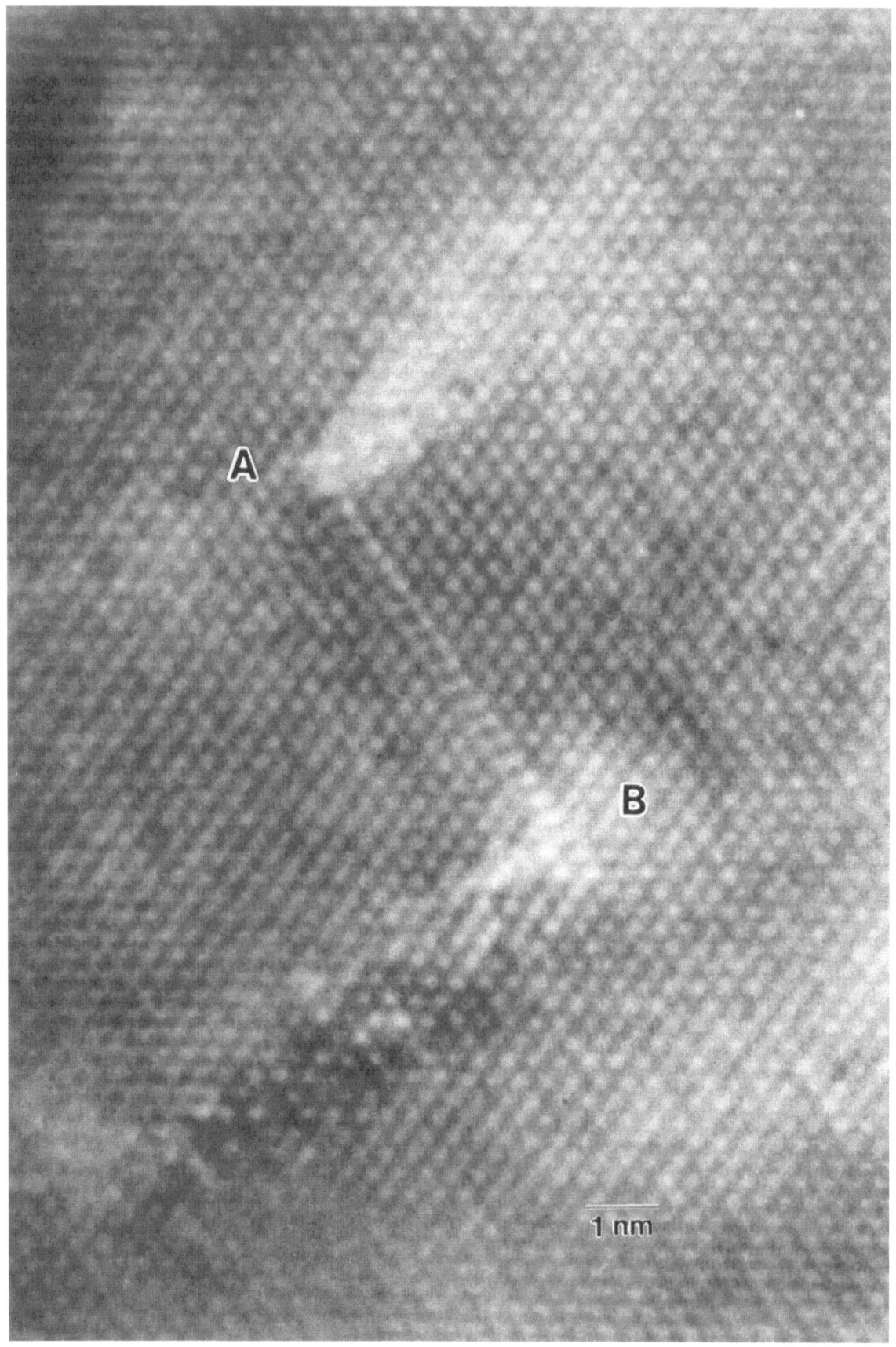

Fig. 3.3. Lattice image of a dissociated dislocation in deformed silicon.
Specimen: Si; **Preparation**: chemical polishing (HNO_3/HF); **Observation**: 1000 kV EM.
Remark: At the ends of the stacking fault there are 30° (*A*) and 90° (*B*) partial dislocations

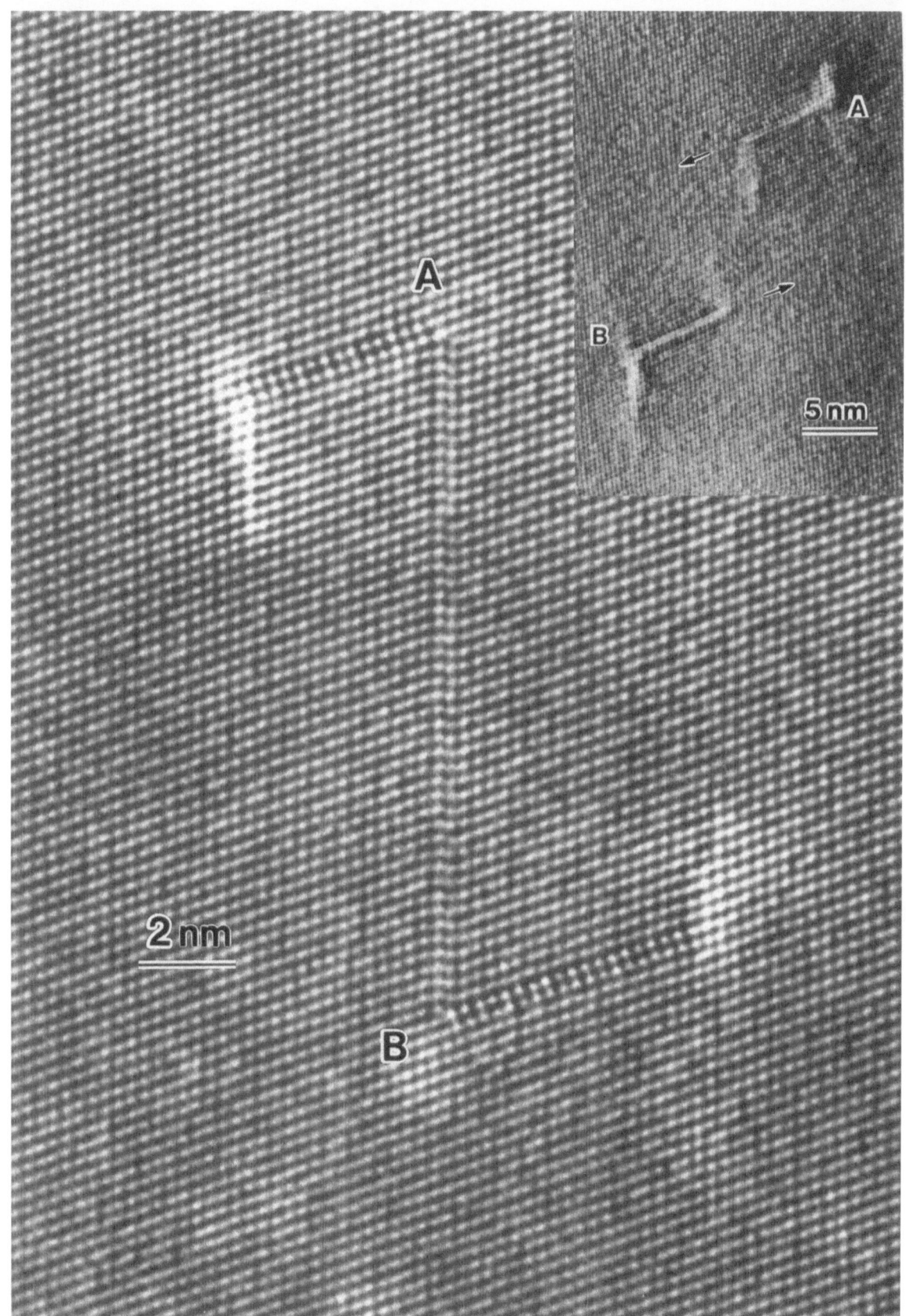

Fig. 3.4. Lattice image of a faulted dipole in silicon.
Specimen: Si; **Preparation**: chemical polishing (HNO₃/HF); **Observation**: 1000 kV EM.
Remark: The Z-type faulted dipole is formed from three stacking faults. This faulted dipole results from the interaction of two dissociated dislocations moving in the crystal, as indicated in the *inset*

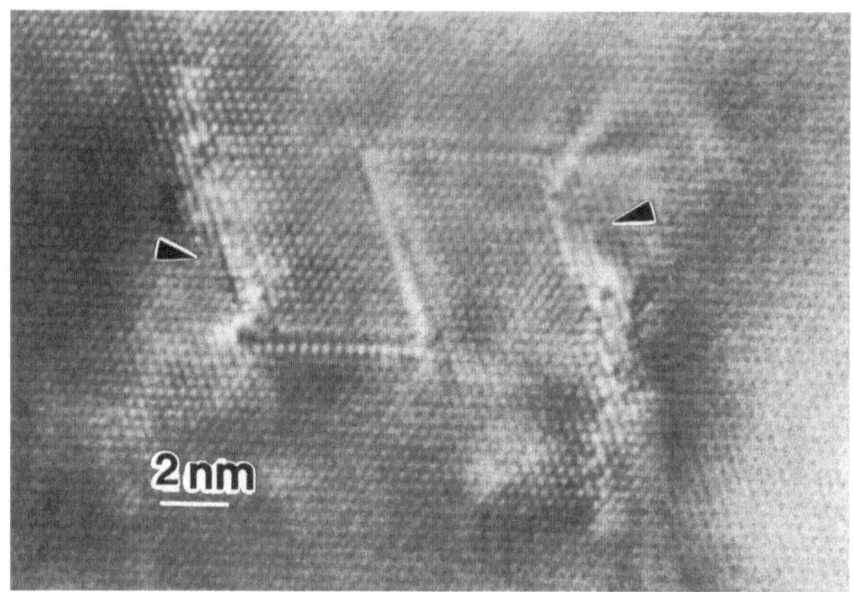

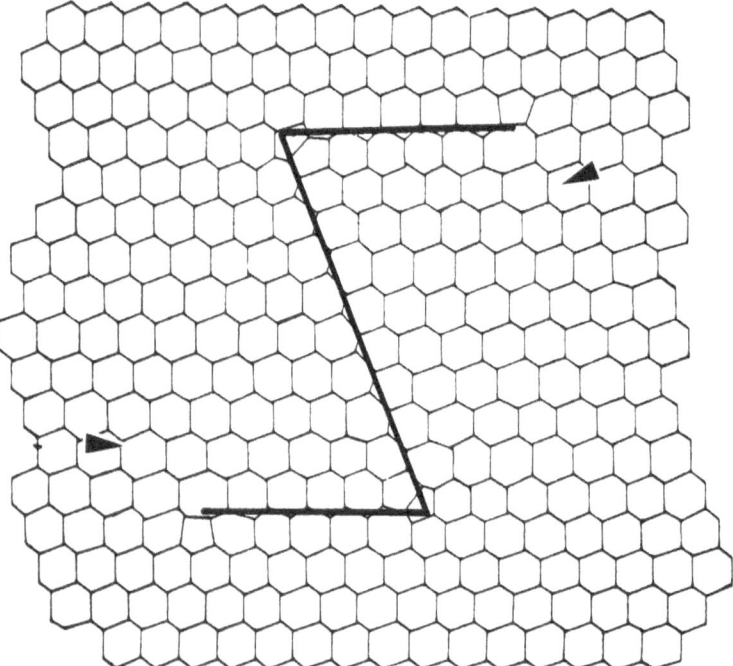

Fig. 3.5. Lattice image of dislocations and irradiation damage in silicon, and an atomic arrangement model. **Specimen**: Si; **Preparation**: chemical polishing (HNO$_3$/HF); **Observation**: 1000 kV EM.
Remark: At the ends of a Z-type faulted dipole there are dislocation loops (see *arrowheads*) caused by electron irradiation. As indicated in the atomic arrangement below, the dislocation loops are located in the expanded lattice regions near the partial dislocations

respond to strain contrast.[2] Thus, it is clear that the dislocation core has slightly dissociated. Figure 3.7b is the processed image obtained by the filtering method, as in Fig. 3.2. It is easier to find the core structure in the processed image than in the original high-resolution image, and the width of the dissociation is estimated to be 1.2 nm. Screw

---

[2] The diffraction contrast due to lattice strain is generally called strain contrast.

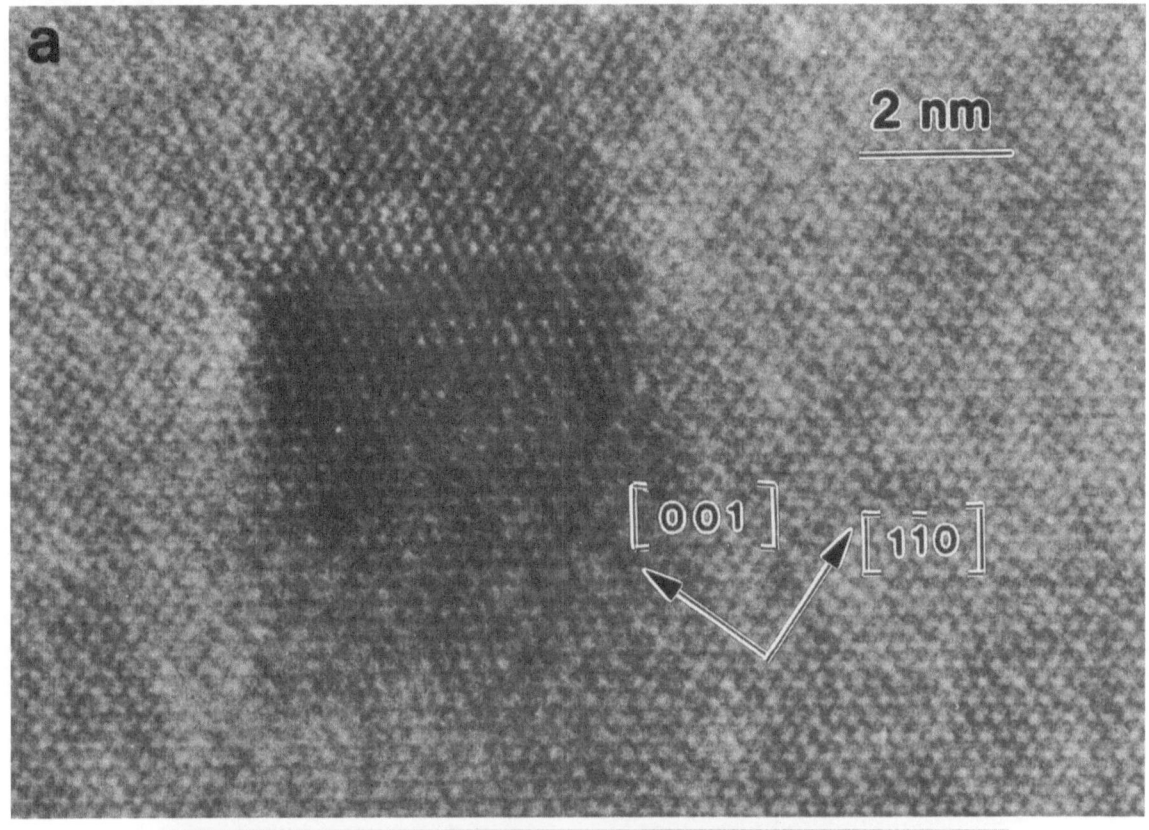

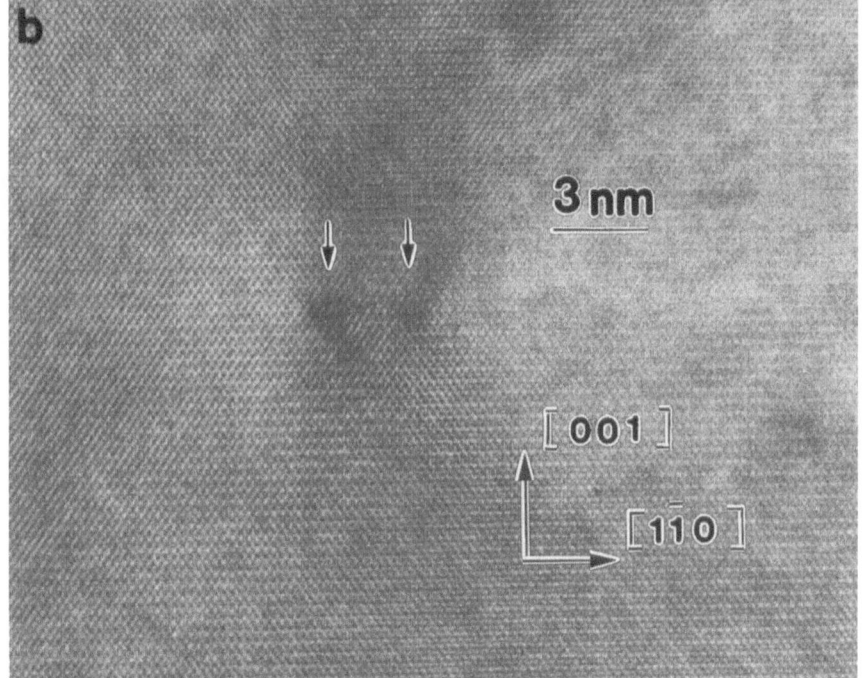

Fig. 3.6. Lattice images of superlattice dislocations in intermetallic compounds (L1$_2$-type).
Specimen: Ni$_3$(Al, Ti); **Preparation**: jet electropolishing (CH$_3$OH:H$_2$SO$_4$ = 4:1); **Observation**: 200 kV EM.
Remark: Superlattice dislocations (screw dislocations) with Burgers vector of a[110] dissociate with **a** a stacking fault, and **b** an APB

Intermetallic compounds—high-temperature structural materials

Alloys where the consitutent elements take regular atomic arrganements under the melting points, and the order–disorder transformation (see Sect. 3.2.1) does not occur, are generally called intermetallic compounds. In most of these materials, the mechanical strength decreases with an increase in temperature, while in some intermetallic compounds such as Ni_3Al and Ni_3 (Al,Ti), the mechanical strength increases with temperature and shows peak strength at high temperatures around 1000°C. Thus, these intermetallic compounds are very useful as high-temperature structural materials, and are actually used for turbine blades in jet planes, for example.

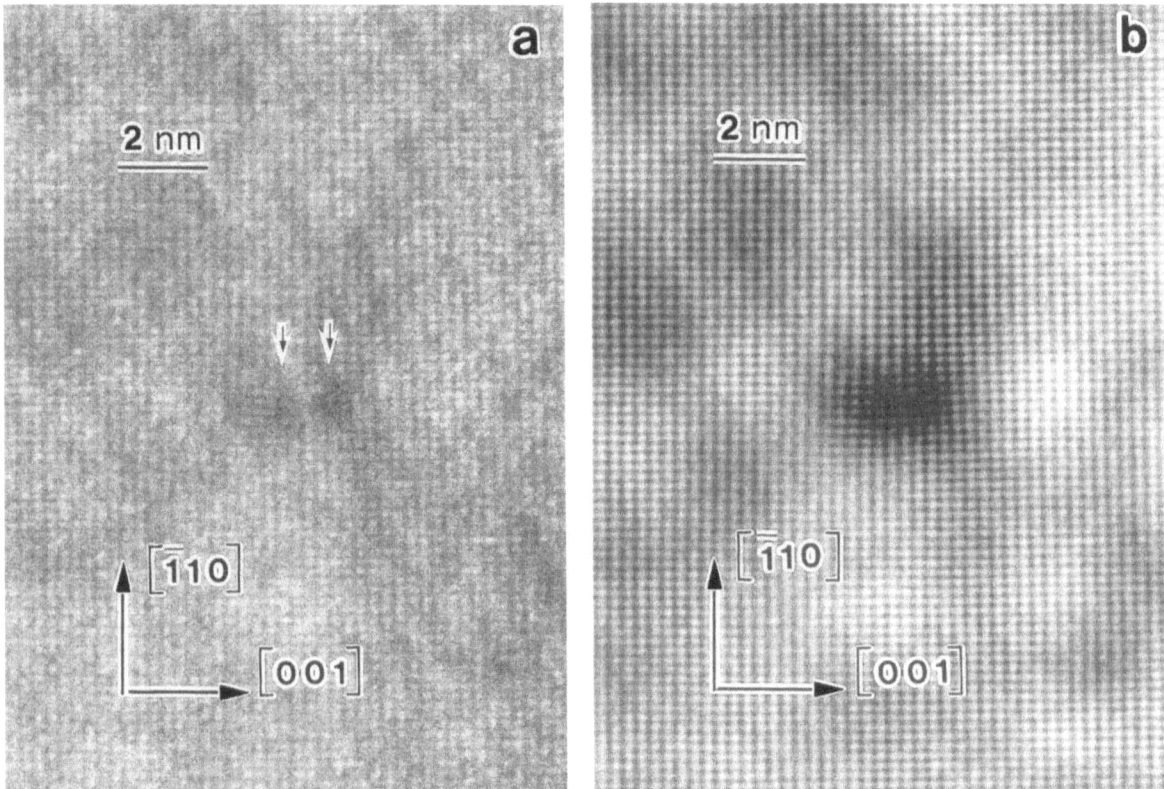

Fig. 3.7. Lattice image of a superlattice dislocation in an intermetallic compound CoTi (B2-type) and its filtered image.
Specimen: CoTi; **Preparation**: jet electropolishing (HClO4 : CH_3COOH_4 = 20 : 80); **Observation**: 400 kV EM.
Remarks: **a** High-resolution image of an edge dislocation. **b** Processed image with noise filtering. This material shows abnormal temperature-dependence for yield stress

dislocations in this compound have also been observed by high-resolution electron microscopy [9], and it was found that superlattice dislocations were dissociated on the {010} plane, which is different from the slip plane, i.e., the {110} plane [9]. It is of interest that no dissociations of superlattice dislocations have been observed in NiAl, which has the same B2 structure as CoTi [10, 11]. Thus, the abnormal temperature-dependence in the mechanical strength of CoTi can be explained in terms of the difference in the core structure of dislocations in these materials [12].

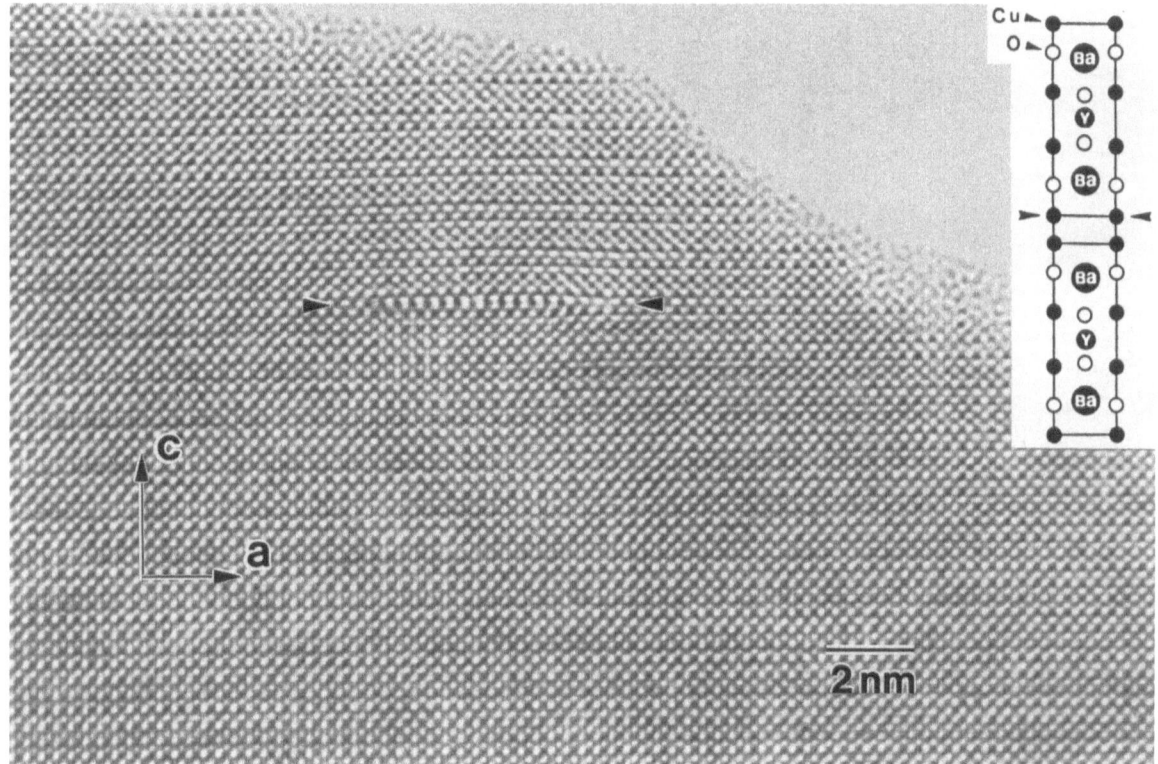

Fig. 3.8. Structure image of a dislocation loop in a high-Tc superconductor.
Specimen: $YBa_2Cu_3O_7$; **Preparation**: crushing; **Observation**: 400 kV EM.
Remark: The dislocation loop was formed with an extra Cu-O plane introduced during its crystal growth

Figure 3.8 shows a high-resolution image of a dislocation loop in a $YBa_2Cu_3O_7$ superconductor. An extra atomic plane exists between the arrowheads. As indicated in the model inset, the extra atomic plane is a Cu–O layer which was probably introduced in the crystal growth process. Thus, this image is interpreted as showing that edge dislocations exist at the arrowheads with Burgers vectors parallel to the c-axis [13].

3.1.1.2 Electron Beam Perpendicular to Dislocation Lines and a Stacking Fault (incident along the b-axis in Fig. 3.1)

This observation mode is appropriate for observing changes in the width of a stacking fault (or an anti-phase boundary), and identifying kinks on the partial dislocations. Figure 3.9 shows an example of high-resolution images of silicon observed in this mode [14]. In the stacking fault between the 30° and 90° partial dislocations, lattice fringes with a spacing of 0.33 nm are seen. In general, these lattice fringes are not observed in a perfect silicon crystal. As illustrated in Fig. 3.10,

the stacking fault corresponds to missing double planes parallel to the $(1\bar{1}1)$ plane, and this defect is considered to produce the lattice fringes. In fact, in the optical diffractogram in Fig. 3.9c, obtained from the stacking fault, forbidden reflections such as $(42\bar{2})/3$ can be seen. Since partial dislocations exist at the ends of the lattice fringes, the positions and density of the kinks can be investigated by tracing the dislocation lines. In Fig. 3.9a, the density of the kinks on the partial dislocations was estimated to be extraordinarily high. This may have been caused by electron irradiation damage. Figure 3.11 shows a high-resolution image of a dislocation in silicon observed by minimizing the electron intensity, and no effects of electron irradiation damage are observed on the dislocation lines. As illustrated in the inset, if the lattice fringes are viewed at an oblique angle, the kink positions (indicated by arrows) can be clearly identified. This observation mode for kinks on dislocation lines has been also applied to GaAs [15], and kink velocity was investigated by *in-situ* observations [16].

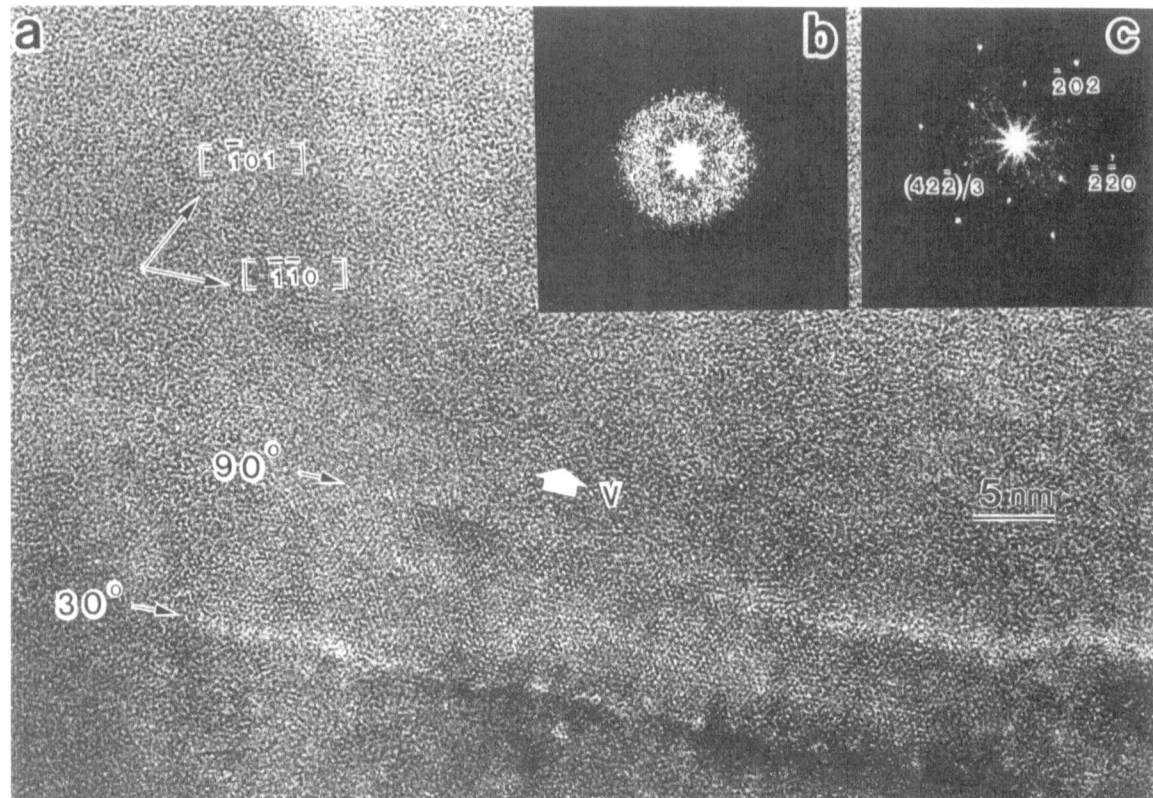

Fig. 3.9. Lattice image of silicon with the incident electron beam perpendicular to both partial dislocation lines, and a stacking fault between them.
Specimen: Si; **Preparation**: chemical polishing (HNO₃/HF); **Observation**: 200 kV EM, [1$\bar{1}$1] incidence.
Remarks: Lattice fringes are observed only at the region corresponding to a stacking fault. The fringes have a spacing of 0.33 nm, corresponding to forbidden reflections. Kinks are located by tracing the edge of the stacking faults. In the optical diffractogram in the *inset* **c**, forbidden reflections from the stacking fault are clearly seen. From the optical diffractogram in *inset* **b**, which was obtained from a high-resolution image of an amorphous region at a crystal edge, the optimum defocus condition (Scherzer focus) can be confirmed. The *white arrow* at *V* indicates the moving direction of the dislocation when the crystal was stressed

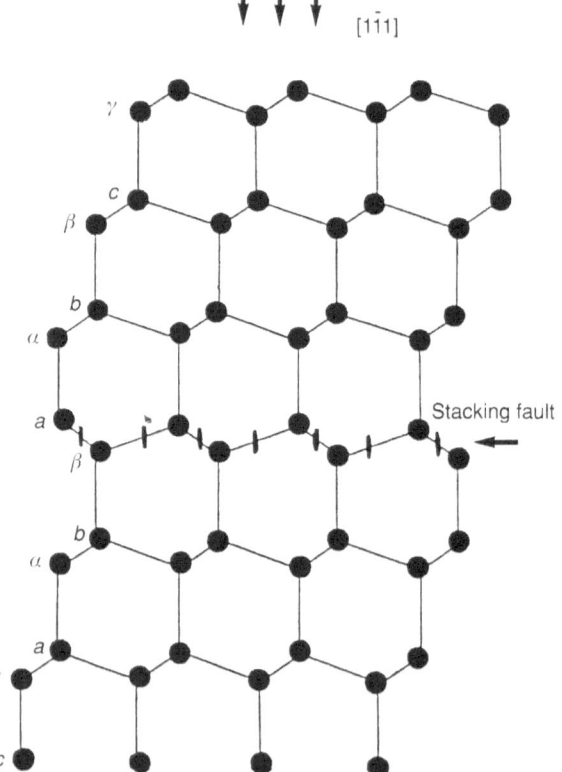

Fig. 3.10. Intrinsic stacking fault between partial dislocations in Si. Note the missing double planes parallel to the (1$\bar{1}$1) plane (indicated by an *arrow*)

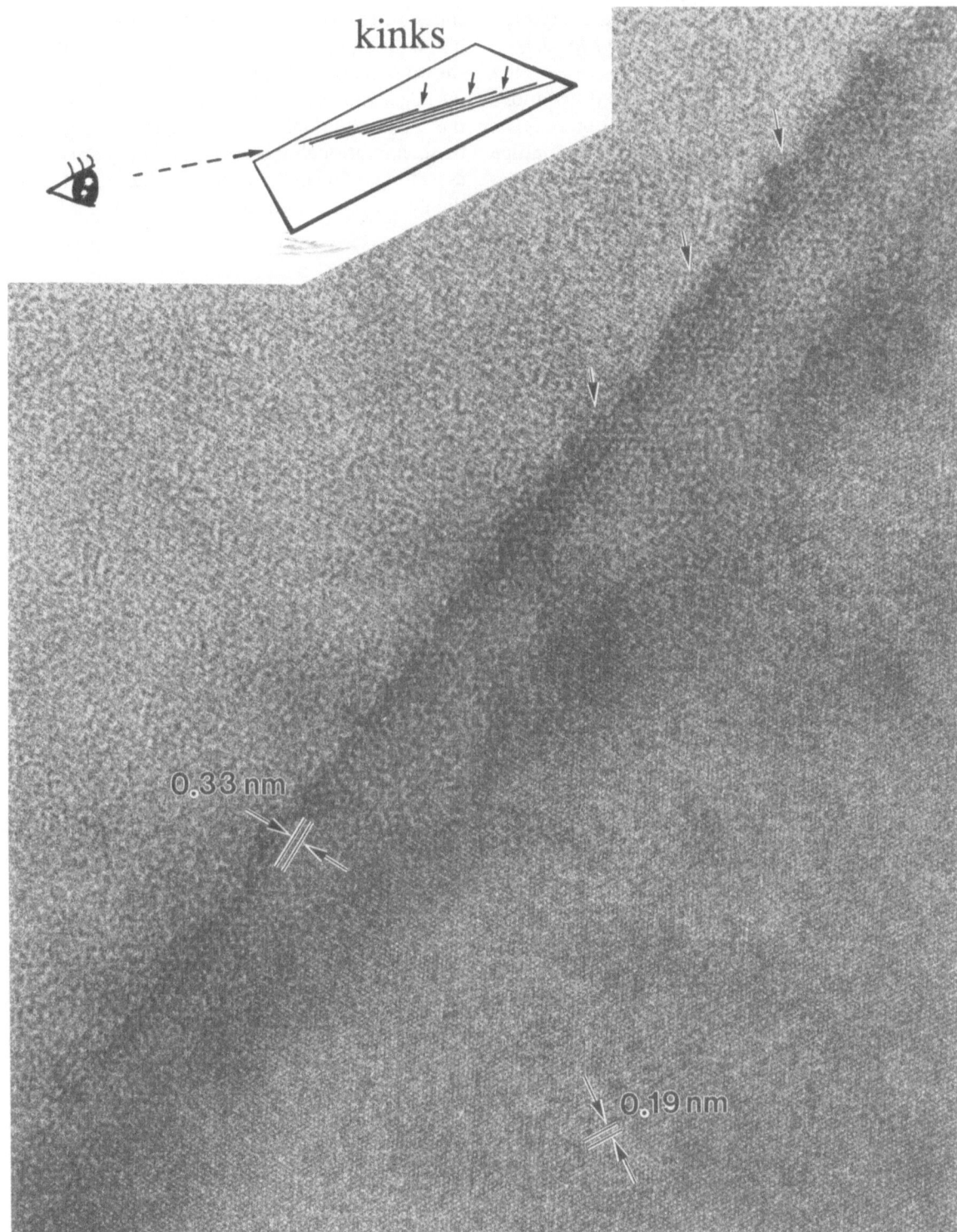

Fig. 3.11. Lattice image of silicon showing the kinks on a dislocation.
Specimen: Si; **Preparation**: chemical polishing (HNO$_3$/HF); **Observation**: 200 kV EM, [1$\bar{1}$1] incidence.
Remarks: Kinks are observed at positions indicated by the *arrows*. As illustrated in the *inset top left*, it is easier to locate the kinks by looking obliquely along the lattice fringes with a spacing of 0.33 nm which correspond to the forbidden reflections. It should be noted that lattice fringes are observed only at the region corresponding to a stacking fault whereas in the other region, lattice fringes corresponding to the fundamental spacing of 0.19 nm can be seen

3.1.1.3 Electron Beam Perpendicular to Dislocation Lines and Parallel to a Stacking Fault (incident along the c-axis in Fig. 3.1)

In this observation mode, jogs on the dislocations and related defects may be clarified. Figure 3.12a shows a high-resolution image of a superlattice partial dislocation in Fe_3Al with a DO_3-type structure. It is well known that a superlattice dislo-

cation in this material dissociates into 4-fold superlattice partial dislocations with wide anti-phase boundaries of about 20 nm in width, as shown schematically in Fig. 3.12b [17]. At the crystal edge, the partial dislocations come out from the surface and only the anti-phase boundary can be seen. Although this anti-phase boundary exists as a straight line in the wide area, there is a step on the boundary, as indicated by D in Fig. 3.12c. The

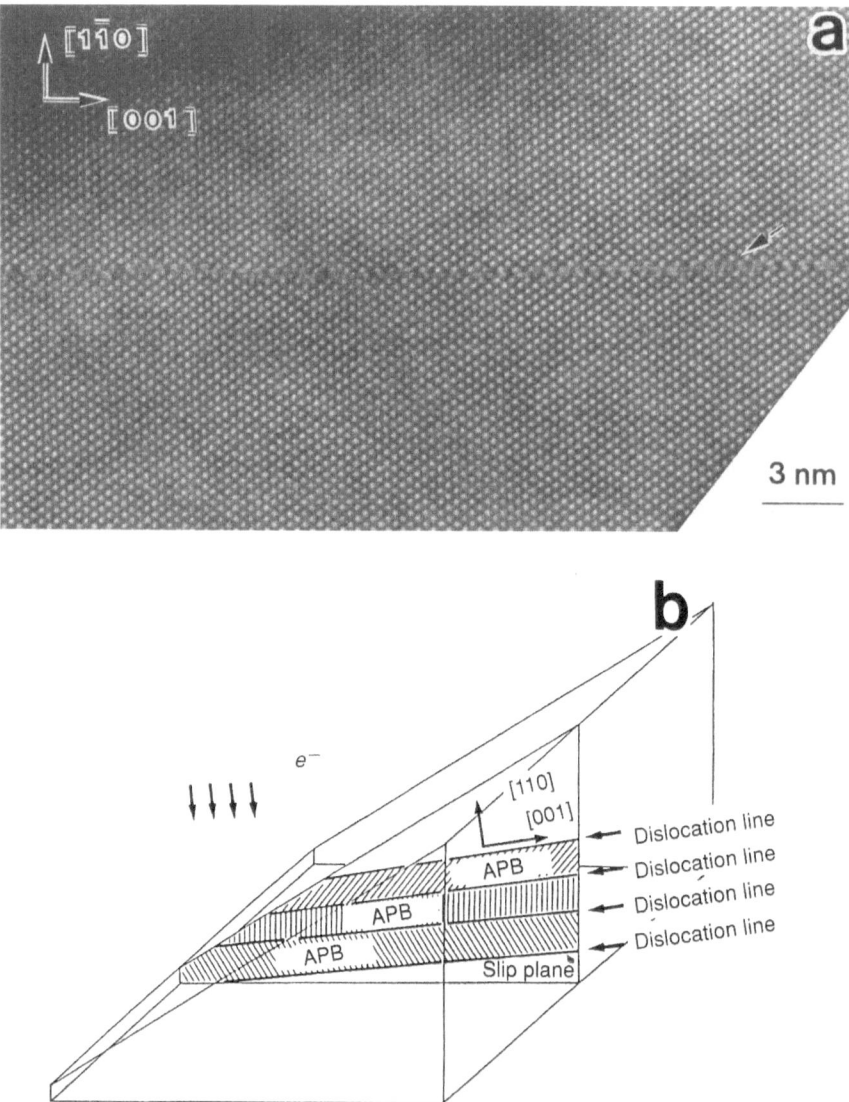

Fig. 3.12. High-resolution images of an APB, and schematic illustrations showing a geometrical configuration of the incident beam, partial dislocation lines, and the APBs between them.
Specimen: Fe_3Al; **Preparation**: jet electropolishing ($CH_3COOH : H_2SO_4 = 95:5$); **Observation**: 200 kV EM.
Remarks: **a** High-resolution image (superstructure image) of an APB which is supposed to be between partial dislocations. **c** Enlarged micrograph of **a**. **b** and **d** are schematic illustrations showing the geometrical configuration of the incident electron beam, partial dislocations, and APBs

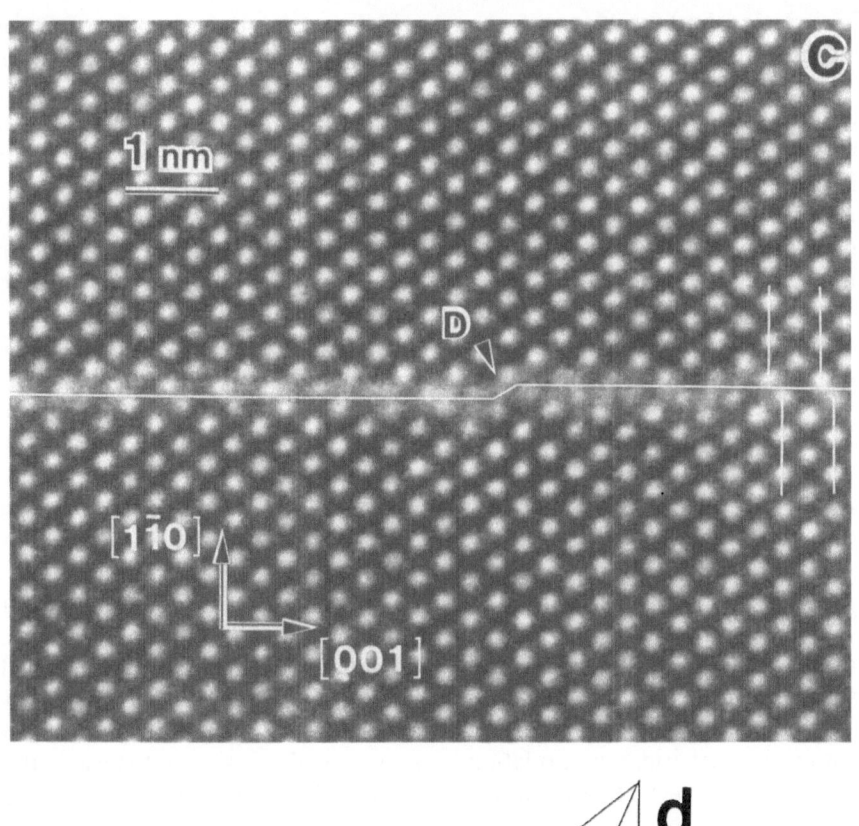

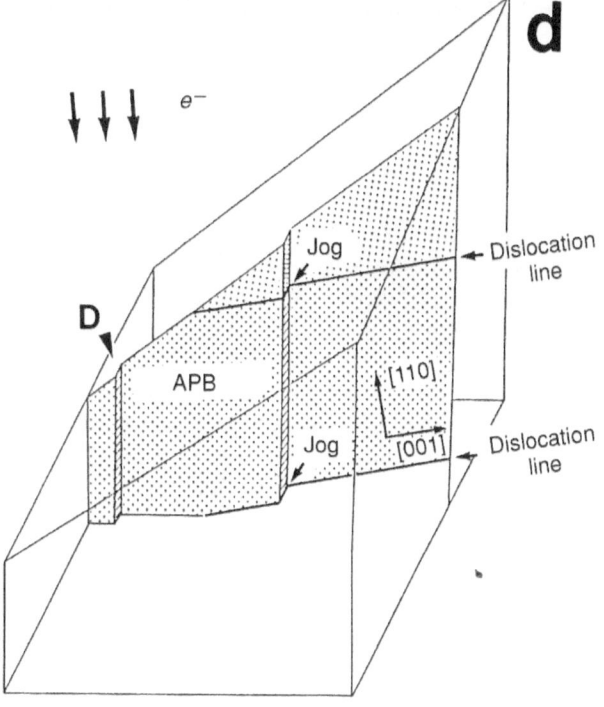

Fig. 3.12. *Continued*

step on the boundary corresponds to an atomic step caused by jogs on the superlattice dislocations, as shown schematically in Fig. 3.12d. In the image, the white spots correspond to Al atoms projected along the incident electron beam. By investigating the distance between the white spots at the anti-phase boundary, the repulsive force between Al–Al atoms in Fe_3Al has been evaluated [4].

3.1.2 Grain Boundaries and Interfaces Between Different Phases

Inorganic materials are usually formed with aggregations of small grains or composites of different phases. Therefore, the properties of such materials are strongly dependent upon the structure of *grain boundaries* and *interfaces (interface boundaries)*. In particular, the mechanical properties of intermetallic compounds and ceramics, which are in demand as structural materials at high temperatures, are sensitive to these structures, so studies on grain boundaries and interfaces are important fields in materials sciences. This section presents some results from the high-resolution electron microscopy of grain boundaries and interfaces.

3.1.2.1 Grain Boundaries

In general, grain boundaries in materials, except for those prepared by special methods, are formed with adjacent grains having random orientations to each other with no special orientation relationships (*random grain boundaries*). High-resolution electron micrographs are images projected along the incident beam, so it is generally difficult to reveal the structures of grain boundaries with random orientations from high-resolution images. However, for intermetallic compounds and ceramics prepared by a sintering method, high-resolution observations have given us valuable information about the existence of grain boundary phases, even though the structures of the grain boundaries and interfaces cannot be determined directly.

Figure 3.13 shows high-resolution images of grain boundaries and triple junctions in Si_3N_4 ceramics prepared by different methods. The images in Fig. 3.13a and b were taken from Si_3N_4 prepared by *hot isostatic pressing* (HIP) without any sintering additives. As can be seen in Fig. 3.13a, a thin layer of an amorphous phase about 1 nm thick exists at the grain boundary, although this specimen was prepared without any additives [18, 19]. This layer is considered to be amorphous SiO_2, which was present on the surface of the raw Si_3N_4 powder. This type of SiO_2 layer becomes liquid at high temperatures and acts as a sintering agent, and consequently a high-density material is produced. Also, the liquid phase makes it possible to grow crystals without restrictions, and consequently they form facets with stable crystal surfaces, as can be seen in Fig. 3.13a.

In general, impurity phases are more usually formed at triple junctions than at ordinary grain boundaries, as seen in Fig. 3.13b. Therefore, even though no impurity phases are observed at grain boundaries, they can sometimes be observed at triple junctions. Thus, the chemical content of the impurity phases appearing at relatively wide areas of the triple junctions can sometimes be investigated by analytical electron microscopy.

It was thought that Si_3N_4 crystals with strong covalent bonding could not be joined directly with grain boundaries, since it was difficult to sinter them without additives, and also impurity phases were always observed at grain boundaries in materials sintered with additives. That consideration was ruled out by high-resolution observations of grain boundaries in Si_3N_4 prepared by the *chemical vapor deposition* (CVD) method, as shown in Fig. 3.13c–e [20, 21]. These images show that impurity phases are not observed either at grain boundaries (e and c) or at a triple junction (d), and that crystals are directly connected at the boundaries in the Si_3N_4 crystal prepared by the CVD method, which is known to be a method of synthesizing high-purity bulk materials. The image of a high-angle grain boundary in Fig. 3.13c shows no deformation of lattice planes near the boundary, whereas that of a small-angle grain boundary in Fig. 3.13e shows periodic strain contrasts corresponding to grain boundary dislocations. These structural features of grain boundaries are similar to those of grain boundaries in metallic alloys.

In order to investigate the existence of thin-grain boundary layers by high-resolution electron microscopy, images must be taken with the incident beam parallel to the planes of the grain boundaries. If the electron beam is inclined to the boundary plane, the overlap of lattice fringes in two grains is observed, like the boundary at the upper left of Fig. 3.13d, and consequently it becomes difficult to clarify the existence of grain boundary layers. It is not easy to set the incident beam exactly parallel to grain boundaries, so it is important to observe high-resolution images of grain boundaries in thin regions, where the effect of the inclined grain boundaries is minimized.

When grain boundaries are formed artificially or introduced with the appearance of twin boundaries, so-called *tilt boundaries* appear. In tilt boundaries, one crystallographic axis is common to the adjacent grains, and thus the crystals in these grains are rotated in relation to each other by some angle around the common axis. In this case, two-dimensional lattice images can be

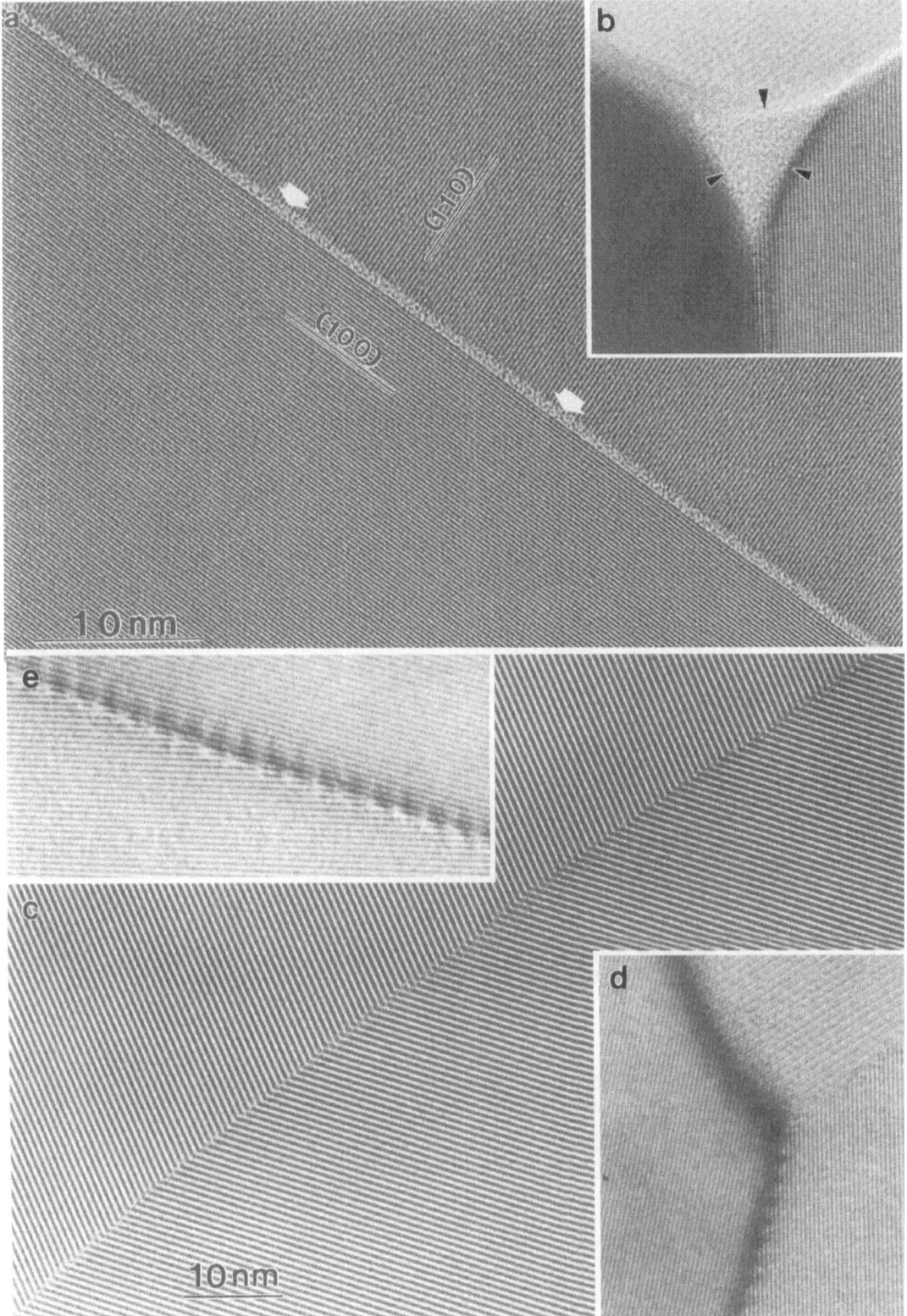

Fig. 3.13. High-resolution images of boundaries and triple junctions in Si_3N_4.
Top: **a** Si_3N_4 grain boundary, and **b** a triple junction in Si_3N_4–SiC sintered by HIP.
Specimen: HIP–Si_3N_4–SiC; **Preparation**: ion milling; **Observation**: 400 kV EM.
Remark: Amorphous SiO_2 is observed at the grain boundary and the triple junction.
Bottom: **c, e** Grain boundaries, and **d** a triple junction in Si_3N_4 prepared by CVD.
Specimen: CVD–Si_3N_4; **Preparation**: ion milling; **Observation**: 400 kV EM.
Remarks: No impurity phases are observed at the triple junction as well as at the grain boundary. The contrast at the small angle boundary in **e** shows a periodic arrangement of dislocations

Fig. 3.14. Lattice image showing a tilt boundary and lattice defects in SiC prepared by CVD.
Specimen: CVD-SiC; **Preparation**: ion milling; **Observation**: 200 kV EM, [110] incidence.
Remarks: Three grains (P, Q, R) and their grain boundaries are identified from the line contrasts of stacking faults and micro-twins. The [110] directions of three grains are parallel

Fig. 3.15. Enlarged micrograph of the regions A, B, and C in Fig. 3.14. **a** Σ9 grain boundary; **b** grain boundary affected by twins and stacking faults; **c** a triple junction

Tilt boundaries and Σ values

Special grain boundaries, which are formed by rotating adjacent grains around the common axis parallel to the boundary plane, are called "tilt boundaries." The relationships of crystallographic orientations between adjacent grains in the tilt boundaries are characterized by Σ values, which indicate the fraction of the *coincident sites* associated with the two grains where the lattices are considered to be interpenetrating. If one of the lattices is then rotated relative to the other around an axis through a common lattice point, for certain rotational angles, it will be seen that a portion of the two sets of lattice points will coincide, and these common lattice points form the so-called coincidence lattice. The Σ value is defined as the reciprocal of the ratio of the coincident lattice sites to the original lattice sites. Hence, a small Σ value shows a high coincidence between the lattice points of two crystals. For example, Σ5 means that for boundaries at an angle θ to each other, one-fifth of the original lattice sites coincide, and this corresponds to boundaries where $\theta = 36.87°$, $53.13°$, $126.87°$, and $143.13°$ for the common axes <100> in a cubic lattice. The grain boundaries indicated by Σ are called coincidence boundaries.

observed, with the incident beam parallel to the rotational (common) axis.

Figure 3.14 shows one example of lattice images of SiC prepared by the CVD method [20, 21]. The many sharp lines in the image correspond to twin boundaries or stacking faults. From the orientations of these lines, it can be seen that three grains exist. The zigzag contrast from the upper left to the lower left, and also from the lower left to the upper right, are tilt boundaries, and enlarged micrographs of the three framed regions A, B, and C are shown in Fig. 3.15.

Figure 3.15a is a coincidence site lattice (CSL) tilt boundary (*coincidence boundary*) described as Σ9. On the boundary, a *symmetric boundary* (although it has no mirror symmetry and there is a slight displacement between the two sides along the boundary) can be seen in the *x–y* region, whereas in the *y–z* region the boundary is formed in a close-packed plane of one crystal (called *close-packed boundaries*), although the boundary is disturbed by the existence of stacking faults. These observations show that boundaries such as symmetric and close-packed boundaries are energetically stable. Figure 3.15b is a boundary disordered by a number of stacking faults and twin planes. In this boundary, there is a good connection of two grains with slight lattice deformation near the boundary. Figure 3.15c is an image of a triple junction, where it can be seen that the grains of R and P are formed as a result of twinning.

The intermetallic compound Ni_3Al has received considerable attention as a potential high-temperature structural material, because its flow strength increases with increasing temperature at high temperatures [22]. Many studies have shown that grain boundaries play a crucial role in the mechanical properties and fracture behavior of polycrystalline Ni_3Al, and its inherent brittleness

at grain boundaries could be overcome by the addition of a small amount of boron [23]. Therefore, many transmission and high-resolution electron microscope studies have been carried out into the grain boundaries of Ni_3Al.

Figure 3.16 shows three types of tilt grain boundary in a $Ni_3(Al,Ti)$ compound [24, 25]. These boundaries were prepared by hot-pressing of bicrystals cut along specific orientations. The adjacent grains across the boundaries are rotated at an angle of (a) 10°, (b) 16.26°, and (c) 36.87° to each other around the common [100] axis. The relationship can be seen from the electron diffraction pattern in the insert in Fig. 3.16b. The grain boundaries of Fig. 3.16b and c are coincidence boundaries described as Σ25 and Σ5, respectively. In the three images, the [100] axes of the adjacent grains are slightly tilted in relation to each other (a few degrees) due to experimental error in preparing the bicrystals, so differences in contrast between adjacent grains and the asymmetry of the image about the boundaries can be noticed. It should be noted that the effect of the misalignment is enhanced in the thicker regions on the right of Fig. 3.16.

In the low-angle grain boundaries of Fig. 3.16a and b, boundary dislocations (indicated by arrowheads) can be seen, whereas in the high-angle boundary of Fig. 3.16c there are no visible dislocations. From close examination of Fig. 3.16a and b, it was found that the boundary dislocations dissociate into two partial dislocations with $a/2[110]$ and $a/2[1\bar{1}0]$ Burgers vectors, and that these dislocations are arrayed along the boundaries [24, 25].

Figure 3.17 shows a structure image of a *twin boundary* observed in a Bi-based high-Tc superconducting oxide [26]. The unusual dark contrast, which is arranged periodically in the image of

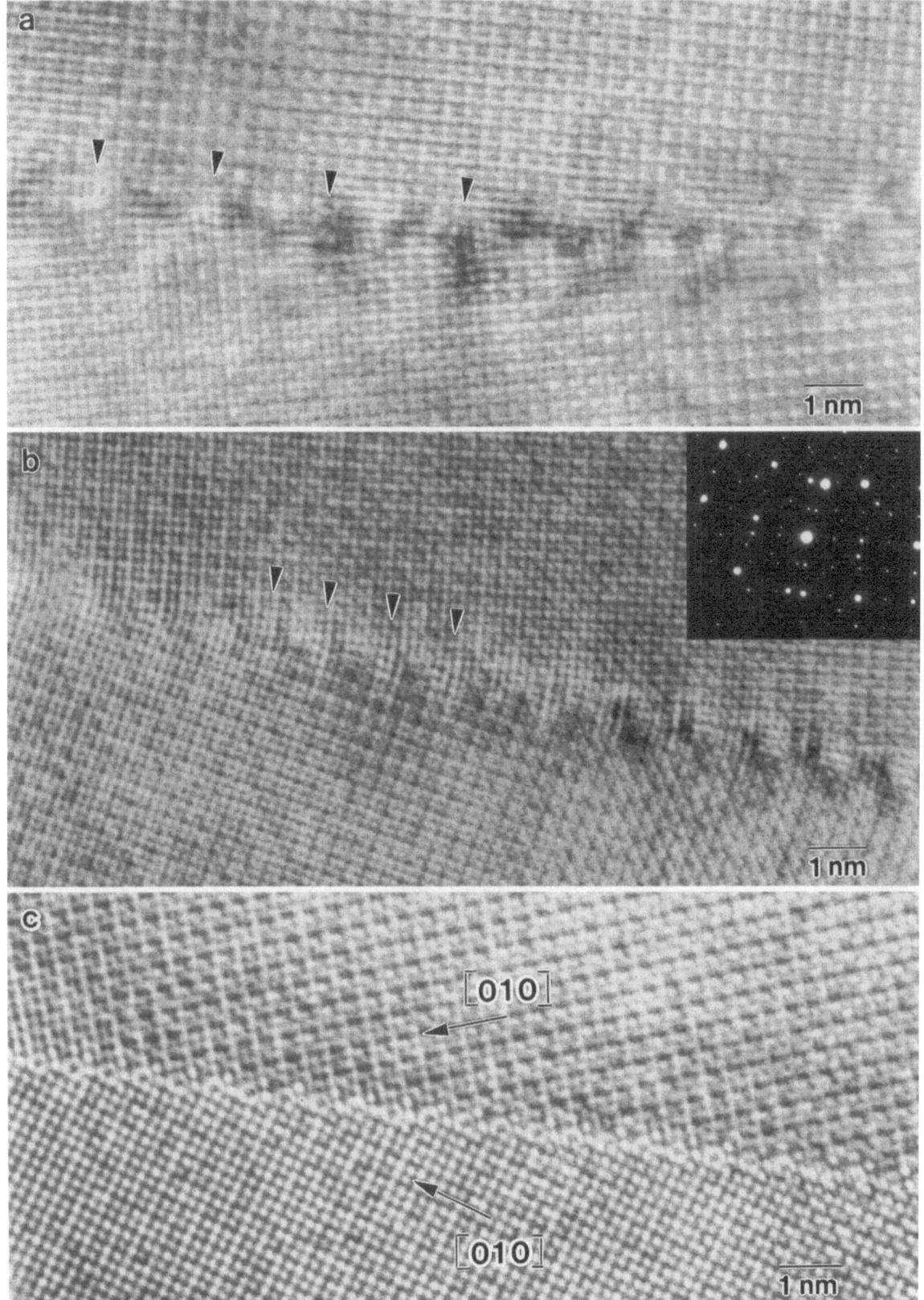

Fig. 3.16. High-resolution images of tilt grain boundaries in a $Ni_3(Al, Ti)$ compound prepared by hot pressing of bicrystals.

Specimen: $Ni_3(Al, Ti)$; **Preparation**: jet polishing ($H_2SO_4 : CH_3OH = 1:9$);

Observation: 400 kV EM, [100] incidence.

Remarks: The adjacent grains across the boundaries are inclined at **a** 10°, **b** 16.26°, and **c** 36.87° to each other around the common [100] axis. **a** and **b** show boundary dislocations, indicated with *arrowheads*. The grain boundaries of **b** and **c** are coincidence boundaries, described as $\Sigma 25$ and $\Sigma 5$, respectively

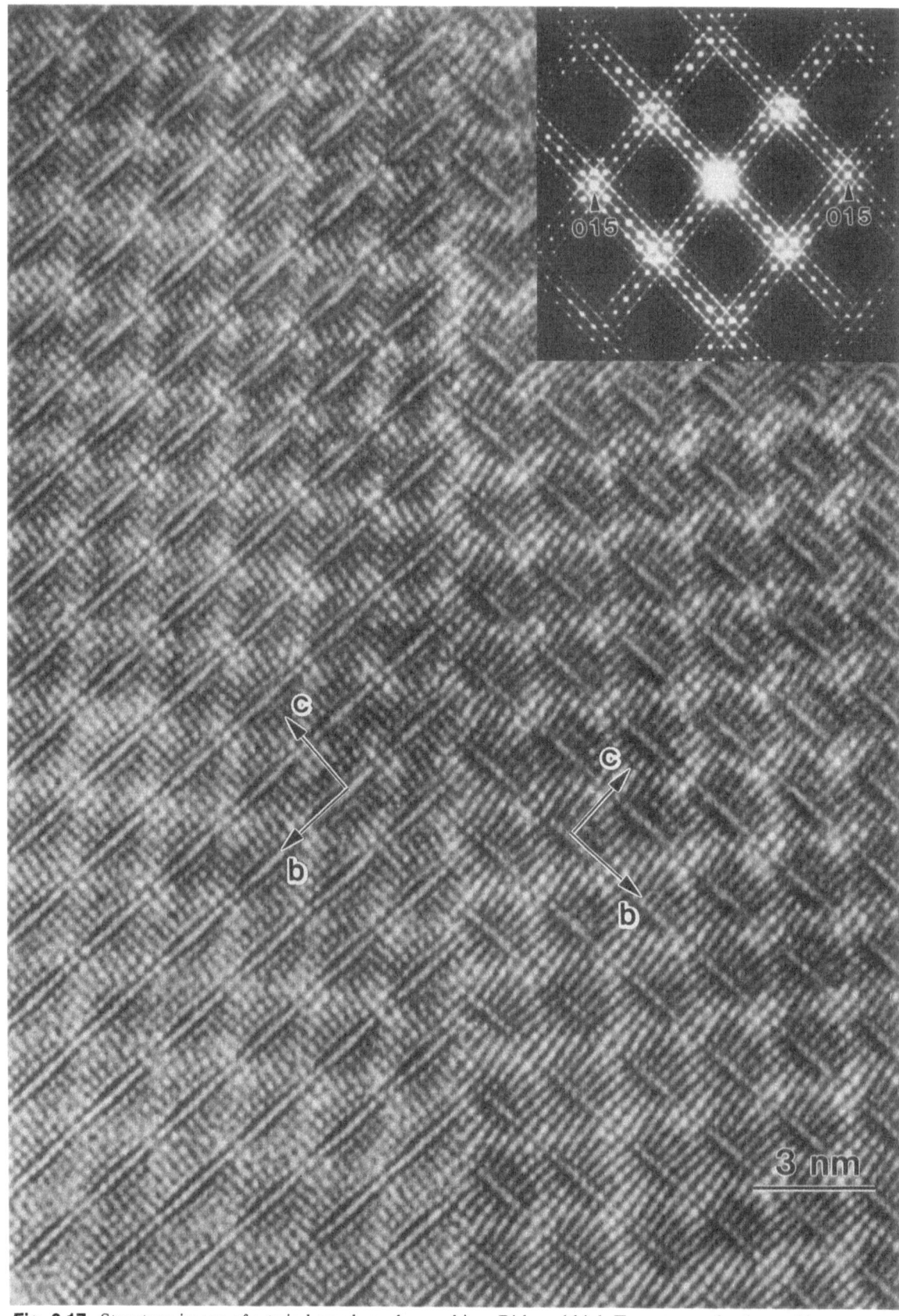

Fig. 3.17. Structure image of a twin boundary observed in a Bi-based high-Tc superconducting oxide.
Specimen: Bi–Sr–Ca–Cu–O; **Preparation**: ion milling; **Observation**: 400 kV EM, [100] incidence.
Remarks: The twin boundary does not have mirror symmetry, and has a relative displacement of half of the period. The two indices in the inset diffraction pattern correspond to the two domains

both grains, results from the modulated structure (lattice modulation), being known to be a characteristic structural feature of this superconductor. As can be seen from the orientations of the two grains, they have a twinning relationship, but there is no mirror symmetry at the boundary. It should be noted that one grain is displaced relative to the other by one-half of the period along the boundary in such a manner that the most lattice-deformed regions with dark contrast in one grain are not adjacent to those in the other grain. It is difficult to tell from the electron diffraction pattern whether the boundary has mirror symmetry or not, but this can easily be seen in the high-resolution image.

3.1.2.2 Interfaces Between Different Phases

There is currently an intensive worldwide research effort devoted to producing composite materials with new and improved properties, which cannot be obtained in single-phase materials. The properties of such composites are strongly dependent upon the structure of the interfaces between grains in different phases (*interphase boundaries*), as well as the heterogeneous structures of the different phases.

The characteristics of permanent magnets are determined by their saturation magnetization and magnetic coercivity. The saturation magnetization is fixed as an inherent property of a crystal, whereas the magnetic coercivity is strongly sensitive to the microstructure of the magnet. At present, magnets are mainly divided into two types: the pinning type and the nucleation type. The former establishes high coercivity by preventing movement of the magnetic domain walls, and the latter does so by inhibiting nucleation of magnetic domains with reversal magnetization. A representative of the pinning-type magnet is a Cu-doped Sm–Co-based magnet in which interfaces between Sm_2Co_{17} and $SmCo_5$ domains act as pinning sites for the magnetic walls. Hence, a Sm–Co-based magnet is also called a precipitation hardened magnet.

Figure 3.18 a and b are transmission and high-resolution electron micrographs, respectively, of a Sm–Co-based magnet after the optimum heat treatment (at 800°C for 4h) to obtain the highest possible coercivity [27]. The transmission electron micrograph in Fig. 3.18a shows strong strain contrasts as well as a complex microstructure, which is caused by the coexistence of two phases.

In the diffraction pattern inserted in Fig. 3.18a, strong reflections, such as 002 and 110 reflections, are common in Sm_2Co_{17} and $SmCo_5$ structures, and the elongated reflections in the vertical direction between the strong reflections are superlattice reflections of the Sm_2Co_{17} structure. That is, the electron diffraction pattern shows that the Sm_2Co_{17} and $SmCo_5$ phases have similar fundamental unit cells and coexist with the same crystal orientations.

Figure 3.18b is a lattice image showing the coexisting microstructure. The Sm_2Co_{17} domains have a micro-twinning structure which gives rise to elongated superlattice reflections of Sm_2Co_{17} in the vertical direction in the diffraction pattern. As can be seen, the $SmCo_5$ domains are precipitated with the definite crystallographic relation among the Sm_2Co_{17} domains. The Sm_2Co_{17} structure is a superlattice structure where special Sm positions (closed circles in Fig. 3.18c) are replaced with pairs of double Co atoms (open circles) in the $SmCo_5$ structure. Therefore, a unit cell of the Sm_2Co_{17} structure is 3×3 times as large as a unit cell of the $SmCo_5$ structure, and consequently lattice fringes showing 3×3 units in the Sm_2Co_{17} domains can be seen. The interface between the Sm_2Co_{17} and $SmCo_5$ domains has a structural relationship, as shown in Fig. 3.18c. The mismatch in the lattice constants of the Sm_2Co_{17} and $SmCo_5$ structures results in lattice distortion at the interfaces, which can be seen at an oblique angle to the horizontal direction, and this lattice distortion leads to the pinning of the magnetic walls at the interfaces.

The coercivity of this magnetic material is greatly reduced by annealing for a long time (at 800°C for 150h). Figure 3.19a is a transmission electron micrograph of the magnetic material with low coercivity. As can be seen from the electron diffraction pattern inserted, the Sm_2Co_{17} and $SmCo_5$ domains have the same crystallographic relationship as in Fig. 3.18. However, the difference in the lattice constants of the Sm_2Co_{17} and $SmCo_5$ phases has increased, and consequently the separation of the strong reflections, such as the 220 and 222 reflections, can be seen. The domains of Sm_2Co_{17} and $SmCo_5$ have also grown larger and interface dislocations between those domains have appeared. For example, periodic arrays of elongated dislocations, which are inclined against the beam direction, are observed in the upper grain of Fig. 3.19a, and also straight arrays of periodic dark spots, corresponding to interface

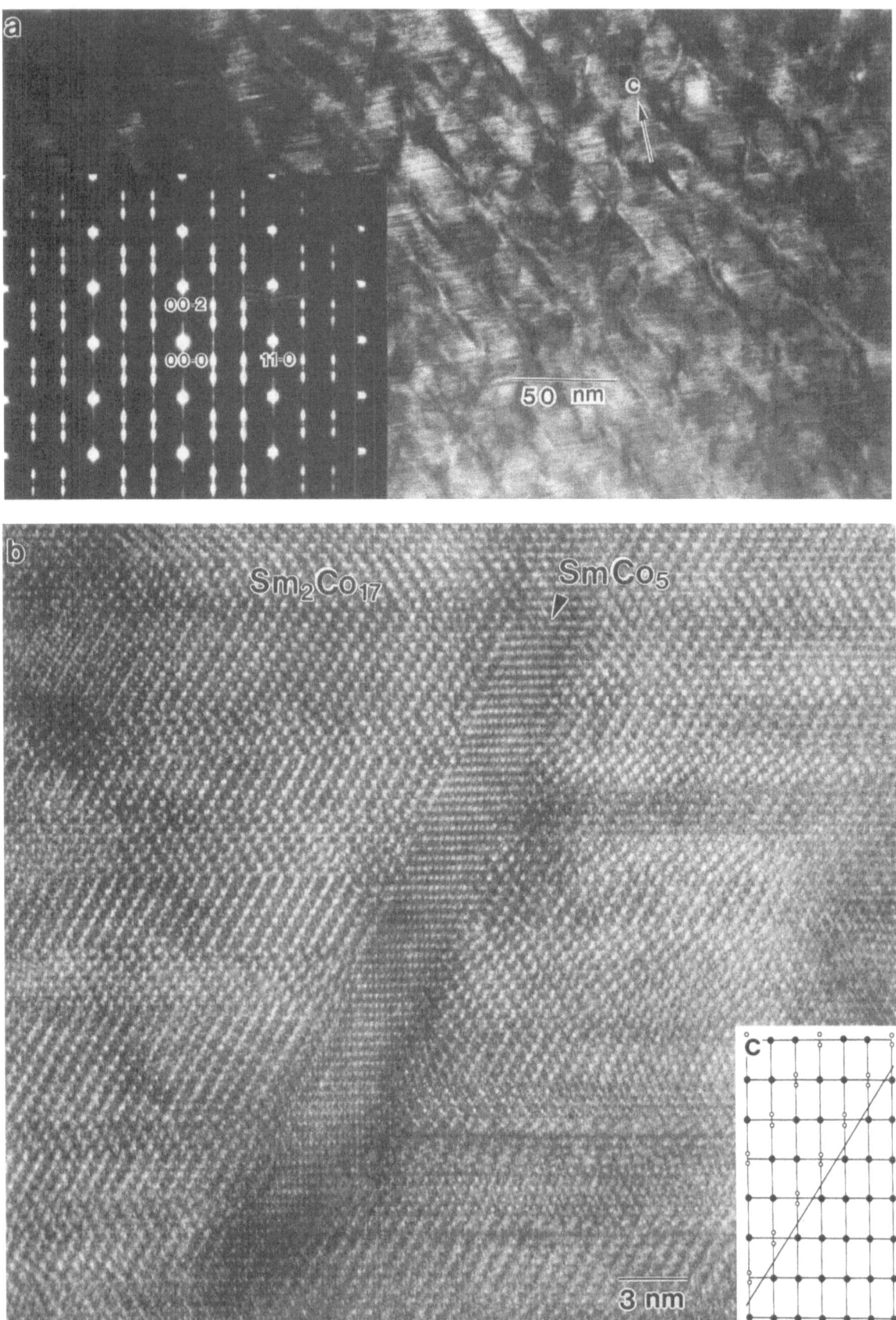

Fig. 3.18. a Bright-field image, and **b** high-resolution image of the Sm–Co system permanent magnet prepared by heat treatment to obtain the highest magnet coercivity.
Specimen: Sm–Co permanent magnet; **Preparation**: jet electropolishing;
Observation: 200 kV EM, [110] incidence for the structure of Sm_2Co_{17}

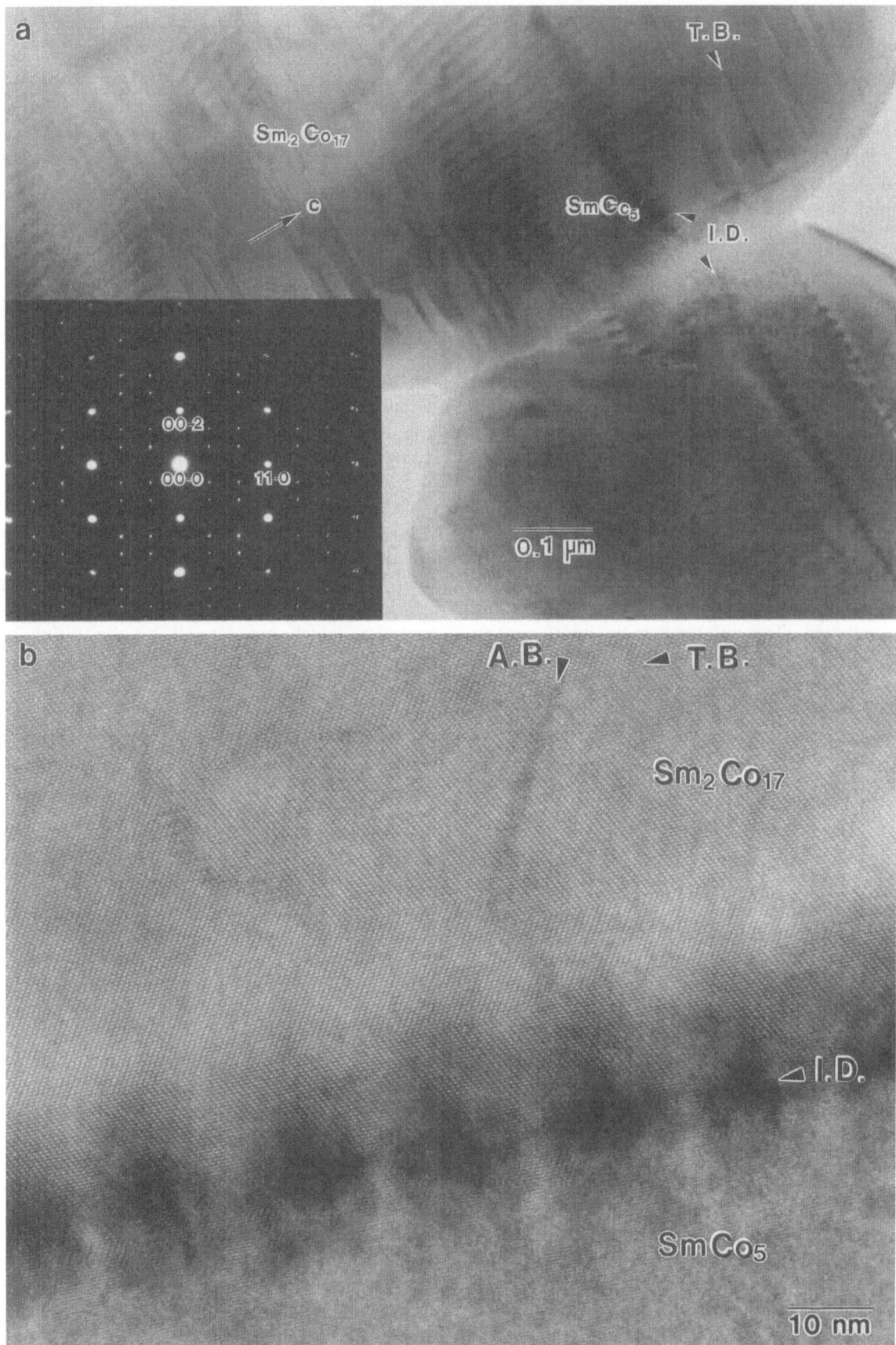

Fig. 3.19. a Bright-field image, and **b** high-resolution image of the Sm–Co system permanent magnet annealed for a long time, which shows low magnetic coercivity.

Specimen: Sm–Co permanent magnet; **Preparation**: jet electropolishing;

Observation: 200 kV EM, [110] incidence for the structure of Sm_2Co_{17}

dislocations which are parallel to the beam direction, can be seen to the right of the lower grain (labeled I.D.).

Figure 3.19b is a lattice image taken with the incident beam parallel to the interface. In the large Sm_2Co_{17} domain, anti-phase boundaries (labeled A.B.) and twin boundaries (T.B.) can be seen, and at the interface between the Sm_2Co_{17} and $SmCo_5$ domains, broad strain fields of interface dislocations (I.D.) are represented as periodic dark regions. From this observation, the interface is seen to be parallel to the (001) plane, and the interface dislocations are located normal to the micrograph.

The interfaces in material with high coercivity are nearly parallel to the c-axis (strictly speaking they are tilted at about 30° (Fig. 3.18c)) and show local lattice distortion. On the other hand, in the specimen with low coercivity, the interfaces are perpendicular to the c-axis, and the misfit of the lattice constants is relaxed by the appearance of interface dislocations, which have long-range strain fields. This difference in the interface structures leads to considerable differences in the magnetic coercivity of the magnets. The magnetization axis of the magnet is parallel to the c-axis, and so the magnetic domain walls tend to run parallel to the c-axis. Thus, the magnetic domain walls in a high-coercivity magnet have surface contact with the interfaces, which are nearly parallel to the c-axis. Conversely, the domain walls in a low-coercivity magnet are touched to the interfaces by a line contact. This difference in contact between the domain walls and the interfaces is believed to cause the different pinning forces in the magnet materials. Also, taking account of the width (a few nanometers) of the domain walls, the long-range strain fields of the dislocations (Fig. 3.19) are considered to produce a weak pinning force compared with the local strain fields of the lattice distortion at the interfaces of Fig. 3.18. It may be concluded that these differences in microstructure produce considerably different magnetic coercivities in the specimens.

Much effort has been devoted to developing composite materials with useful properties which the matrix material does not have. For example, in a Si–Al composite material, small Si crystals precipitate in the Al matrix and a highly wear-resistant material is obtained.

Figure 3.20 shows electron micrographs of rapidly solidified Al–Si alloy powders prepared by a gas-atomization method [28]. The powders, with grain sizes of a few tens to about 100 μm (Fig. 3.20b), are used as bulk specimens by pressing and sintering them. In the powder specimens, small Si crystals are dispersed in the Al matrix, and dislocations twisted around the Si crystals can be seen (Fig. 3.20c).

The Si crystals precipitated in the Al matrix have a definite crystallographic relation ($[110]_{Al}$ // $[110]_{Si}$). This relationship can be seen in the high-resolution image in Fig. 3.20a. The image was taken with the incident beam parallel to the [110] direction of Al and Si crystals, and so two-dimensional lattice images of both the Al and Si crystals can be seen. The Si crystal is composed of five domains divided by five twin planes with a multiple-twinning relation, which is often observed in material particles with a face-centered cubic (fcc) or diamond structure. The multiple-twinned crystal formed with five $(1\bar{1}1)$ twin planes has a misfit of about 7° from 360°, and consequently the [110] axes in the five domains are slightly inclined to each other to relax the misfit. The inclination effect can be seen as slightly different image contrasts, which are formed by the tilting of the crystals from the [110] axis, in the five domains.

Interface structures between the Al matrix and the Si crystal can clearly be seen in the lattice image in Fig. 3.20a, where the interface between the Si crystal and the Al matrix is exactly parallel to the incident beam. At the boundaries between the Al matrix and the A and E domains of Si, one can see semi-coincidence boundaries formed with a definite boundary plane of $(1\bar{1}1)$. A semi-coincidence boundary results from the small misfit between four times the lattice spacing of $(1\bar{1}1)_{Al}$ (0.936 nm), and three times that of $(1\bar{1}1)_{Si}$ (0.939 nm). The relationship can be seen as a bright fringe at every third fringe near the boundary of the E domain. However, there are no definite crystallographic relations between the Al matrix and the B and C domains, and deformed regions with no lattice fringes, like amorphous regions, are observed along the interfaces.

Interfaces between Si_3N_4 with high-covalent bonding and TiN with metallic bonding are now shown [29]. In a Si_3N_4–TiN composite prepared by the CVD method, TiN fibers extend along the [001] direction of the β-Si_3N_4 matrix, and can be seen as moiré fringes in Fig. 3.21a [30]. Figure 3.21b is a lattice image showing cross sections of the TiN fibers embedded in the [001] projection of β-Si_3N_4. The bright dots in the matrix align in a hexagonal lattice with a spacing of 0.76 nm, which corresponds to the lattice constant of β-Si_3N_4. In

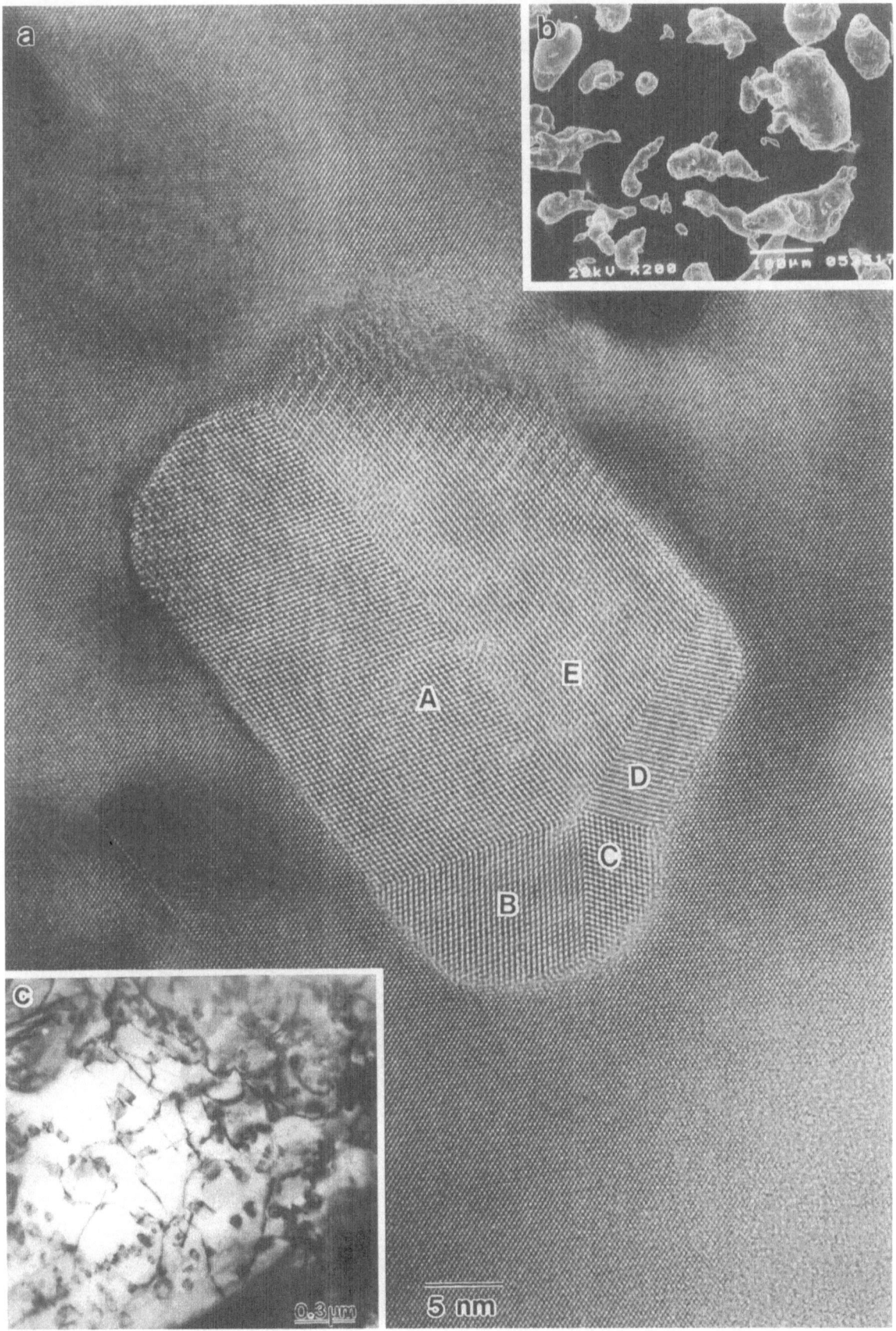

Fig. 3.20. a High-resolution image, **b** scanning electron microscope image, and **c** bright-field image of gas-atomized Al–Si alloy powders.
Specimen: Al–20wt%Si–1wt%Ni alloy powders; **Preparation**: ion milling;
Observation: 400kV EM, [110]$_{Al}$ incidence

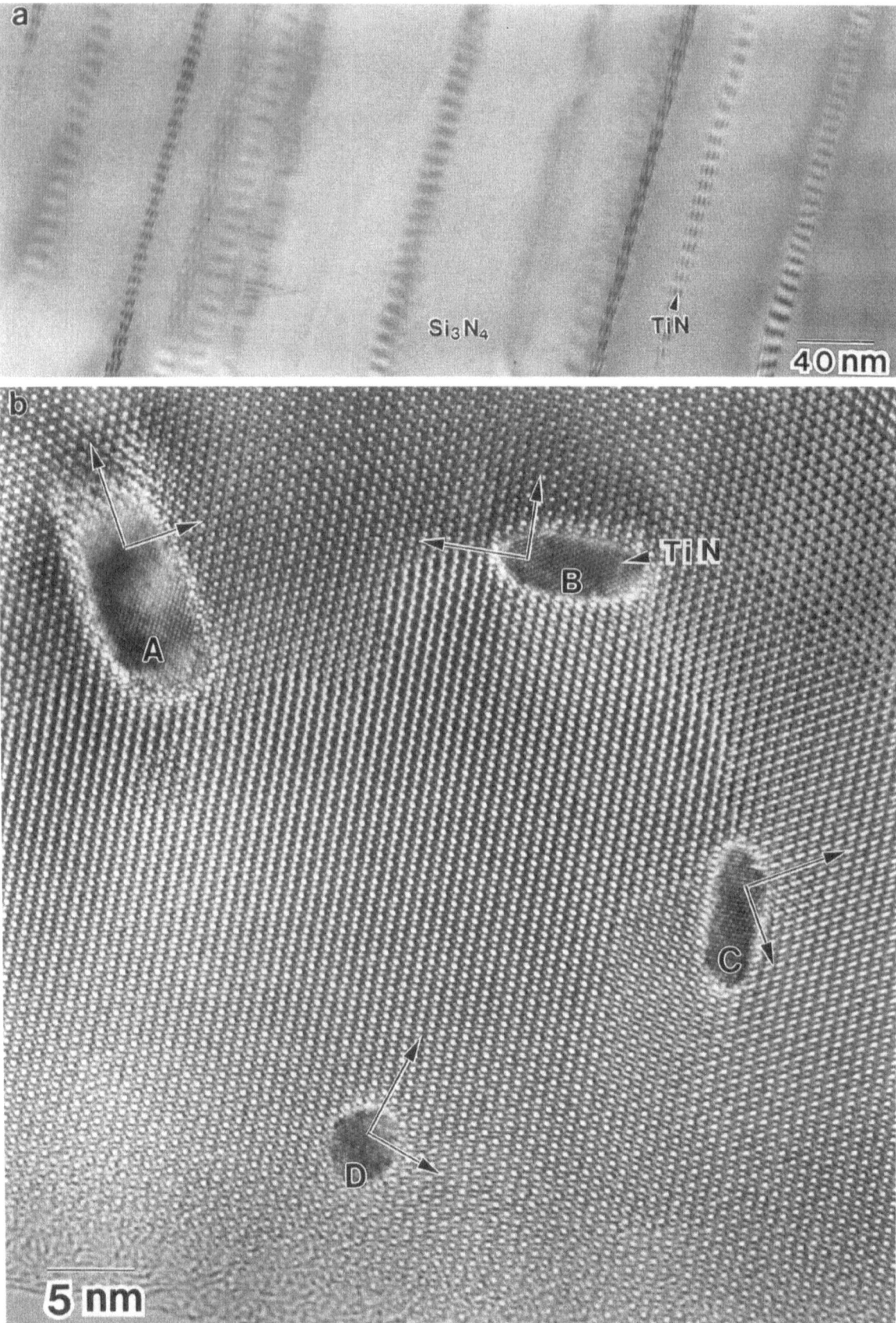

Fig. 3.21. **a** Bright-field image, and **b** high-resolution image of Si_3N_4–TiN composite prepared by CVD.
Specimen: CVD-Si_3N_4–TiN; **Preparation**: crushing;
Observation: 1000 kV EM, incidences of **a** perpendicular and **b** parallel to the [001] direction for β-Si_3N_4.
Remark: The *long* and *short arrows* in **b** correspond to the [001] and [$\bar{1}$10] directions of TiN crystals, respectively

the cross sections of the TiN crystals, small bright dots align in a rectangular lattice corresponding to the [110] projection of an NaCl-type structure. From the high-resolution image in Fig. 3.21b, some crystallographic relationships between the TiN fibers and the β-Si$_3$N$_4$ matrix can be deduced.

The fiber axes of TiN are always $\langle 110 \rangle$, being parallel to the [001] axis of β-Si$_3$N$_4$. From this relationship, it can be understood that the moiré fringes of all the TiN fibers shown in Fig. 3.21a have the same intervals. The moiré fringes result from the difference between the lattice spacing of the (110) planes of TiN and the (001) planes of β-Si$_3$N$_4$.

Also, from Fig. 3.21b, the following relationships can be seen for the TiN crystals A, B, C, and D:

A, B: [001] TiN // [120] Si$_3$N$_4$;
C: [001] TiN // [010] Si$_3$N$_4$;
D: no definite orientation relationship.

The D crystal, with no definite orientation relationship, has a circular cross section, whereas the A, B, and C crystals, with definite relationships, have cross sections with elliptic or rectangular shapes extending in one direction. These crystallographic orientation relationships between TiN and β-Si$_3$N$_4$, and the shapes of the TiN fibers, can be understood from the coincidence of the corresponding lattice spacings of TiN and β-Si$_3$N$_4$, as listed in Table 3.2. That is, the lattice spacing of (110) TiN along the fiber axis is very similar to that of (001) Si$_3$N$_4$, and the spacings of (200) and (222) TiN along the long axis of the elliptic cross sections are similar to those of (300) and (330) Si$_3$N$_4$, respectively.

In an α-Si$_3$N$_4$–TiN composite prepared by the same (CVD) method, it was found that TiN crystals precipitate with no special orientation relationships and have no special shape. This can be considered to be due to the poor coincidence of lattice spacings between TiN and α-Si$_3$N$_4$ compared with those of TiN and β-Si$_3$N$_4$.

Sometimes a reaction at an interface plays an important role in the mechanical properties of a material, and the reaction depends on the fabrication process of that material. It is well known that Si$_3$N$_4$-whisker(Si$_3$N$_{4w}$)-reinforced Al–Mg–Si alloy (Al alloy 6061) composites exhibit superplastic behavior at high strain rates above $10^{-2}\,\text{s}^{-1}$ [31]. However, this superplastic behavior is also known to be very sensitive to the fabrication process of the composites. For example, the composite prepared by hot-press sintering at 873 K at a pressure

Table 3.2. Distances of lattice planes in TiN, β-Si$_3$N$_4$, and α-Si$_3$N$_4$

TiN		β-Si$_3$N$_4$		α-Si$_3$N$_4$	
hkl	d (nm)	hkl	d (nm)	hkl	d (nm)
100	0.424	100	0.663	100	0.669
		110	0.382	110	0.388
		200	0.331	200	0.337
110	0.299	001	0.291	002	0.281
111	0.244	210	0.249	210	0.255
200	0.212	300	0.219	303	0.223
211	0.173	220	0.191	220	0.194
222	0.122	330	0.127	330	0.129

of 390 Mpa for 1.2 ks showed a maximum 75% elongation at a high initial strain-rate of $2 \times 10^{-1}\,\text{s}^{-1}$ at 818 K, whereas the composite prepared by hot extrusion at 773 K with a reduction ratio of 100:1 after hot-press sintering exhibited a large elongation of about 600% in the same conditions [31]. This remarkable difference in superplastic behavior was found to be due to the difference in the interfacial reaction between Si$_3$N$_{4w}$ and the Al matrix. Figure 3.22 shows microstructures of interfaces in the two composites prepared with the different fabrication processes [32]. In Fig. 3.22a, which is a typical transmission electron micrograph of a Si$_3$N$_{4w}$ crystal embedded in the Al matrix in the as-sintered specimen, the cross section of the Si$_3$N$_{4w}$ crystal can be seen to be formed mainly with flat planes parallel to {100} and partially with planes parallel to {110}. Figure 3.22b shows a high-resolution image of a Si$_3$N$_{4w}$–Al-matrix interface. In the image, the Si$_3$N$_{4w}$ crystal, in which lattice fringes corresponding to hexagonal unit cells on the basal plane of β-Si$_3$N$_4$ can be seen, shows an atomistically flat edge parallel to the (100) plane of Si$_3$N$_4$ covered by a thin amorphous layer about 1 nm in thickness, as shown by the pair of arrowheads. The amorphous layer appears to be SiO$_2$, which was present on the surfaces of the raw Si$_3$N$_4$ whiskers. Figure 3.22c is a typical transmission electron micrograph of a Si$_3$N$_{4w}$ crystal embedded in the Al matrix in the extruded specimen. In the image, the edges of the Si$_3$N$_{4w}$ crystal are seen to be slightly deformed, and some precipitate particles appear around the crystal. Figure 3.22d is a high-resolution image of the enclosed rectangular region in Fig. 3.22c. This shows the wavy surface of the Si$_3$N$_{4w}$ crystal, and also reaction phases. It can be concluded that the low melting points of the reaction phases produce partially liquid regions at the interfaces at the tensile-testing temperature of 818 K, and cause high-strain-rate superplasticity [32].

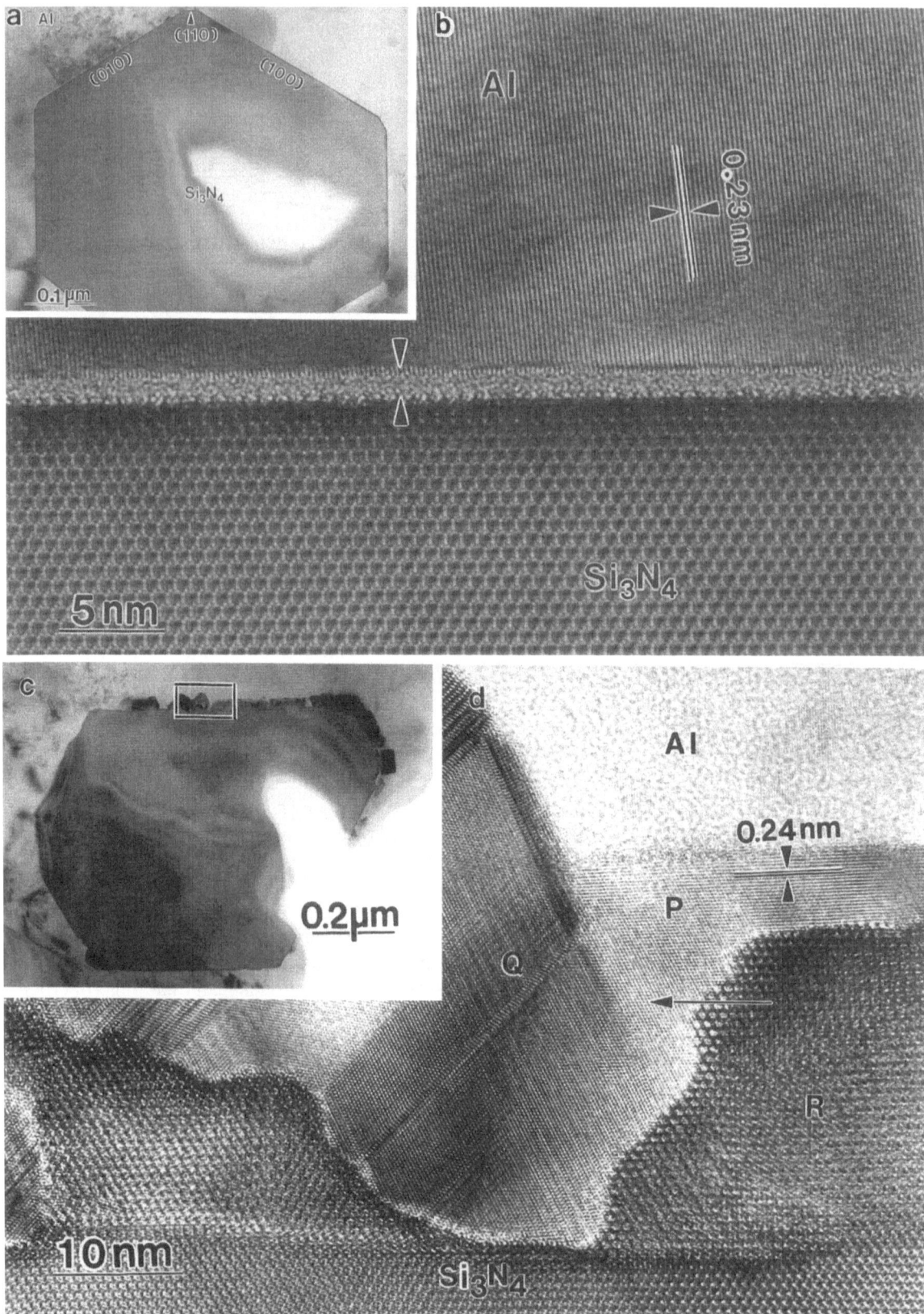

Fig. 3.22. a, **c** Bright-field images of Si_3N_4 crystals embedded in the Al-matrix of as-sintered and hot-extruded specimens, respectively. **b**, **d** High-resolution images of interfaces between the Al-matrix and Si_3N_4 in the as-sintered and hot-extruded specimens.

Specimen: Si_3N_4-whiskers/Al-alloy composite; **Preparation**: ion milling; **Observation**: 400 kV EM

Remarks: In **b**, the interface is covered by a SiO_2 amorphous layer, showing no interfacial reaction. In **c** and **d**, the interfacial reaction between the Al-matrix and Si_3N_4 can be seen

3.1.3 Surfaces

Internal structure with some periodicity terminates at the surface. Thus, the surface of a material may be considered in some senses to be a planar defect. At the surface, an atomic arrangement which is different from the internal structure often appears. For example, the so-called 7×7 structure forms at the surface of the (111) plane in silicon, and this has been extensively studied by LEED (low-energy electron diffraction) and RHEED (reflection high-energy electron diffraction). With transmission electron microscopes, the surface has been studied by conventional electron diffraction [33] and also by REM (reflection electron microscopy) [34]. Recently, the surface structure has been investigated even more extensively by STM (scanning tunneling microscopy).

In transmission electron microscopy, there are two ways to observe the surface of a material. One is using an incident electron beam which is almost perpendicular to both top and bottom surfaces, such as region P in Fig. 3.23. This observation mode is called a *plan view*. In this observation mode, the surface structure can be observed directly, but the surface structures at the top and the bottom may overlap, and these surface structures may also overlap with the internal structure. The signal obtained from the surface is much smaller than that from the internal structure, and thus analysis of the surface structure is generally difficult. Nevertheless, the structure on the (111) surface in silicon has recently been investigated by high-resolution electron microscopy with an ultrahigh vacuum electron microscope [35]. The other observation mode is illustrated as region E in Fig. 3.23, where the incident electron beam is parallel to the surface. This observation mode is called an *edge-on view* or a *profile view*. In this mode, the cross section of the surface structure is investigated and a relatively strong image signal can be obtained comparable to that from the internal

structure. High-resolution images obtained using an edge-on view will now be considered.

In the study of surface structures by transmission electron microscopy, the degree of vacuum in the specimen chamber should generally be 10^{-7}–10^{-8} Pa. However, in the study of special materials such as oxides, which have stable surfaces and are not oxidized even in air, their surface structure can be studied by conventional electron microscopes. The high-resolution images presented below were obtained using conventional electron microscopes.

As will be noted in Sect. 4.6.1, superconducting oxides prepared by sintering are crushed to get cleaved surfaces for electron microscopy. In the cleaved surface obtained, the characteristic features of the surface structure of these superconductors can be observed. Figure 3.24a and b show structure images of $TlBa_2CaCu_2O_7$ and $YBa_2Cu_3O_7$, respectively [13]. In both images, surfaces cleaved almost parallel to the *c*-plane are observed, and cations projected along the incident electron beam appear as dark dots. Thus, from the arrangements of dark dots around the surface, one can specify the atomic plane cleaved. In the image of $TlBa_2CaCu_2O_7$ (Fig. 3.24a), cleavages occur between the Tl-plane and the Ba-plane, as shown in the inset, while in $YBa_2Cu_3O_7$ (Fig. 3.24b), cleavages occur between the Ba-plane and the Cu-plane, as illustrated in the inset.

However, different features are observed in surfaces cleaved at the plane almost perpendicular to the *c*-plane. Figure 3.25a and b show structure images of $TlBa_2Ca_3Cu_4O_{11}$ and $Pb_2Sr_2Y_{0.5}Ca_{0.5}Cu_3O_8$, respectively [36]. In both cases, irregularities in the atomic arrangements or surface reconstructions can be seen. In Fig. 3.25b in particular, it is clear that the Pb layers are preferentially concave, resulting in surface reconstruction. It is not possible to determine whether the surface reconstruction was originally caused in the air or in the organic solvent, but it can be said that the cleaved surface, being parallel to the *c*-plane, is unstable compared with the one normal to the *c*-plane.

Figure 3.26 shows high-resolution images of the surface structure in iron oxides (hematite: α-Fe_2O_3). Thin specimens were prepared by *ultramicrotomy* (see Sect. 4.6.4). Due to different growth conditions, particles have (Fig. 3.26a) pseudocubic [37], (Fig. 3.26b) peanut [38], or (Fig. 3.26c) hexagonal platelet shapes. Pseudocubic and peanut-shaped particles consist of small single crystals. In the former case, all the particle surfaces consist of

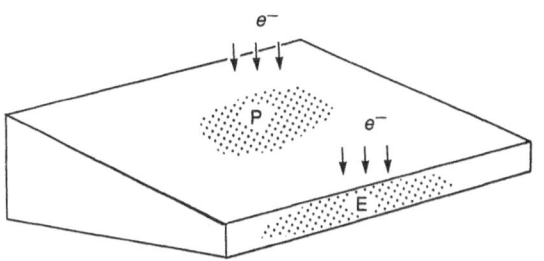

Fig. 3.23. Two observation modes for crystal surfaces

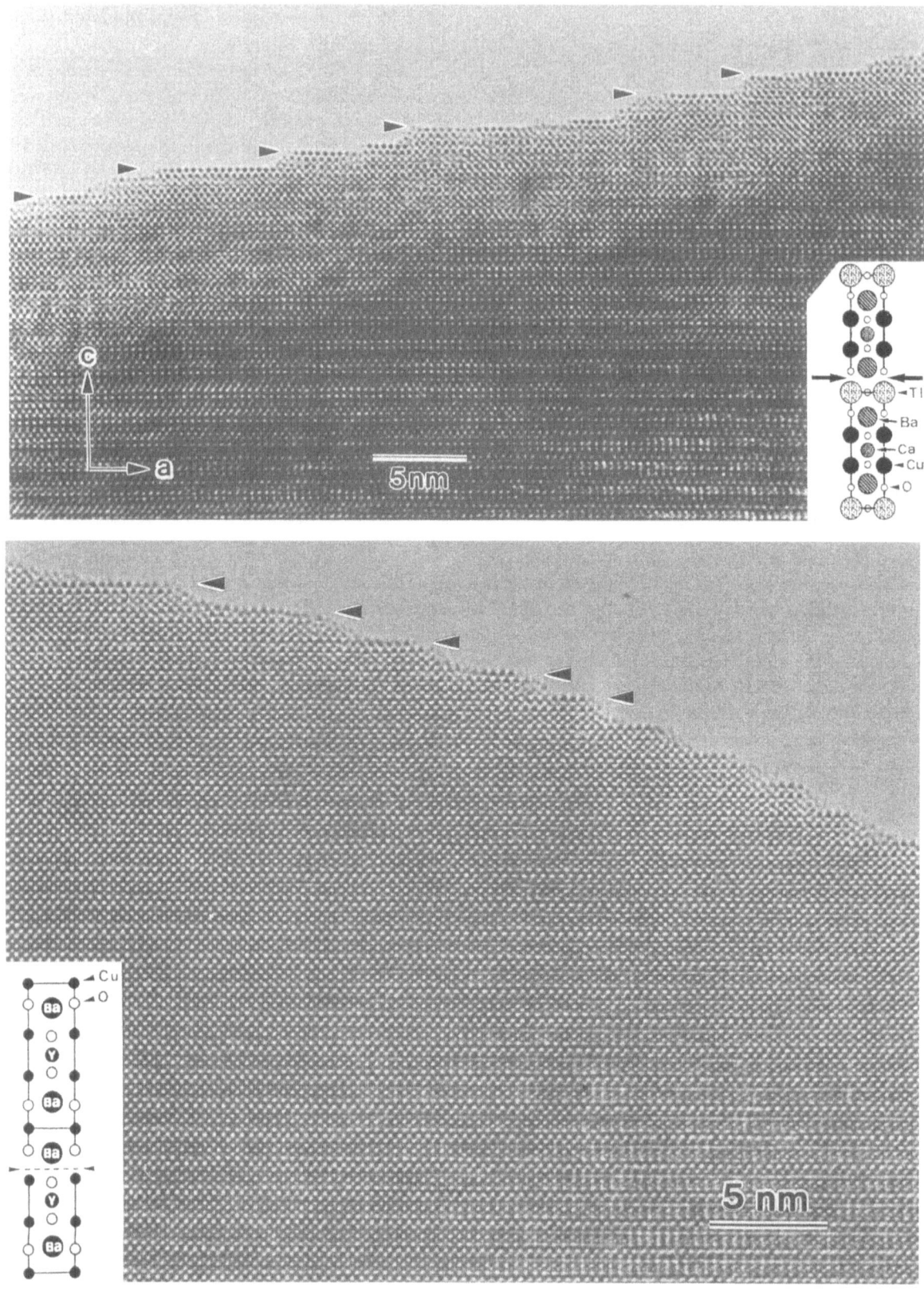

Fig. 3.24. Structure images showing the cleavage surface structure of superconducting oxides **a** $TlBa_2CaCu_2O_7$ and **b** $YBa_2Cu_3O_7$.
Specimen: **a** $TlBa_2CaCu_2O_7$ and **b** $YBa_2Cu_3O_7$ superconducting oxides; **Preparation**: crushing;
Observation: 400 kV EM.
Remark: Cleavages happen at the places indicated by the arrows in the inset structure models

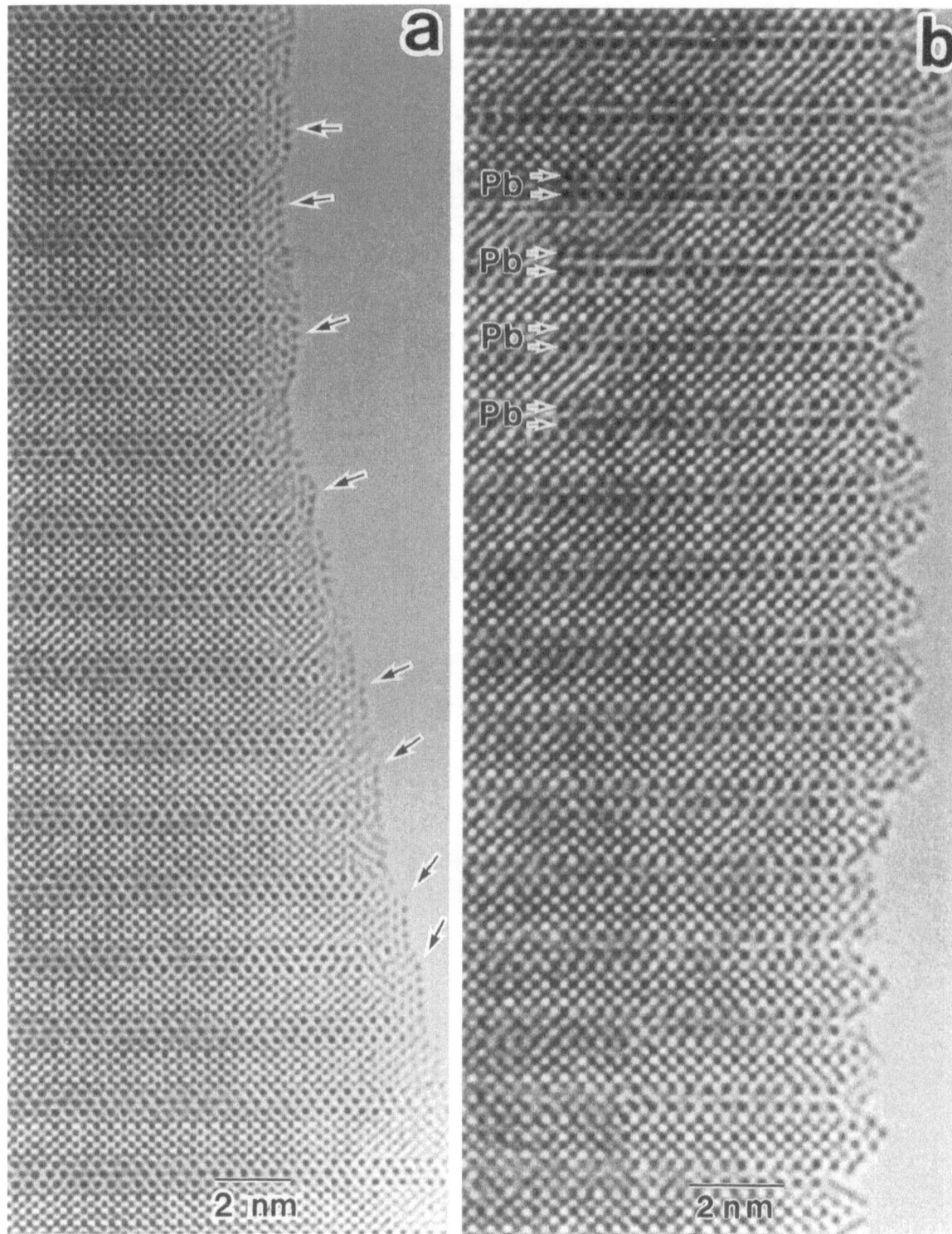

Fig. 3.25. Structure images showing the surface structure of superconducting oxides **a** $TlBa_2Ca_3Cu_4O_{11}$ and **b** $Pb_2Sr_2Y_{0.5}Ca_{0.5}Cu_3O_8$.

Specimen: **a** $TlBa_2Ca_3Cu_4O_{11}$ and **b** $Pb_2Sr_2Y_{0.5}Ca_{0.5}Cu_3O_8$; **Preparation**: crushing; **Observation**: 400 kV EM.

Remark: Reconstruction can be seen near the surface. Especially surface reconstruction is clearly observed at the regions indicated by *arrows* in **a**

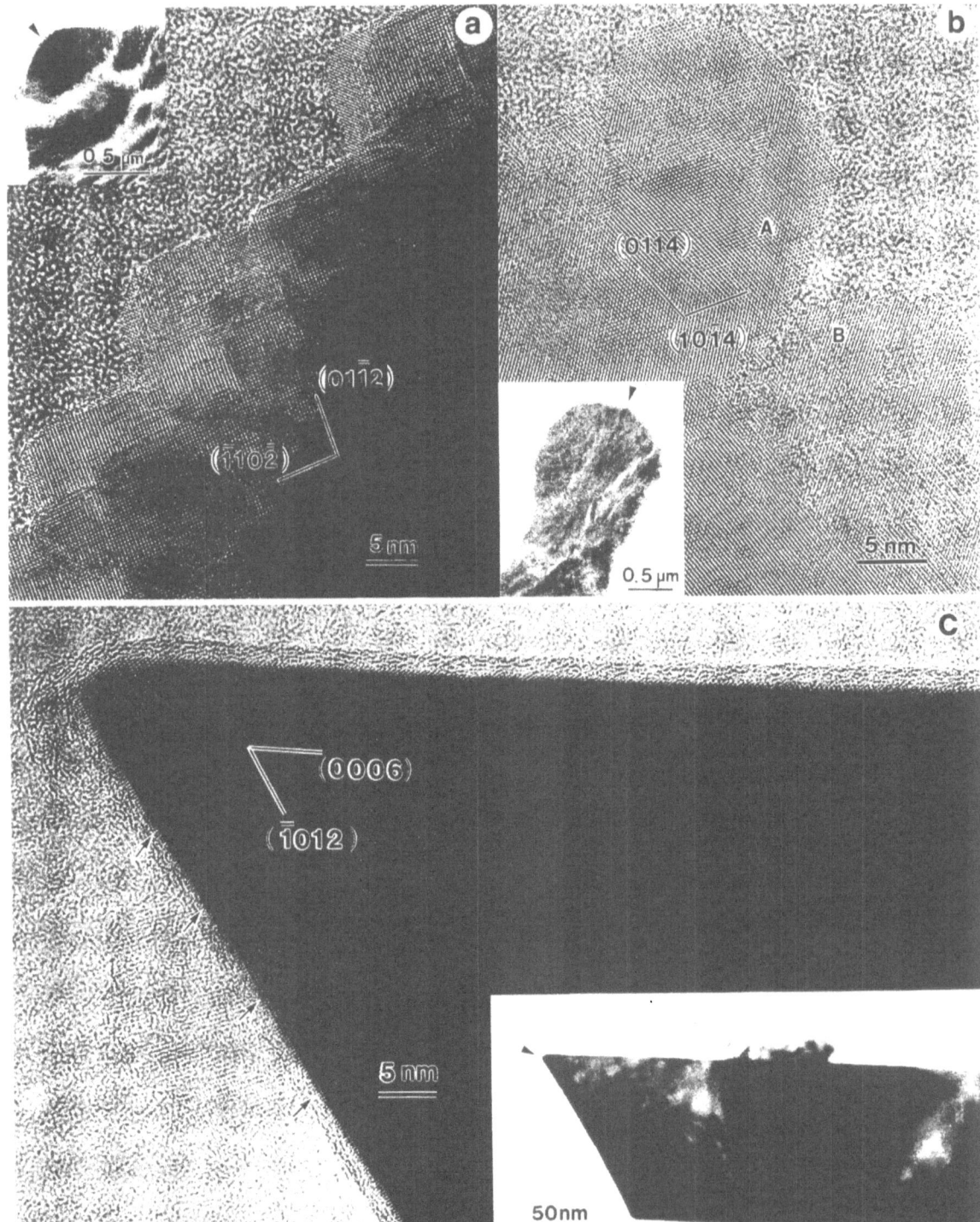

Fig. 3.26. Surface structures of hematite particles: **a** pseudocubic; **b** peanut-type; **c** platelet-type.
Specimen: α-Fe$_2$O$_3$; **Preparation**: ultramicrotomy; **Observation**: 1250 kV EM.
Remarks: The surface of a pseudocubic particle consists of the {01$\bar{1}$2} plane, and that of a platelet-type particle consists of the (0001) and ($\bar{1}$012) planes, while no specific crystal planes are observed in the surface of a peanut-type particle

{10$\bar{1}$2} surfaces[3] of small single crystals, while in the latter case, small single crystals are round, and therefore special crystal planes do not form the surface. However, in the cross section of a hexagonal platelet particle shown in Fig. 3.36c, it is seen that the particle is a single crystal and the surfaces in the basal and side planes consist of the (0001)

___

[3] Hematite has a corundum structure like Al_2O_3, and their crystal planes or reflections are indexed on the basis of the hexagonal system. As shown in Fig. 3.26, indices based on a system with four axes are sometimes used. The relationships of crystal directions and crystal planes between systems with three and four axes are:

$$a' = (2a - b)/3, \qquad b' = (2b - a)/3,$$
$$d' = -(a + b), \qquad c' = c, \qquad h' = h, \qquad k' = k,$$
$$i' = -(h + k), \qquad l' = l$$

where the crystal direction and plane in three-axis notation are [$a\ b\ c$] and ($h\ k\ l$), respectively, while those in four-axis notation are [$a'\ b'\ d'\ c'$] and ($h'\ k'\ i'\ l'$), respectively.

and ($\bar{1}$012) planes, respectively. Small steps on the surface are also seen, as indicated by the arrows in the side plane. The difference in surface structures caused by their growth conditions results in these characteristic shapes.

Figure 3.27a is an edge-on view of the growth surface of EMT zeolite. In the high-resolution image, a step consisting of a structural unit is seen on the (100) surface. This is important structural information in attempts to understand its crystal growth mechanism. Figure 3.27b shows a high-resolution image of Pt particles sputtered on the surface of LTL zeolite [39]. As indicated by the arrows, Pt clusters of about 1 nm in size line up with the same periodicity as the LTL crystal. Thus, it is seen that the Pt clusters are situated in special positions in the LTL crystal, probably in the wide open channel positions. Zeolite *containing* Pt clusters are quite important industrially as a catalyst, and thus these images provide important information for understanding the action mechanism of catalysts.

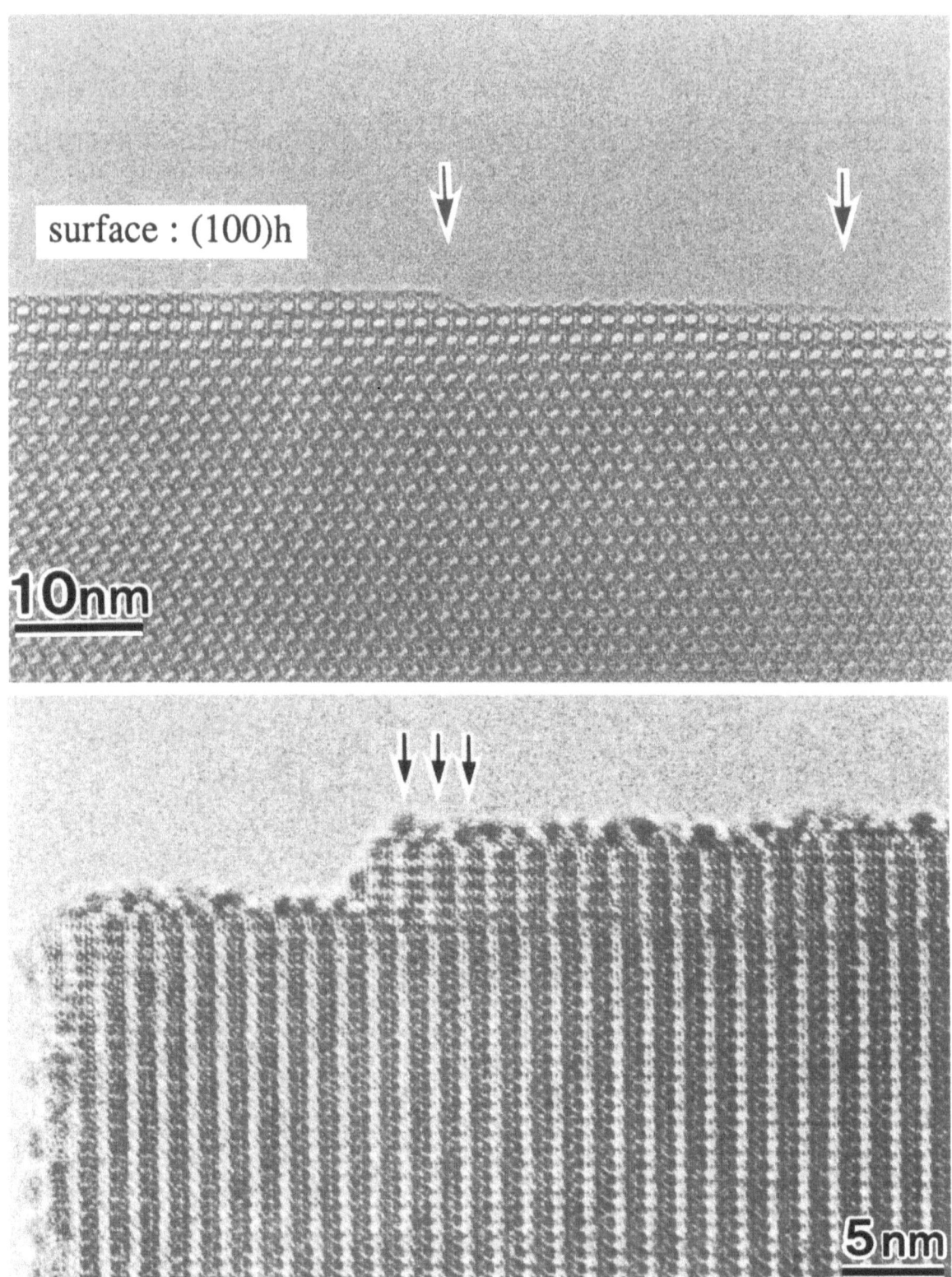

Fig. 3.27. **a** High-resolution image of the crystal growth surface of an EMT zeolite, and **b** high-resolution image of Pt clusters (*arrows*) on the surface of an LTL zeolite.

Specimen: **a** EMT zeolite, and **b** LTL zeolite with Pt clusters prepared by the sputtering method;

Preparation: dispersed on microgrids by using an organic solvent; **Observation**: 400 kV EM.

Remarks: **a** Steps (*arrows*), each consisting of one structural unit produced during crystal growth, are observed on the (100)h plane of the hexagonal structure. **b** Pt clusters (*arrows*) arranged with a periodicity corresponding to the zeolite surface structure. These images were kindly provided by Drs. O. Terasaki and T. Ohsuna, Tohoku University

3.1.4 Other Structural Defects

As well as the lattice defects mentioned above, there are a variety of other structural defects in crystals. These defects are divided into two types: thermodynamically stable and thermodynamically unstable. Unstable defects, like lattice defects such as dislocations and stacking faults, can often be removed by heat treatment. On the other hand, stable defects are such structural defects introduced from nonstoichiometric compositions. This type of defect is important when attempting to understand the structures and properties of nonstoichiometric compounds. This section considers the structural defects in nonstoichiometric oxides.

The metal oxides MoO_3, ReO_3, and WO_3 have the highest oxygen content. The structure of these oxides, known as a ReO_3 structure, is based on a fundamental structural unit of a RO_6 octahedron composed of six oxygen atoms at the vertices and a metal atom R (R = Mo, Re, W) at the center. They form a cubic arrangement of the octahedra by vertex-sharing, as shown in Fig. 3.28. By increasing the content of metal elements from a stoichiometric composition of RO_3 to $R_{1+x}O_3$, a variety of structural defects are formed to accommodate nonstoichiometric compositions. Repre-

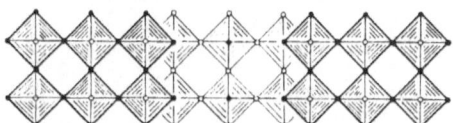

Fig. 3.29. CS planes of edge-sharing formed by shifting RO_3 blocks relative to each other

sentative of these defects is a *Perovskite structure*, in which extra metal atoms occupy the body-centered position in the ReO_3 structure. The structures of most of the high-Tc superconducting oxides are based on this structure (Sect. 3.2.2). Some, special structural defects which accommodate nonstoichiometric compositions through changes of framework structure are now presented.

The simplest way to make nonstoichiometric compositions of $R_{1+x}O_3$ is to change some of the vertex-sharing of the RO_6 octahedra to edge-sharing, as shown Fig. 3.29. The change can be established by shifting RO_3 blocks relative to each other. The boundaries at which the shifting happens are called crystallographic share planes (*CS planes*). The structure of a Nb_2O_5 oxide is formed by two-dimensional CS planes [40].

Figure 3.30 is a high-resolution structure image of a Nb–O–F compound. In this image, bright spots correspond to channels in the arrangement of the octahedra, and dark spots to positions of the octahedra. The cubic arrangements of the octahedra (blocks of the ReO_3 structure) are divided by two-dimensional CS planes, which are seen as dark regions. The CS planes are arrayed in straight lines in the vertical direction, but in the horizontal direction they are on zig-zag lines. The modulation of the CS planes in the horizontal direction causes diffuse scattering in the horizontal direction in the diffraction pattern inserted in Fig. 3.30.

Figure 3.31 is a structure image showing a shock-loading structure change in a single crystal of a Nb_2O_5 compound at a pressure of 54 Gpa [41]. Although there are some deformed areas of image contrast due to the heterogeneity of the specimen thickness, and some disturbance of structures along the incident beam, one can clearly see characteristic structure changes by the shock wave. That is, the ordered arrangement of ReO_3 blocks in the Nb_2O_5 compound undergoes considerable changes with regard to the sizes and

Fig. 3.28. ReO_3 structure based on the fundamental structure unit of a RO_6 octahedron composed with six oxygen atoms at the vertices and a metal atom at the center

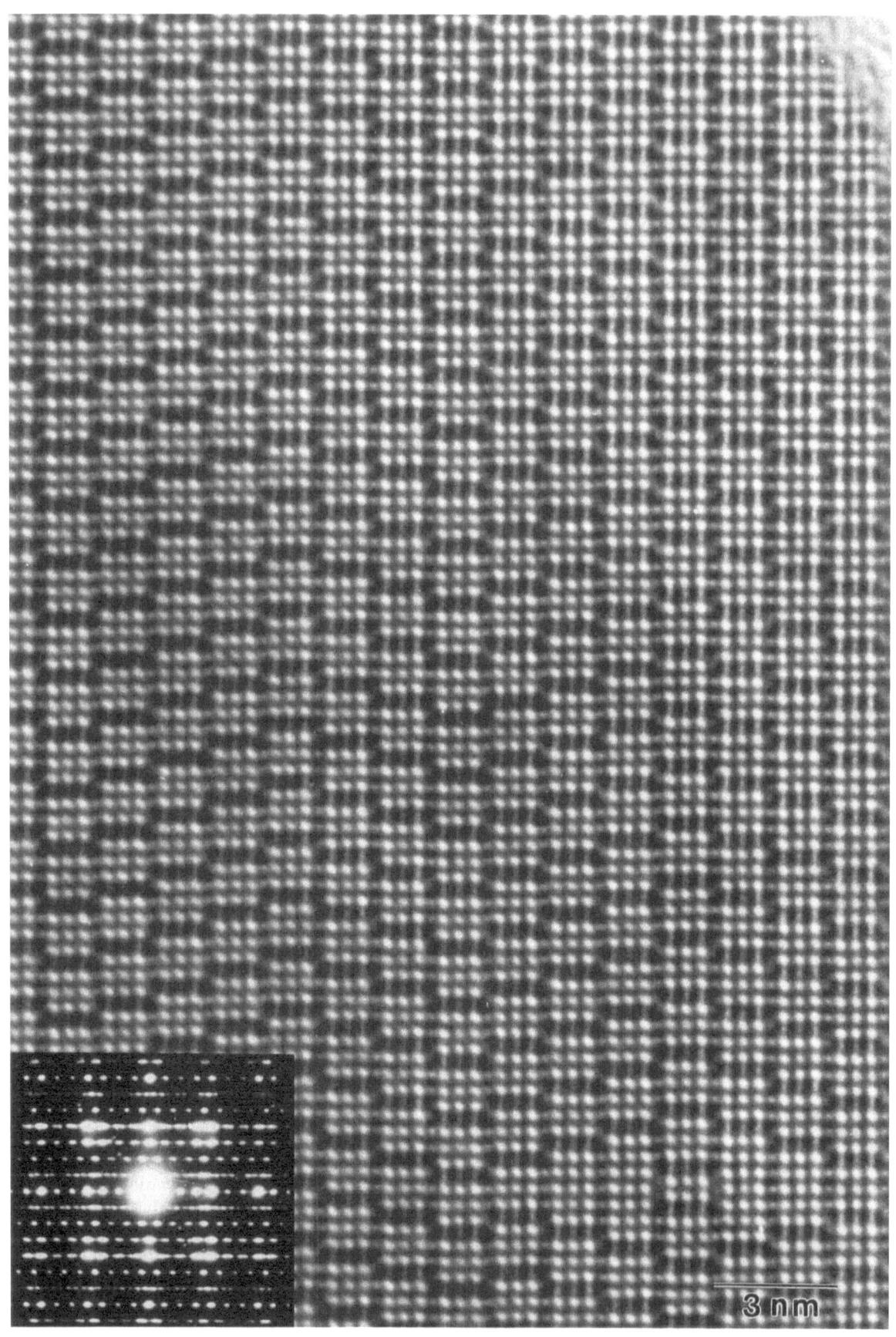

Fig. 3.30. Structure image and electron diffraction pattern of a Nb–O–F compound.
Specimen: Nb–O–F; **Preparation**: crushing; **Observation**: 1000 kV EM

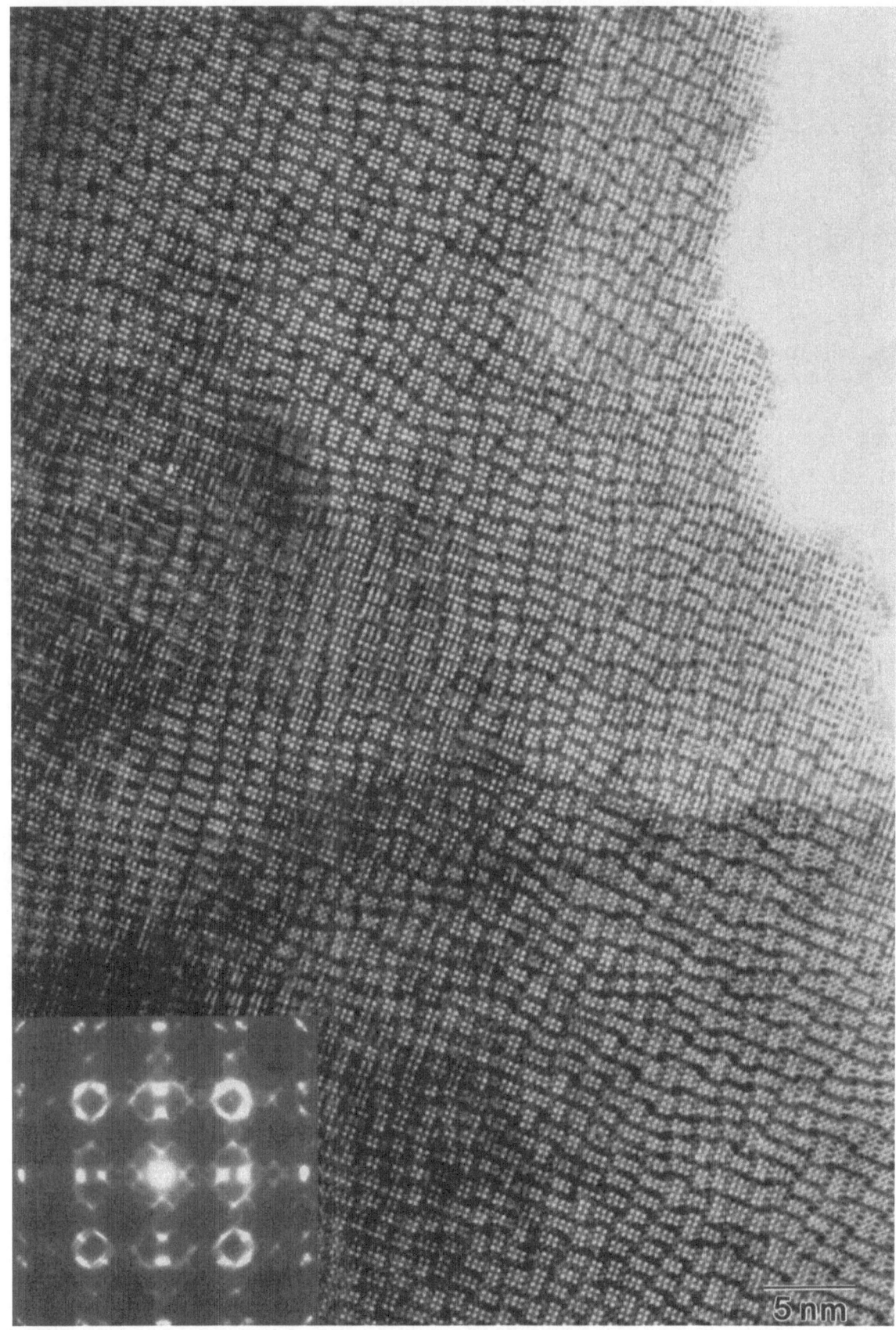

Fig. 3.31. Structure image and electron diffraction pattern of a Nb_2O_5 compound loaded with a shock pressure of 50 GPa.

Specimen: Nb_2O_5; **Preparation**: crushing; **Observation**: 200 kV EM

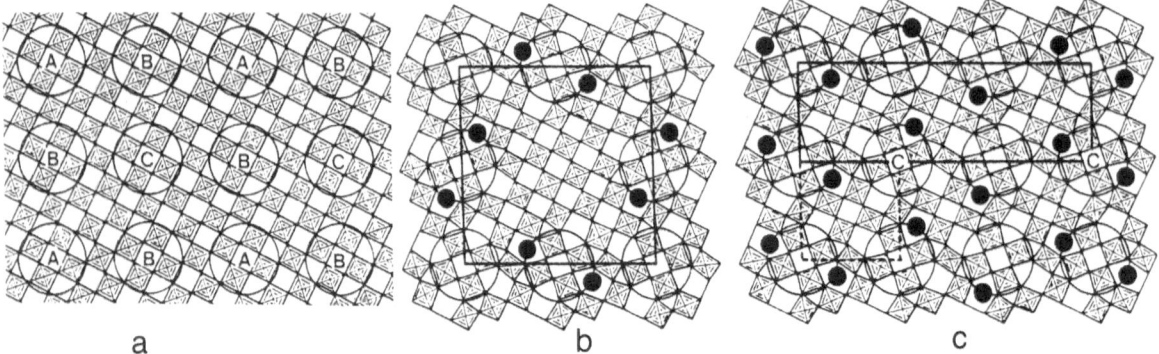

Fig. 3.32. Structures formed by a rotation fault (RF). **a** ReO_3 structure. **b** $7WO_3 \cdot 2Ta_2O_3$ formed by the rotation of A and B in **a**. **c** $9WO_3 \cdot 4Ta_2O_3$ formed by the rotation of A, B, and C

shapes of the ReO_3 blocks, although the overall structure of blocks is basically unchanged. This observation shows that only the arrangement of ReO_3 blocks is modulated by the shock wave while the ReO_3 structure units are not destroyed. This modulation causes characteristic diffuse scattering in the electron diffraction pattern inserted in Fig. 3.31.

Now we consider a different defect, which undergoes structural change by the modulation of the framework structure. By rotating the columns of a 2×2 rhombohedral block (indicated by circles A, B, and C in Fig. 3.32a) through 45°, pentagonal columns are produced which are larger than the cubic columns, as shown in Fig. 3.32b and c [40]. Parts of the pentagonal columns are occupied by extra metal atoms with ordered arrangements, so new structures with $R_{1+x}O_3$ nonstoichiometric compositions are formed. This type of defect is hereafter called a *rotation fault* (RF). It is well known that extra metal atoms are located at two diagonal channels in the four pentagonal channels produced by a RF. By adding a small amount of Ta_2O_5 to WO_3, arrays of RFs along one direction are formed (Fig. 3.33). Figure 3.33 shows arrays of RFs (indicated by black circles) at almost constant distances, and twin boundaries (TB). The bright and dark spots in the image correspond to channels without atoms and atom columns, respectively.

By increasing the Ta_2O_5 content, a periodic arrangement of one-dimensional arrays of RFs is produced, and then a two-dimensional structure (Fig. 3.32b), formed by the rotation of A and B in Fig. 3.32a, appears. Figure 3.34 is a structure image of a WO_3–$1/4Ta_2O_5$ specimen [42]. In the image, a periodic arrangement of one-dimensional arrays of RFs can be seen at the bottom, with local two-dimensional arrangements of RFs at the top. The large square outlined at the top corresponds to a unit cell of Fig. 3.32b, and the small square outlined at the bottom to a unit cell of the structure formed by the rotation of B only in Fig. 3.32a. In Fig. 3.34, it should be noted that dark spots corresponding to metal atom positions in a thin region at the right-hand side are extended in the horizontal direction in a thick region at the left-hand side. This feature results from the misalignment of crystal tilting, and this effect is enhanced at the thick region (see Sect. 2.2.3).

With increasing Ta_2O_5 content, the rotation of the C blocks (Fig. 3.32a) occurs in addition to that of the A and B blocks to accommodate the large number of metal atoms, as can be seen in Fig. 3.35. In the image, one can see a structure formed by the rotation of A and B (Fig. 3.32b) towards the top, and a structure formed by the rotation of A, B, and C (similar to the structure of Fig. 3.32c) towards the bottom.

The specimens in Figs. 3.33–3.35 were prepared by quenching them in water from liquid phases, and so they have imperfectly ordered structures. In the images of those specimens, one can see RFs consisting not only of the rotations of A and B, but also that of C, accommodating extra metal atoms with increasing Ta_2O_5 content.

Figure 3.35 was taken with a 400 kV electron microscope with a resolution of 0.17 nm [43]. One can see the difference in resolution between Fig. 3.35 and Figs. 3.33 and 3.34, which were taken with a 200 kV electron microscope with a resolution of 0.23 nm. Although in Fig. 3.35 the effect of the

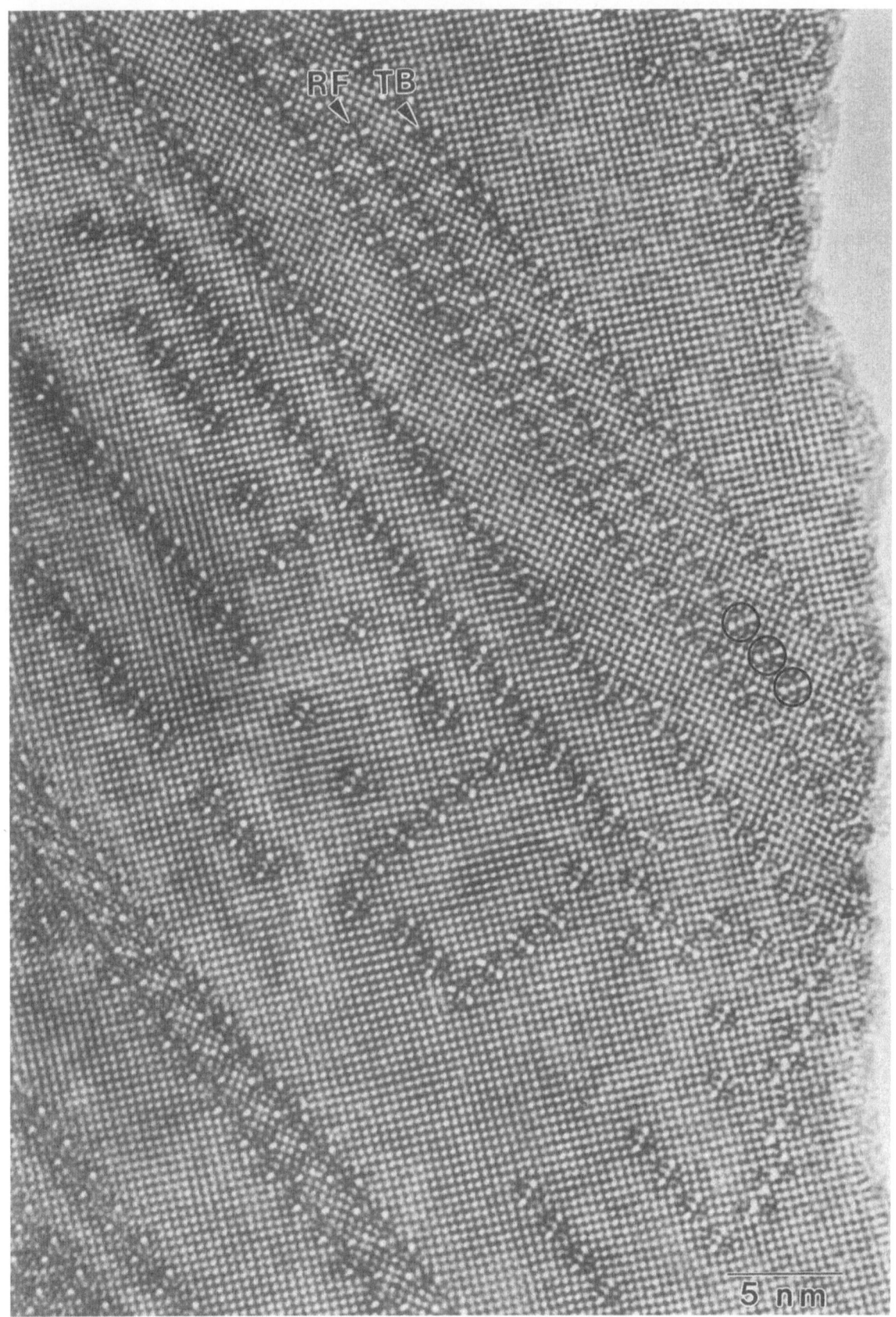

Fig. 3.33. Structure image showing the one-dimensional arrangements of rotational faults in a WO_3–Ta_2O_3 oxide.
Specimen: WO_3–$1/8Ta_2O_3$; **Preparation**: crushing; **Observation**: 200 kV EM.
Remark: Twin boundaries (TB) as well as the one-dimensional arrangements of rotational faults (RF) can be seen

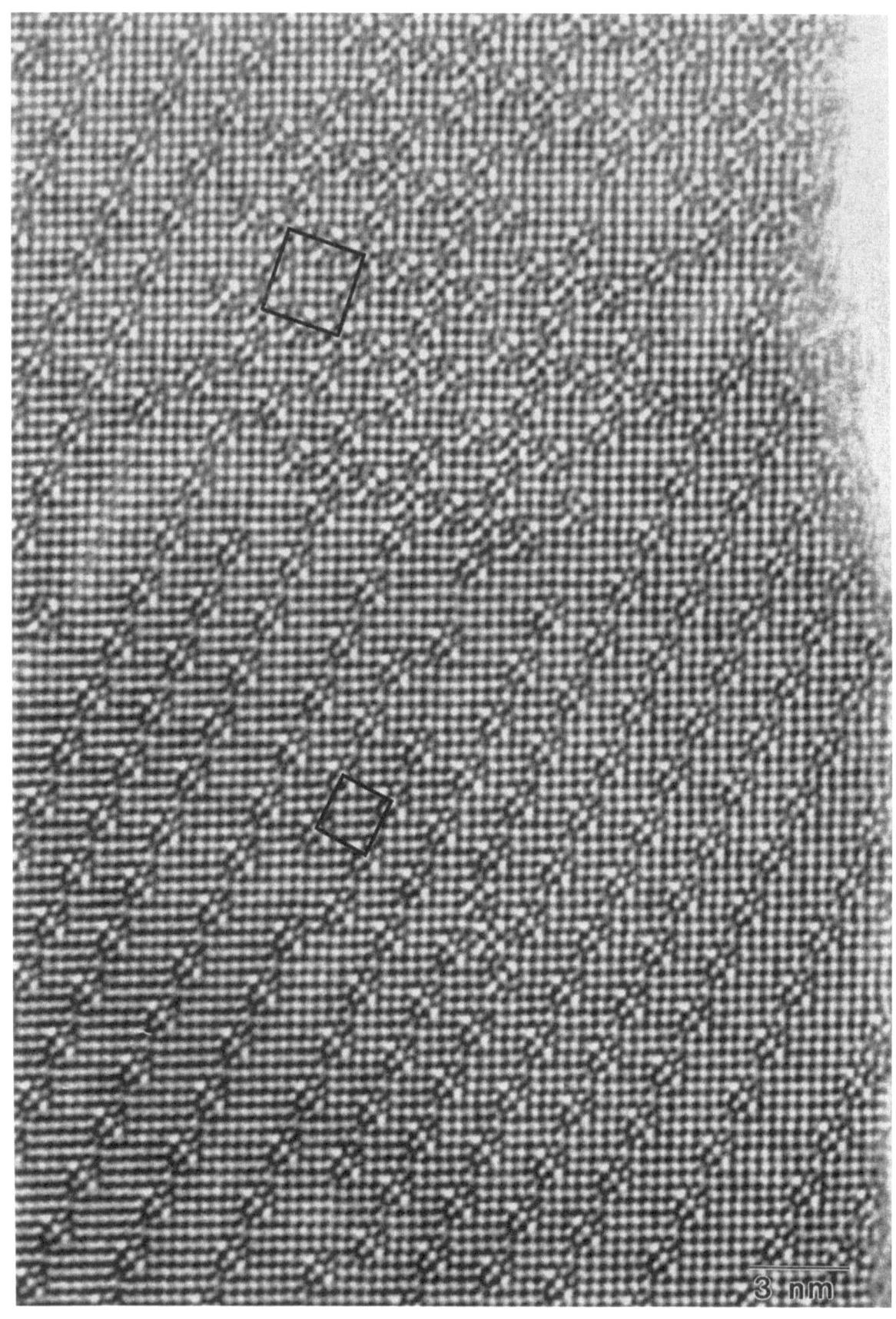

Fig. 3.34. Structure image showing one- and two-dimensional arrangements of rotational faults in a WO_3–Ta_2O_3 oxide.
Specimen: WO_3–$1/4Ta_2O_3$; **Preparation**: crushing; **Observation**: 200 kV EM

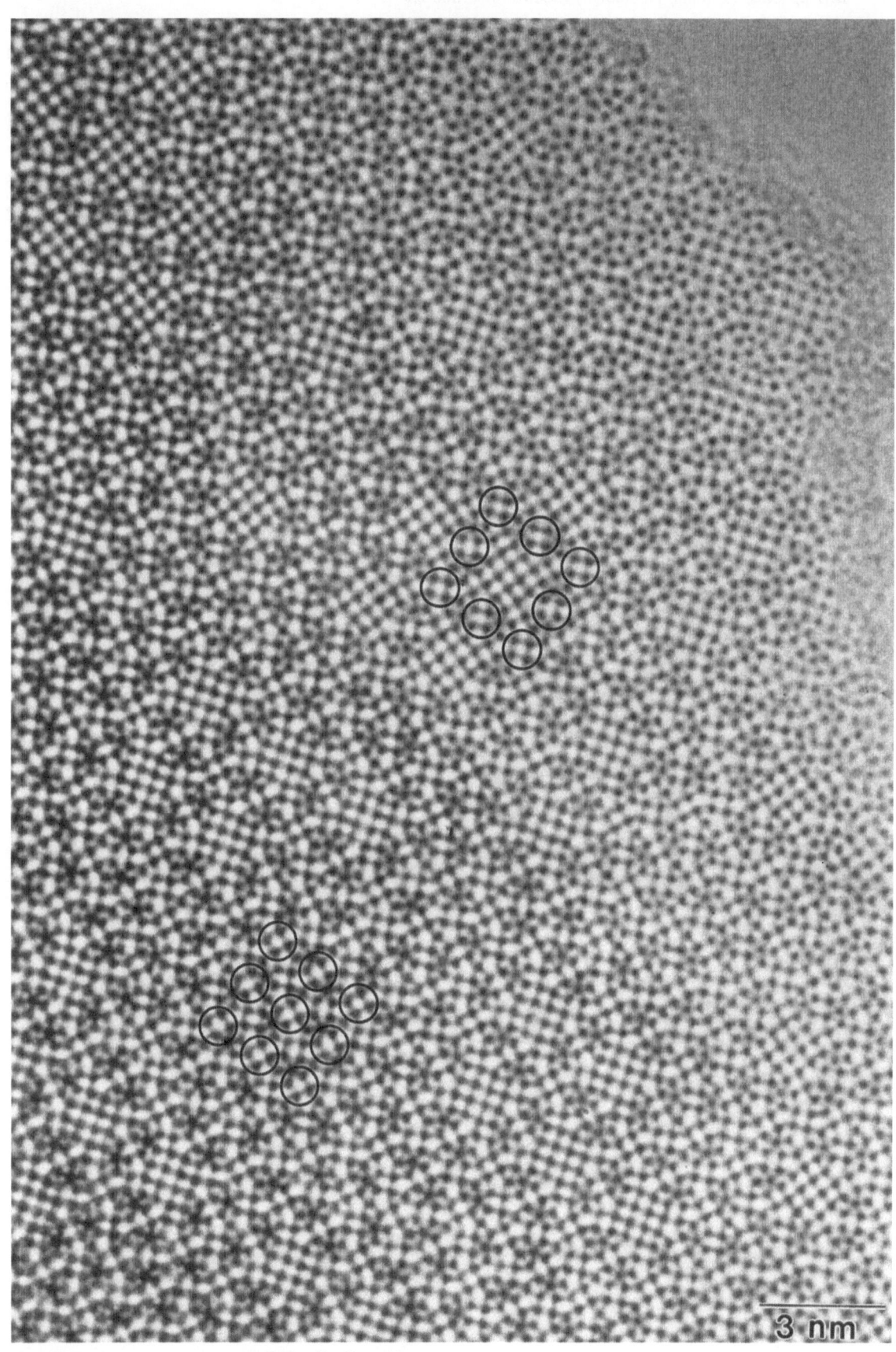

Fig. 3.35. Structure image of a WO_3–Ta_2O_3 oxide.
Specimen: WO_3–$1/4Ta_2O_3$; **Preparation**: crushing; **Observation**: 400 kV EM.
Remark: The black circles indicate rotational faults

misalignment of crystal tilting can be seen as asymmetric arrangements of bright and dark spots in a thick region at the left-hand side, the dark spots in a thin region at the right-hand side faithfully reflect the positions of metal atoms. Therefore, it is possible to determine the positions of metal atoms from the image in that thin region. However, it is hard to get information about the positions of metal atoms from the images in Figs. 3.33 and 3.34, although information about the characteristic features of the defects can be obtained.

Figures 3.30, 3.31, 3.33–3.35 show aperiodic structures. High-resolution electron microscopy is an indispensable tool for studying such structures.

3.2 High-Resolution Images of Various Materials

3.2.1 Ceramics

Ceramics such as SiC, Si_3N_4, and Al_2O_3 have been studied extensively as high-temperature structural materials to try to overcome their severe inherent brittleness. These studies have involved attempts to control the microstructure to prevent the propagation of cracks.

In electron microscopic studies of structural ceramics, it is important to investigate their microstructures, particularly the structures of grain boundaries and interfaces, and also to reveal the relationships between the microstructure and crack propagation, i.e., the fracture mechanisms. Since the interface structures of ceramics are explored in Sect. 3.1.2, only studies concerned with fracture toughness will be presented in this section. Some mechanisms to improve fracture toughening of ceramics have been proposed, and fracture toughening by crack deflection and plastic deformation is now considered.

To observe microcracks in ceramics, the specimens are prepared as follows [44]. Thin slices are mechanically polished to a thickness of about $150\,\mu m$, and disks 3 mm in diameter are obtained using an ultrasonic cutter. These are polished to get a mirror plane. About 20 point indents are made on the mirror plane by a Vickers indentor with a load of 50 g for 15 s. The back-side of the indented plane is further polished to about $30\,\mu m$ thickness with a dimple grinder, and then milled with Ar ions at an accelerating voltage of about 3 kV at a glancing angle of 25°. Finally, both sides are milled with Ar ions for about 30 min and then coated with carbon to prevent charge-up effects during observations.

In efforts to design highly toughened ceramics, whisker-reinforced and platelet-reinforced processing has received much attention. The aim is to reduce fracturing by strong crack deflection. Figure 3.36 shows (a) transmission and (b) high-resolution electron micrographs of crack propagation in a Si_3N_4–SiC (whisker) composite [44]. In Fig. 3.36a, an indent is observed on the right-hand side labeled "I", and cracks starting from this indent are propagated and deflected along grain boundaries and interfaces, as shown by the arrows. Together with the deflection and branching of the main cracks shown in Fig. 3.36a, the high-resolution image in Fig. 3.36b shows that the crack

propagating along the arrows loses a large amount of its driving force of propagation, accompanied by lattice deformation in the crystal. The lattice deformation, which can be seen by viewing obliquely along the white arrows in Fig. 3.36b, is produced in front of the turning point of the crack propagating along the arrows (the amorphous-like contrast is due to the carbon coating). That is, crack deflection not only turns the crack, but also reduces the propagation energy by lattice deformation inside the grains around the crack. This result is very important in understanding the mechanism of fracture toughening by crack deflection.

Figure 3.37 is (a) a transmission electron micrograph and (b) a lattice image showing crack propagation in a SiC (platelet)-reinforced Si_3N_4 matrix composite [19]. In Fig. 3.37a, it can be seen that cracks propagate along interfaces between SiC and Si_3N_4 grains and along the grain boundaries of Si_3N_4, and also inside the large SiC grain, and many microcracks are generated from the main cracks. The branching of cracks and the generation of microcracks play an important role in fracture toughening. The crack inside the SiC grain propagates along a close-packed plane of SiC, as can be seen in Fig. 3.37b.

Ceramics are generally fractured by their lack of plastic deformation. Therefore, if crystals exhibiting plastic deformation are dispersed in the ceramics, the composites are expected to show high fracture toughening. Figure 3.38 shows a lattice image of a raw powder of ZrO_2 [45]. This image shows a simple twinning structure formed in the transformation process from a tetragonal structure in the high-temperature phase to a monoclinic structure in the low-temperature phase without any restrictions. However, in an Al_2O_3–ZrO_2 composite, the ZrO_2 crystal undergoes the tetragonal–monoclinic transformation under strong restrictions from the surrounding crystal grains, and consequently complex twinning structures are observed [44].

It was found that the transformation of ZrO_2 grains in an Al_2O_3–24vol%ZrO_2 composite produced microcracks at ZrO_2/Al_2O_3 interfaces as well as complex structures of ZrO_2 grains [46]. It was also found that the ZrO_2 grains exhibited plastic deformation around areas of crack propagation. In the transmission electron micrograph in Fig. 3.39a, it can be seen that crack A, which propagates transgranularly in a ZrO_2 grain, is remarkably short compared with crack B, which is

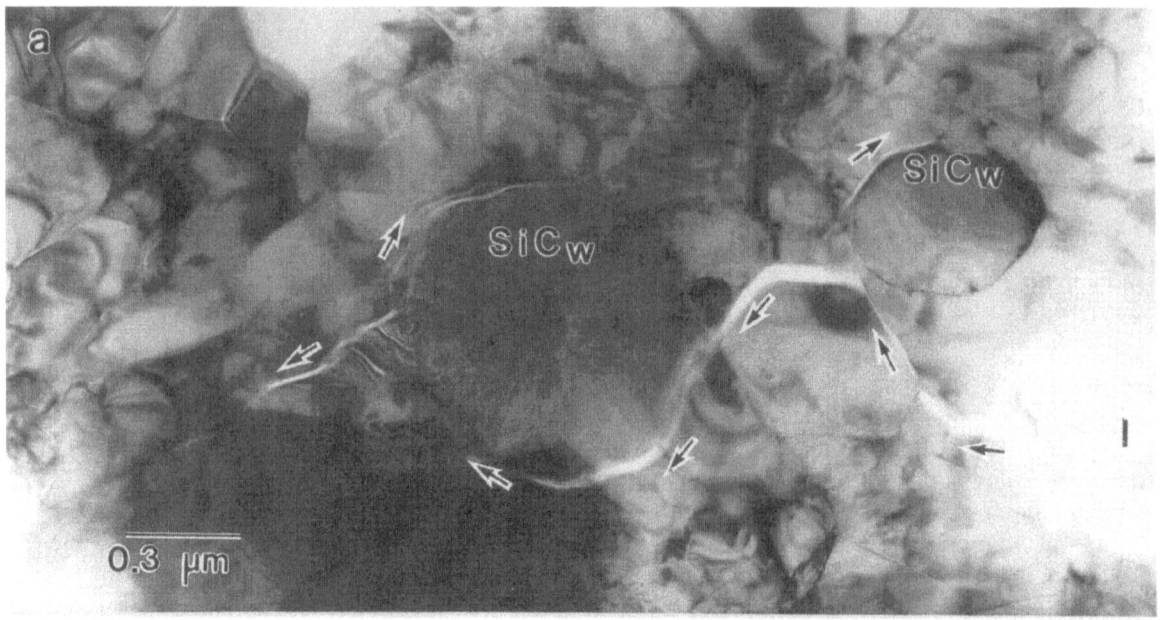

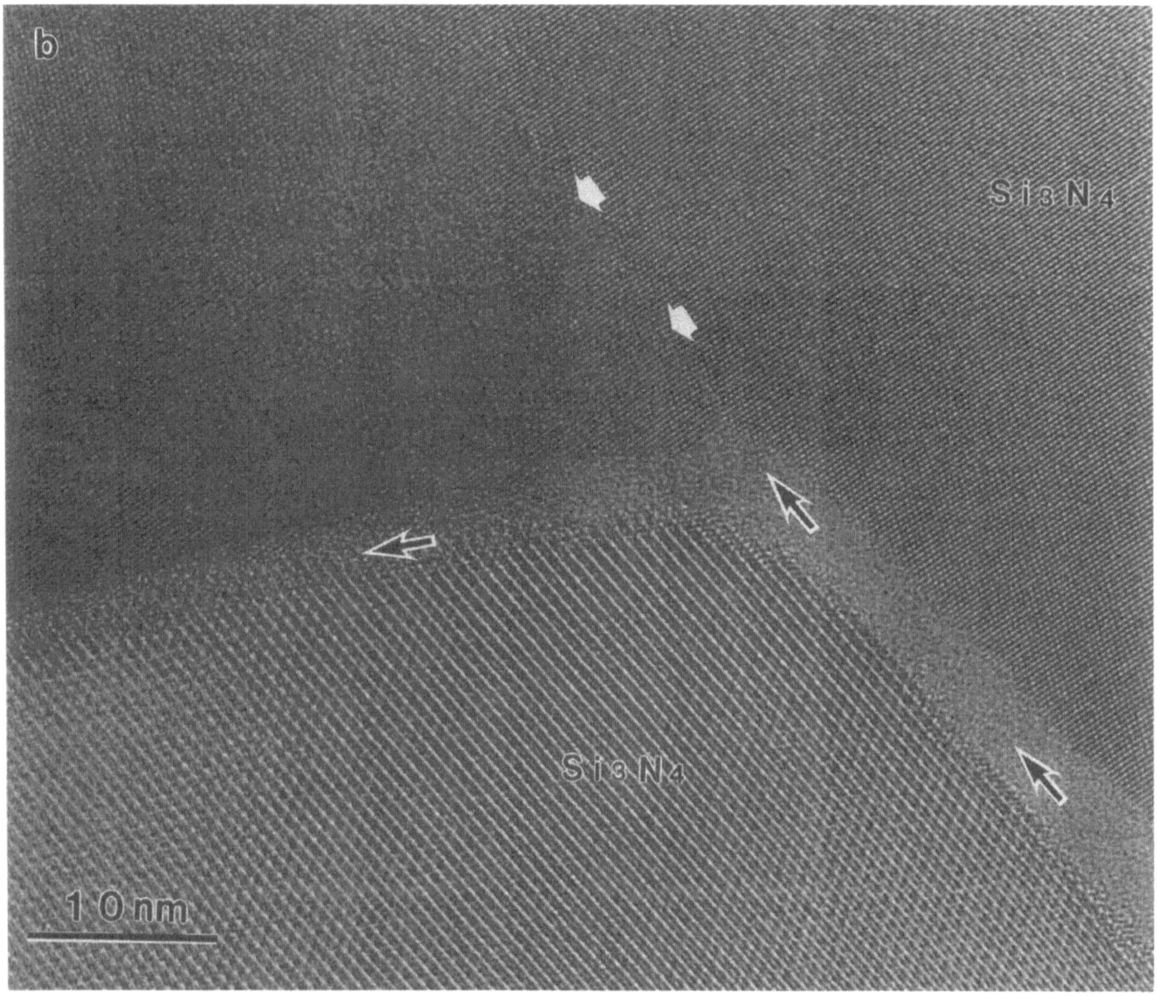

Fig. 3.36. a Bright-field image, and **b** high-resolution image showing crack propagation in a Si_3N_4–SiC whisker composite.

Specimen: Si_3N_4–SiC (whisker); **Preparation**: ion milling; **Observation**: 400 kV EM.

Remarks: **a** Cracks propagate along grain boundaries and interfaces, as indicated by the *arrows*. SiC_W shows a SiC whisker. "*I*" indicates the indentation center. **b** Lattice deformation in the Si_3N_4 crystal located at the front of the turning point of crack propagation

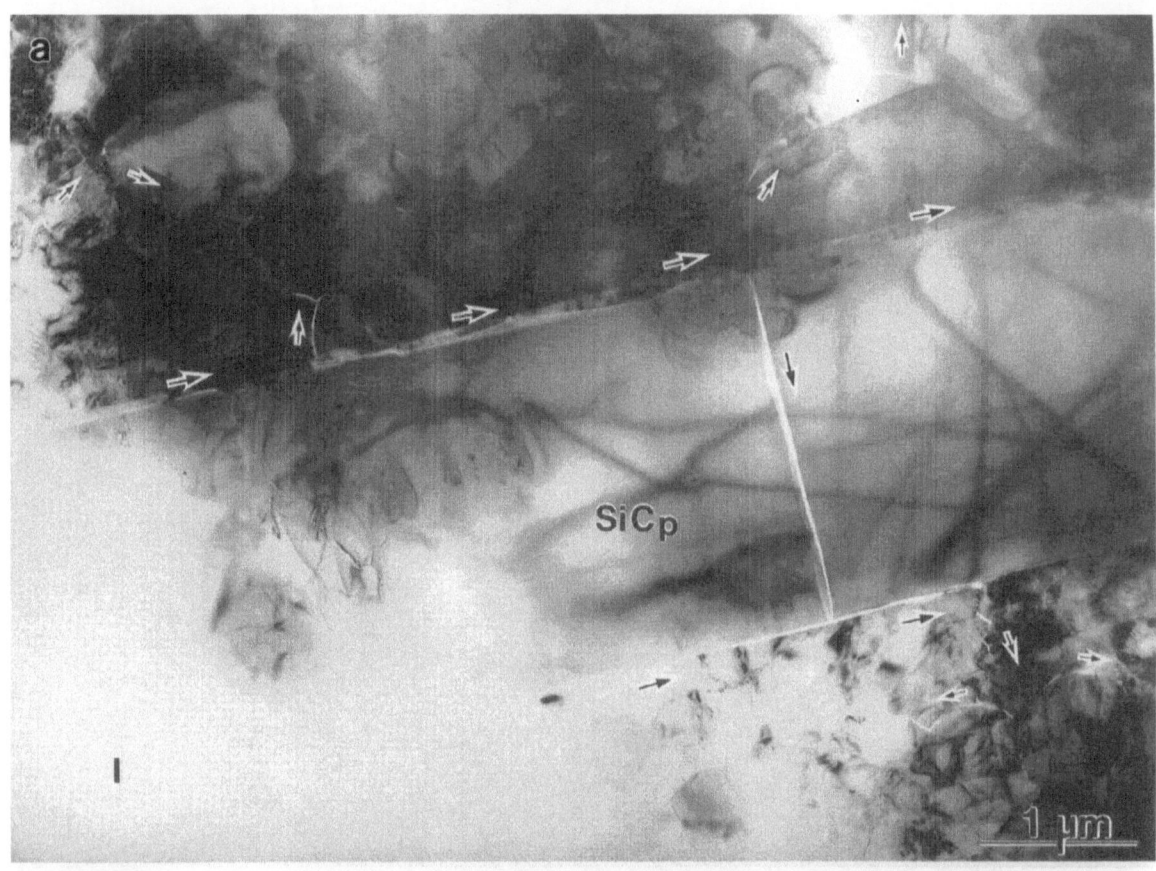

Fig. 3.37. **a** Bright-field image, and **b** high-resolution image showing crack propagation in a Si_3N_4–SiC platelet composite.

Specimen: Si_3N_4–SiC platelet; **Preparation**: ion milling; **Observation**: 400 kV EM.

Remarks: **a** Cracks propagate along grain boundaries and interfaces, and also propagate inside grains, as indicated by the *arrows*. SiCp shows a SiC platelet. "*I*" indicates the indentation center. **b** A fracture plane inside the grain is parallel to the closed-packed plane

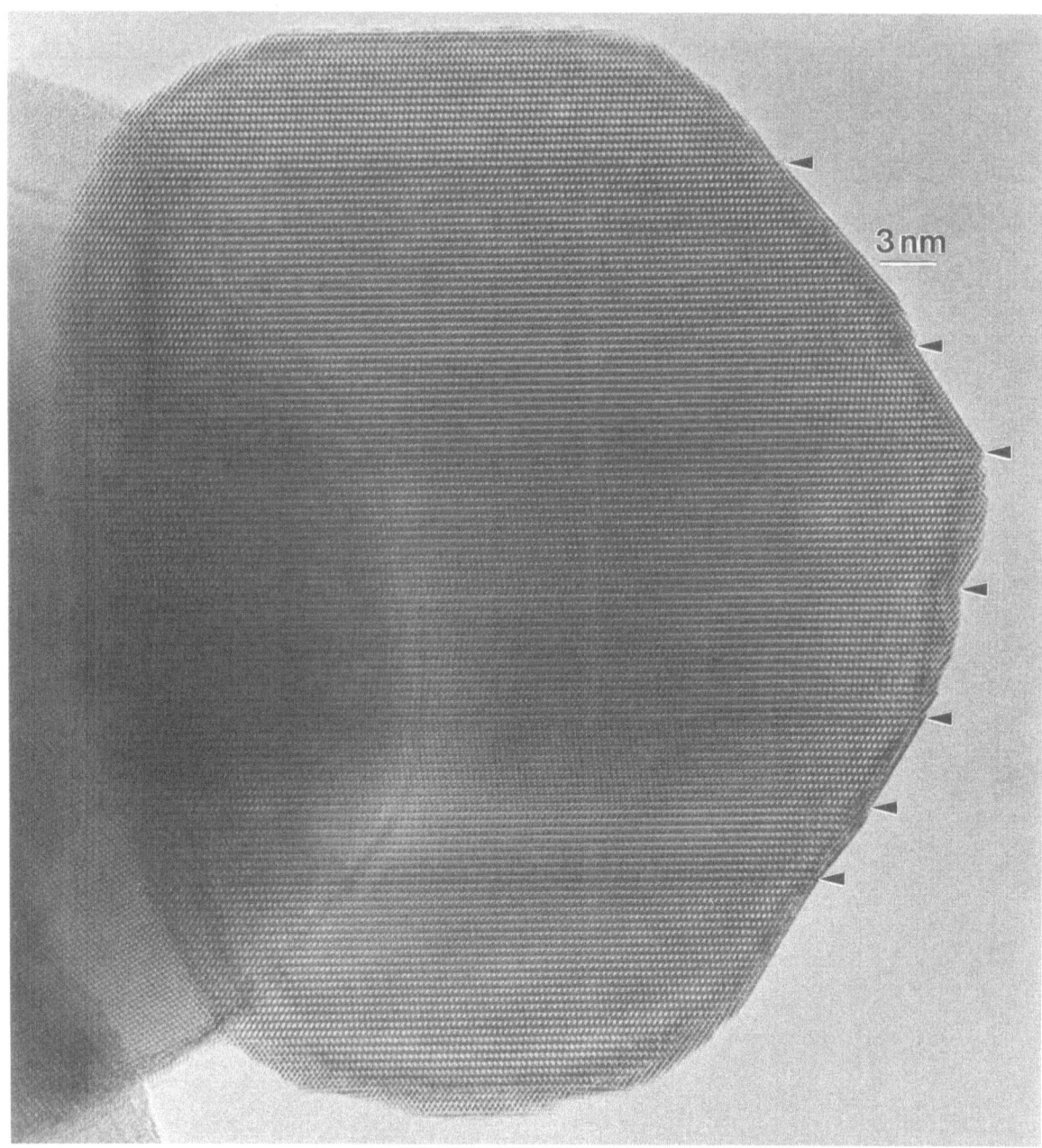

Fig. 3.38. High-resolution image of a ZrO_2 particle.
Specimen: ZrO_2; **Preparation**: ZrO_2 particles were dispersed on a microgrid by using an organic solvent;
Observation: 400 kV EM, [0$\bar{1}$1] incidence.
Remark: Twin planes are indicated by the *arrowheads*

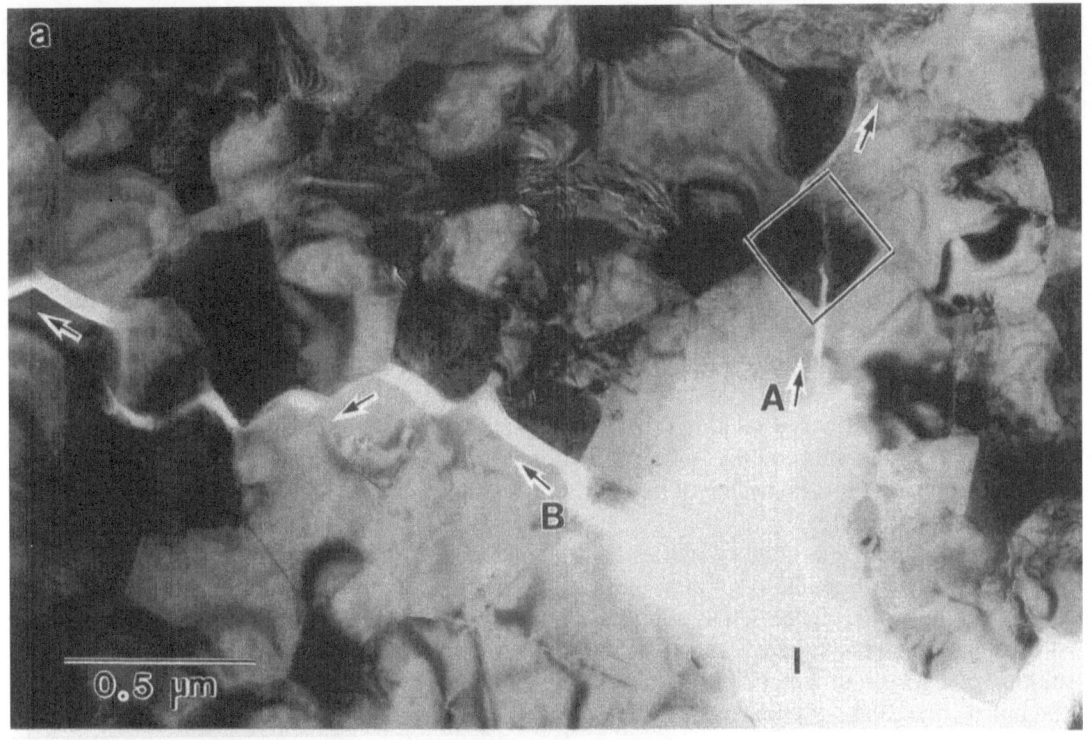

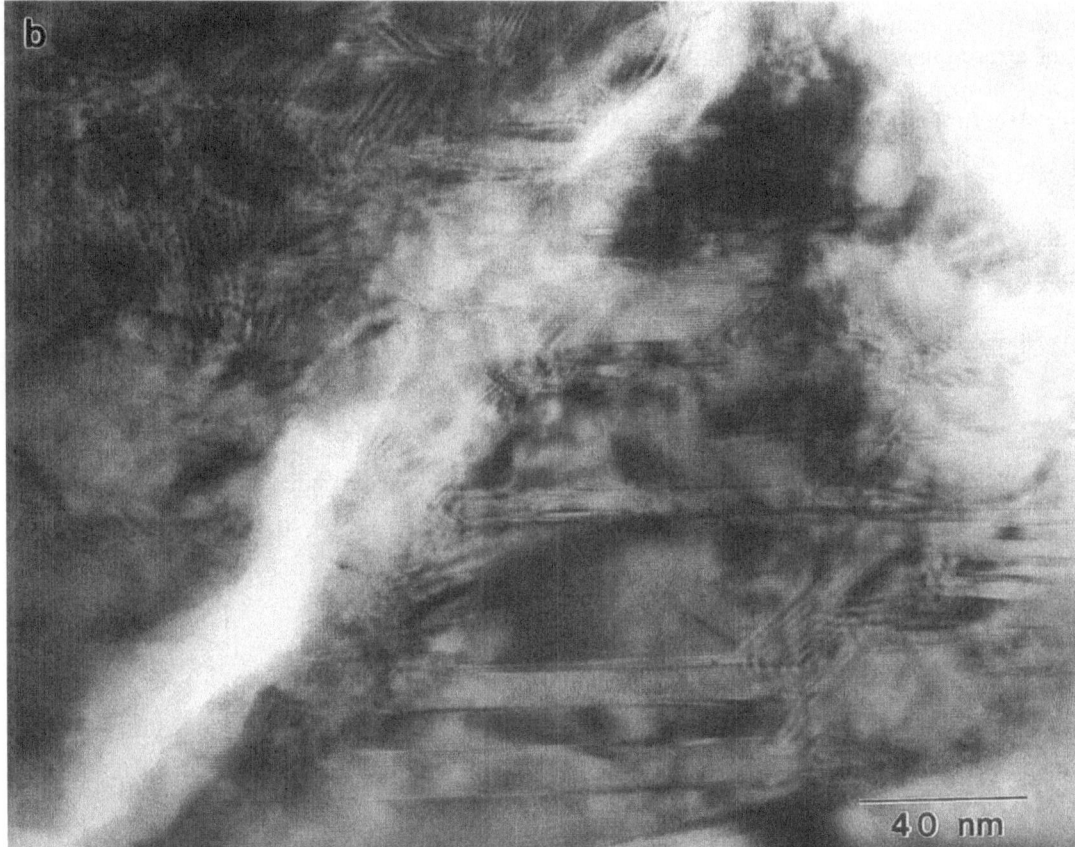

Fig. 3.39. a Bright-field image, and **b** high-resolution image showing crack propagation in an Al_2O_3–ZrO_2 composite.

Specimen: Al_2O_3–24vol%ZrO_2; **Preparation**: ion milling; **Observation**: 400 kV EM.

Remarks: In **a**, the crack "*A*" propagating through the monoclinic ZrO_2 grain loses a large amount of the driving force of propagation. "*I*" indicates the indentation center. **b** A lattice image of the monoclinic ZrO_2 grain, in which the deformation of the lattice due to crack propagation can be seen propagating intergranularly. This observation suggests that the crack A loses a large amount of its driving force of propagation during its passage through the ZrO_2 grain. This is caused by the plastic deformation of the ZrO_2 grain, and the plastic deformation can clearly be seen in a magnified image (Fig. 3.39b) of the region enclosed by the square in Fig. 3.39a. Figure 3.39b, shows a strain field from the plastic deformation as well as a complex twinning structure. This plastic deformation in ZrO_2 grains is considered to play a vital role in the fracture toughening of this material, and the result is important for the design of highly fracture toughened composites.

It has been known that stress-induced martensitic transformation of ZrO_2 leads to great toughness in ceramic composites. Figure 3.40 is a lattice image showing the stress-induced martensitic transformation of ZrO_2 around crack propagation [47]. In the image, one can see the partial transformation from tetragonal to monoclinic structure in a ZrO_2 grain. Monoclinic grains nucleate from the matrix of a tetragonal structure at the top of the figure, and propagate toward the bottom with a twinning structure which is shaped like the feathers of an arrow. The volume change caused by the transformation increases the resistance to crack propagation.

Ceramics, which are brittle at room temperature, show plastic deformation at high temperatures, and their hardness decreases and their fracture toughness increases rapidly with increasing temperature above 1100°C. Figure 3.41b is a transmission electron micrograph of an area around an indentation (I) introduced at 1200°C into an Al_2O_3–24vol%Al_2O_3 composite [46]. No cracks are observed around the indentation. Figure 3. 41a is a lattice image of the region A in Fig. 3.41b. The image shows subgrain boundaries formed in a ZrO_2 grain and their sliding (seen from the rotation of the twin boundaries). Disturbed layers with amorphous-like contrast can be seen at the subgrain boundaries and triple junctions. This result shows that the deformation of ZrO_2 at high temperatures mainly occurs through grain-boundary sliding rather than the generation and movement of dislocations. However, many dislocations were observed in Al_2O_3 grains in this composite. The plastic deformation of the Al_2O_3 grains was produced by the introduction of dislocations.

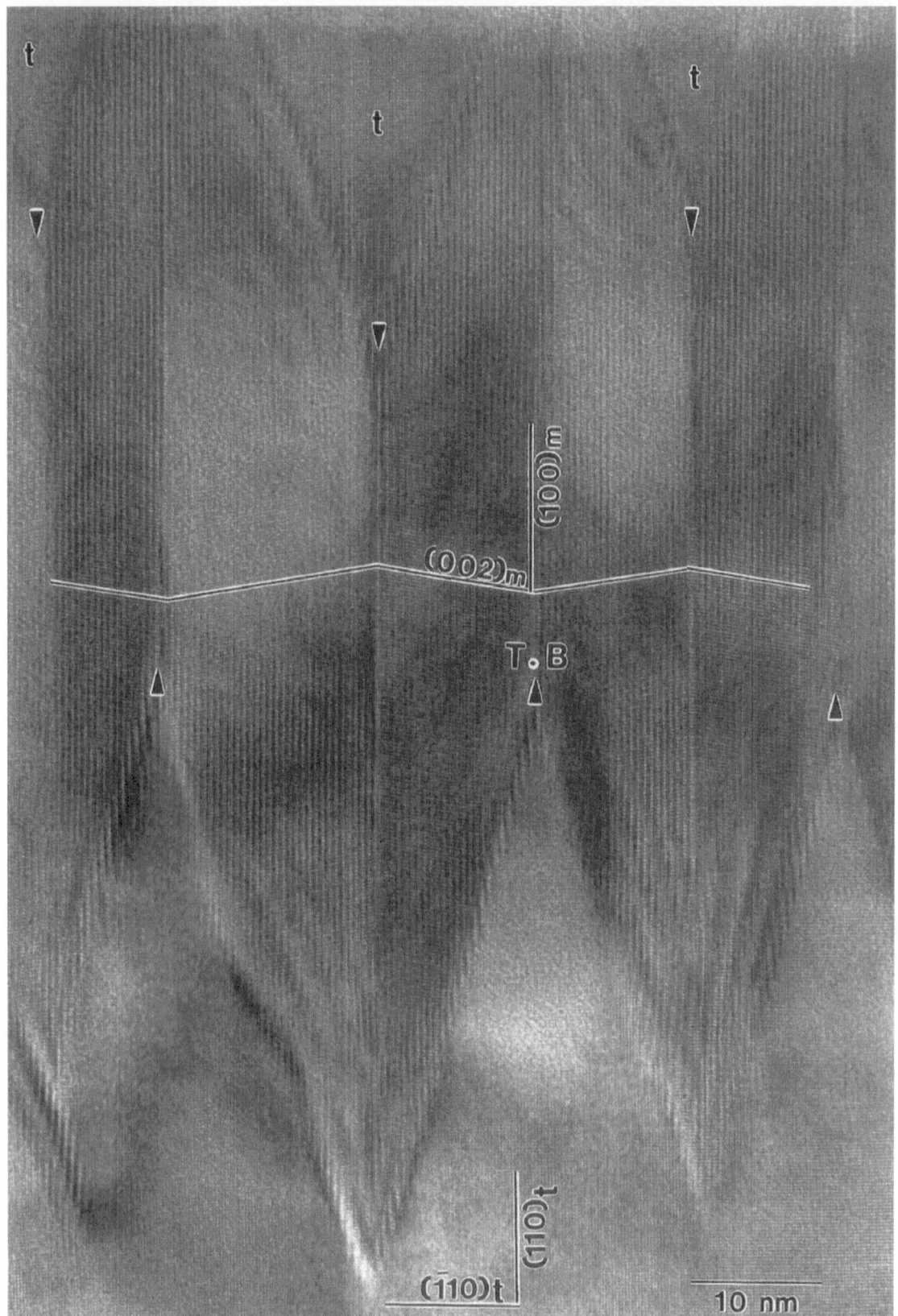

Fig. 3.40. High-resolution image showing the stress-induced martensitic transformation of ZrO_2 around a crack.
Specimen: ZrO_2–24vol%Al_2O_3; **Preparation**: ion milling; **Observation**: 400 kV EM.
Remarks: Monoclinic grains (*m*) nucleate from the matrix of a tetragonal structure (*t*) in the *upper region* and propagate toward the *bottom* with a twinning structure shaped like the feathers of an arrow. *Arrowheads* indicate twin boundaries (*T.B*)

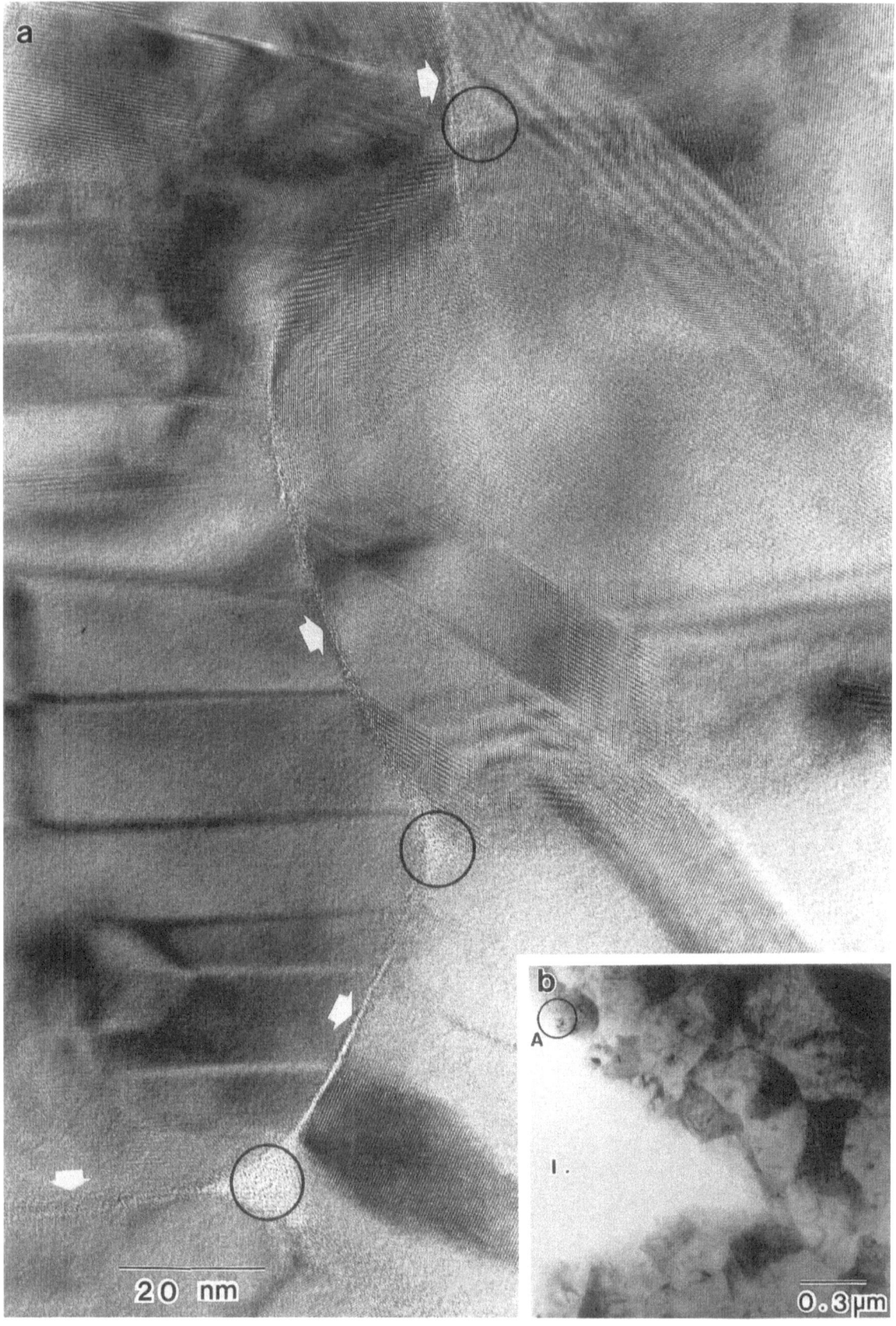

Fig. 3.41. a High-resolution image of the ZrO₂ grain (*A*) near the indentation (*I.*) in **b**. **b** Bright-field image showing no cracks near the indentation which was made at 1200°C.
Specimen: Al₂O₃–24vol%ZrO₂; **Preparation**: ion milling; **Observation**: 400 kV EM.
Remarks: **a** Note the formation of sub-grain boundaries in the ZrO₂ crystal, as indicated by the *arrows*. Along the sub-grain boundaries and triple junctions, disturbed layers with amorphous contrast, which result from grain boundary sliding, are seen. **b** No cracks indicate plastic deformation around the indentation (*I.*).

3.2.2 Superconducting Oxides

The study of superconducting oxides started in 1986 [48], and the discovery of an $YBa_2Cu_3O_7$ superconductor [49] with a superconduction temperature higher than the temperature of liquid nitrogen stimulated this work. Bednorz and Müller, who discovered of a high-Tc superconducting oxide, received a Nobel prize in 1987, but much fundamental and applied research still continues.

In this field, high-resolution electron microscopy has contributed greatly to crystal structure analysis of a group of new materials, because it has made it possible to analyze structures for specimens with several phases, and also in a short period of time.

The structures of high-Tc superconducting oxides are mainly based on a Perovskite-type structure, which is known to be the structure of barium titanates being ferroelectrics, and are formed by stacking blocks of the Perovskite-type structure with other structures. The Perovskite structure consists of A atoms, which are located at the centers of octahedra of oxygen atoms, and a B atom at the body-centered position, as shown in Fig. 3.42. In most superconducting oxides, the A atoms are Cu atoms and the B atoms are alkaline-earth metals or rare-earth metals.

Since high-Tc superconducting oxides have simple structures, and also thin specimens are easily obtained using a crushing method, they are very suitable materials for high-resolution electron microscopy, and their structures have been determined by high-resolution observations.

Figure 3.43 is a structure image of a $YBa_2Cu_3O_7$ superconductor taken with a 400 kV electron microscope with a resolution of 0.17 nm [50, 51]. The image was taken with the incident beam parallel to the [010] axis, and individual cation positions are clearly represented as dark spots. In this type of structure image taken from thin regions, as mentioned in Sect. 1.3.1, the projected potential is reflected directly on the observed images, and the darkness and/or size of the dark spots is proportional to an atomic number of cations. For example, as can be seen in the inserted structure model, Cu ions with a small atomic number are represented as small dark spots, and they can be clearly distinguished from Y and Ba ions with large atomic numbers. Also, the positions of the cations are accurately reflected by the positions of the dark spots of the image, and consequently

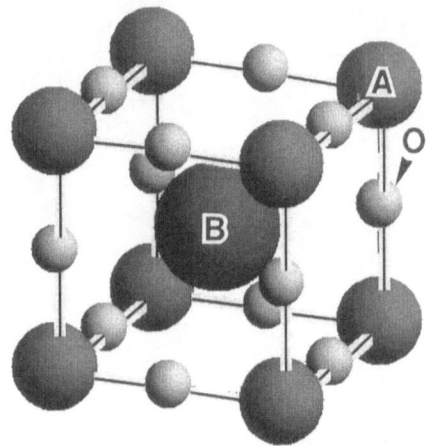

Fig. 3.42. Perovskite structure. A atoms are located at the center of octahedra consisting of oxygen atoms, and a B atom is located at the body-centered position of the structure

coordinates of the cations can be determined from the positions of the dark spots, as will be explained below. Oxygen atoms and oxygen vacancies (indicated by arrowheads) located at bright spots in the image can also be distinguished from the brightness of the bright regions. Thus, this type of structure image includes valuable information concerning not only the arrangements of cations, but also the positions of oxygen atoms.

Figure 3.44 is a structure image of a $Tl_2Ba_2CuO_6$ superconductor [52]. This contains much clearer information about the crystal structure than the image of $YBa_2Cu_3O_7$ in Fig. 3.43. In the image, Tl and Ba atoms with large atomic numbers appear as large dark spots, whereas Ca and Cu atoms with small atomic numbers appear as small dark spots. From a close examination of the image, Tl and Ba, and Ca and Cu atoms can be distinguished. With the aid of information obtained by other methods, such as chemical analysis, X-ray diffraction, and so on, the arrangements of cations can be determined unequivocally from this type of structure image.

Close examination of Fig. 3.44 shows systematic displacements of the dark spots from the positions of an ideal Perovskite structure. By comparing the average positions of about ten dark spots with atomic coordinates determined by X-ray diffraction with a single crystal, it was found that those positions agreed well with each other within an experimental error of 0.01 nm. That is, the result

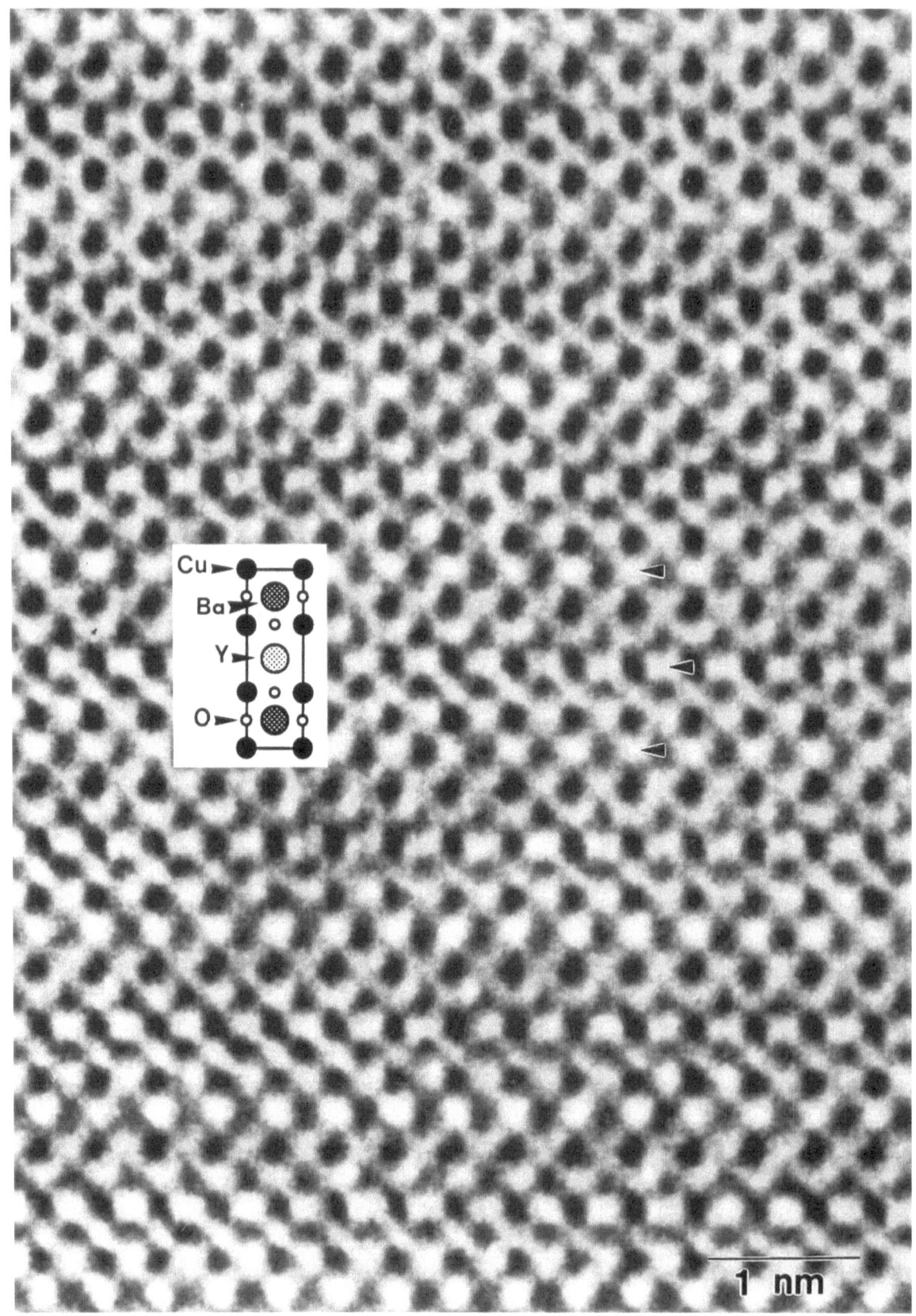

Fig. 3.43. Structure image of a $YBa_2Cu_3O_7$ superconductor.
Specimen: $YBa_2Cu_3O_7$; **Preparation**: crushing; **Observation**: 400 kV EM, [010] incidence.
Remark: Cations can be seen as *dark spots* by a comparison with the structure model inserted. *Arrowheads* indicate vacant positions of oxygen atoms

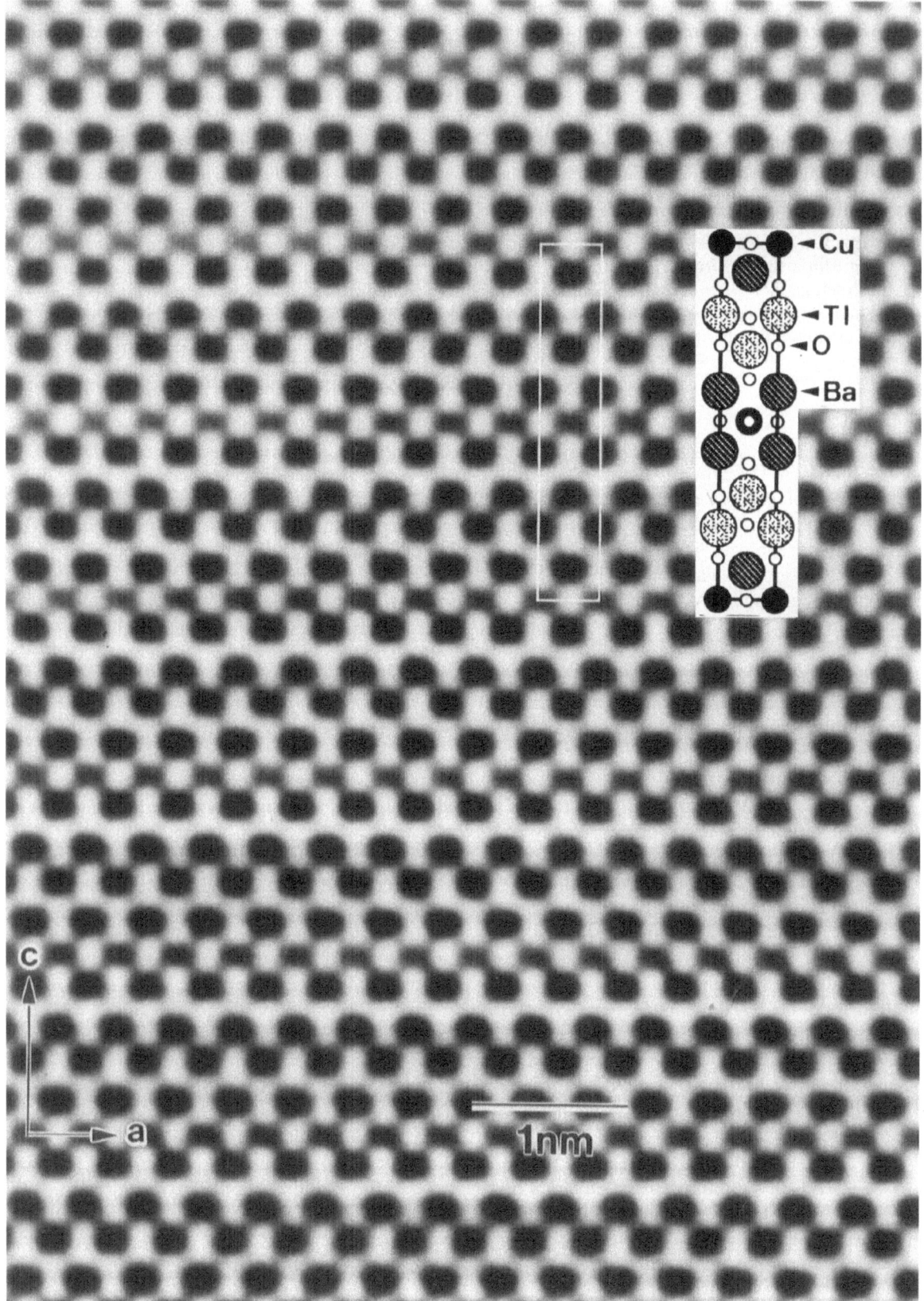

Fig. 3.44. Structure image of a $Tl_2Ba_2CuO_6$ superconductor.
Specimen: $Tl_2Ba_2CuO_6$; **Preparation**: crushing; **Observation**: 400 kV EM, [010] incidence.
Remarks: There is good correspondence between an arrangement of dark spots in the image and that of cations in the structure model inserted. Note the systematic displacements of dark spots from the ideal positions of a Perovskite structure

---

Reminiscence: High-resolution observation of a $YBa_2Cu_3O_7$ superconductor

After the discovery of high-Tc superconductors, there was fierce competition to produce structure analyses using X-ray diffraction, neutron diffraction, and electron microscopy. The photograph in Fig. 3.43 is a structure image of a $YBa_2Cu_3O_7$ superconductor which was observed at an early stage, just after the discovery of this material. Although the experimental conditions for high-resolution observations were not perfect, it was considered to be one of the best images of this material at that time, and has been cited in various newspapers and many books. The image was actually observed at midnight using a 400 kV electron microscope with a support from T. Honda at JEOL (Tokyo, Japan). Since the electron microscope was under construction, there were several problems, especially from contamination, and thus it was necessary to take the high-resolution images in a very short time. Since then, high-resolution electron microscopy has contributed greatly to the structure analyses of the high-Tc superconductors which have since been discovered.

---

shows that from this type of structure image, the coordinates of cations can be determined with accuracy of 0.01 nm [53, 54].

The structure images like those in Figs. 1.7 and 3.43 can also give us valuable information concerning the arrangement of oxygen atoms, although the oxygen atoms can not be seen as dark spots, like cations. The structures of superconducting oxides are characterized by the appearance of oxygen vacancies compared with an ideal Perovskite, and the arrangements of oxygen vacancies are closely related to the mechanism for the occurrence of high-Tc superconductivity. Therefore, the determination of the arrangements of oxygen vacancies is an important subject. Most oxygen atoms in the structures of superconducting oxides are located between cations and at the bright regions between the dark spots which correspond to cations in the images in Figs. 1.7 and 3.43. Thus, the bright regions in the structure images should be investigated in detail to get information about the oxygen atoms. Figure 3.45 shows calculated images of two models, where it is assumed that the oxygen atom positions on the Ca layers are either fully occupied by oxygen atoms (a) or vacant (b). The images were calculated with an underfocus value of 45 nm (near the Scherzer focus) and a specimen thickness of about 2 nm, with a 400 kV electron microscope having a resolution of 0.17 nm. The calculated images reveal that the bright spots on the Ca layer in Fig. 3.45b, which correspond to the vacant oxygen positions indicated by Ov, are brighter than those in the same positions in Fig. 3.45a. This result confirms that ordered arrangements of oxygen vacancies can be seen from the close examination of bright regions in this type of structure image [53, 54].

(One can also see brighter vacant oxygen positions in Figs. 1.7 and 3.43.)

We have seen that it is possible to determine the arrangements and coordinates of cations, and the arrangements of oxygen atoms, by observing structure images. Figure 3.46 shows two examples of the crystal structure analysis of complex structures. From these images, and with the aid of information from chemical analysis, the structure models inserted in the images were proposed.

The final example gives the positions of oxygen atoms which were determined from a structure image. Figure 3.47 shows simulated images of two models in which oxygen atoms are located (a) at the centers of tetrahedra formed by Sm atoms, and (b) at the normal oxygen positions in a Perovskite structure. The simulated images were calculated with an underfocus of 45 nm at a variety of thicknesses. The different features shown by contrasts within the bright regions can be clearly seen. These differences are enhanced with increasing thickness. Figure 3.48 shows an observed structure image [55]. In this image, the specimen thickness increases from upper left to lower right. The change of contrast in the image shows good correspondence with the change of image with increasing thickness in Fig. 3.47a. In this way, we can get information about the positions of oxygen atoms from observed structure images with the aid of computer simulation.

As can be seen, high-resolution electron microscopy is a powerful tool in the structure analysis of superconducting oxides with relatively simple structures, and valuable information concerning crystal structures can be obtained with the aid of computer simulation.

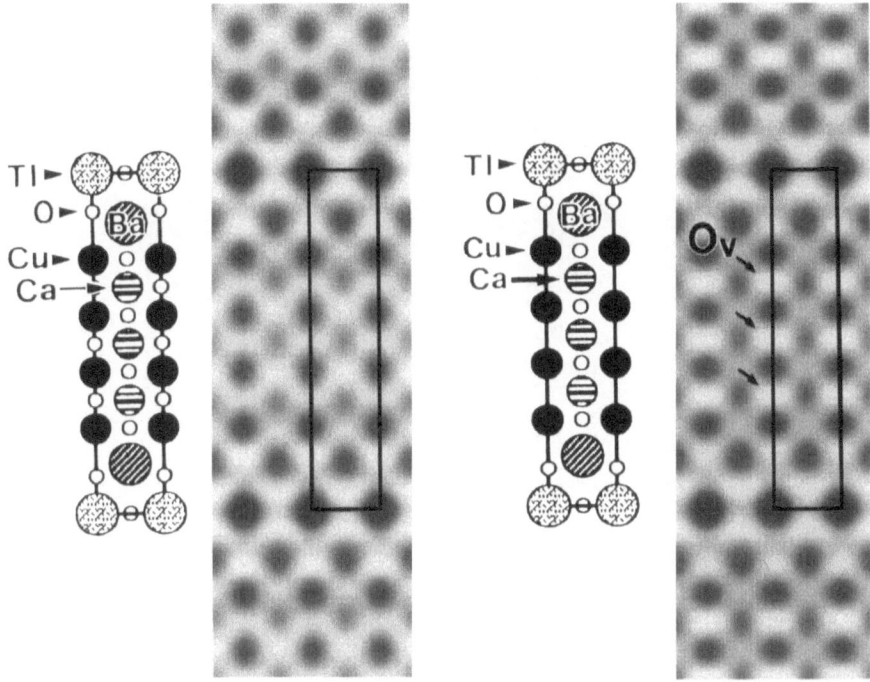

Fig. 3.45. Simulated images of two models (TlBa$_2$Ca$_3$Cu$_4$O$_x$), where the oxygen atom positions on the Ca layers are fully occupied by oxygen atoms **a**, and vacant **b**. Note that vacant oxygen positions (O_v) appear bright in **b**

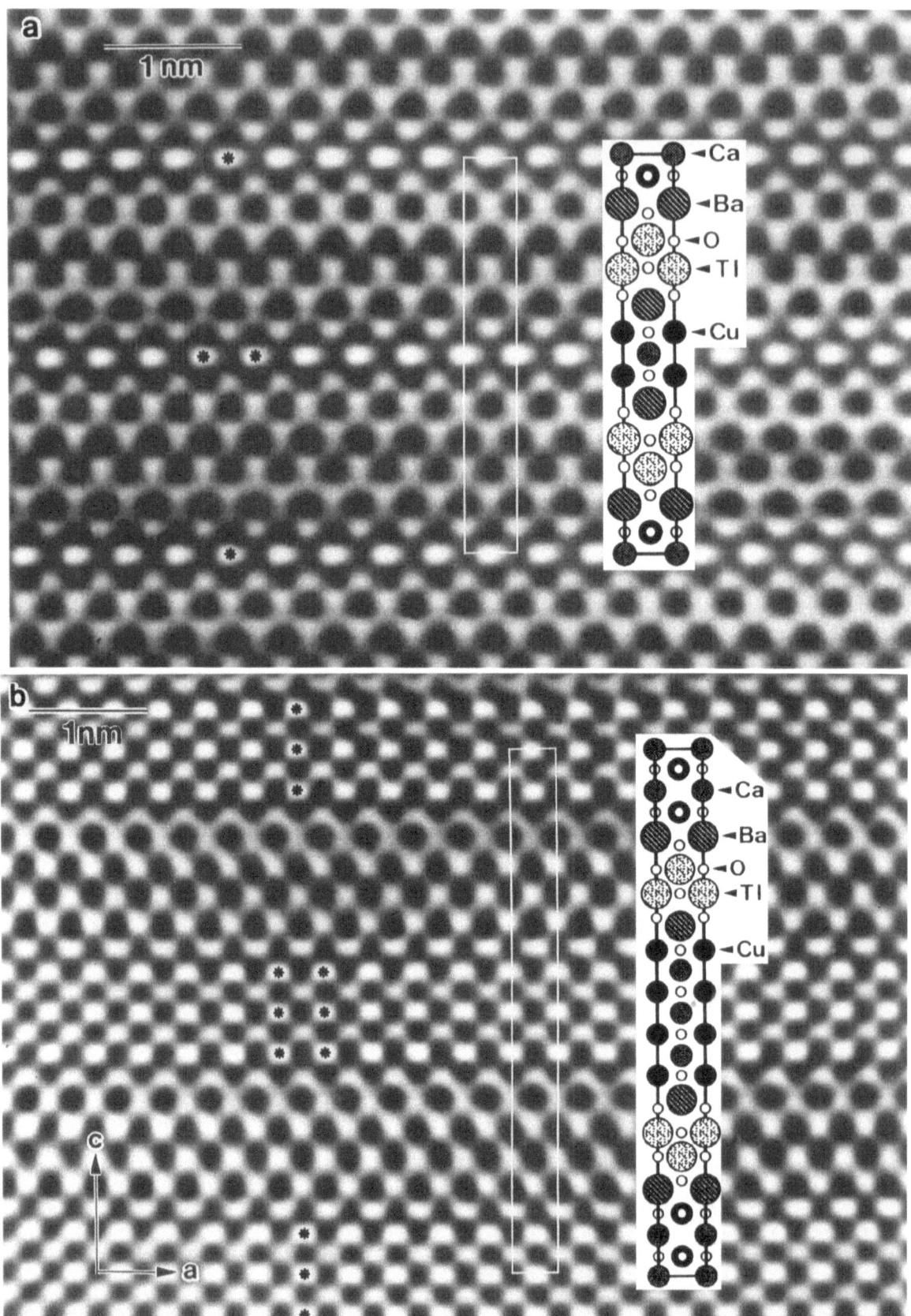

Fig. 3.46. Structure images of **a** $Tl_2Ba_2CaCu_2O_8$ and **b** $Tl_2Ba_2Ca_3Cu_4O_{12}$ superconductors.
Specimen: **a** $Tl_2Ba_2CaCu_2O_8$ and **b** $Tl_2Ba_2Ca_3Cu_4O_{12}$; **Preparation**: crushing;
Observation: 400 kV EM, [010] incidence.
Remark: *Stars* show vacant oxygen positions

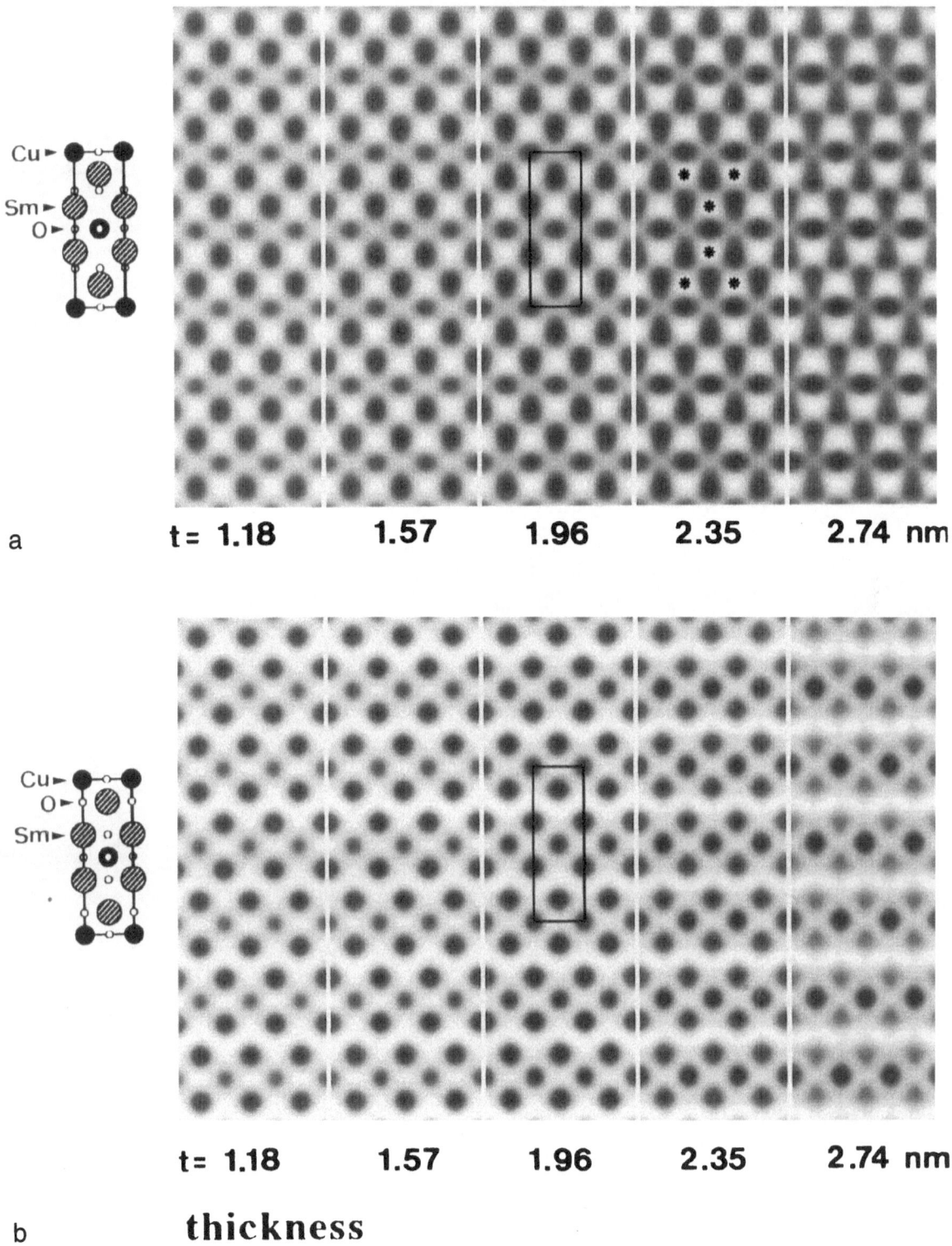

Fig. 3.47. Thickness-dependency of structure images of two models in Sm-based superconductors

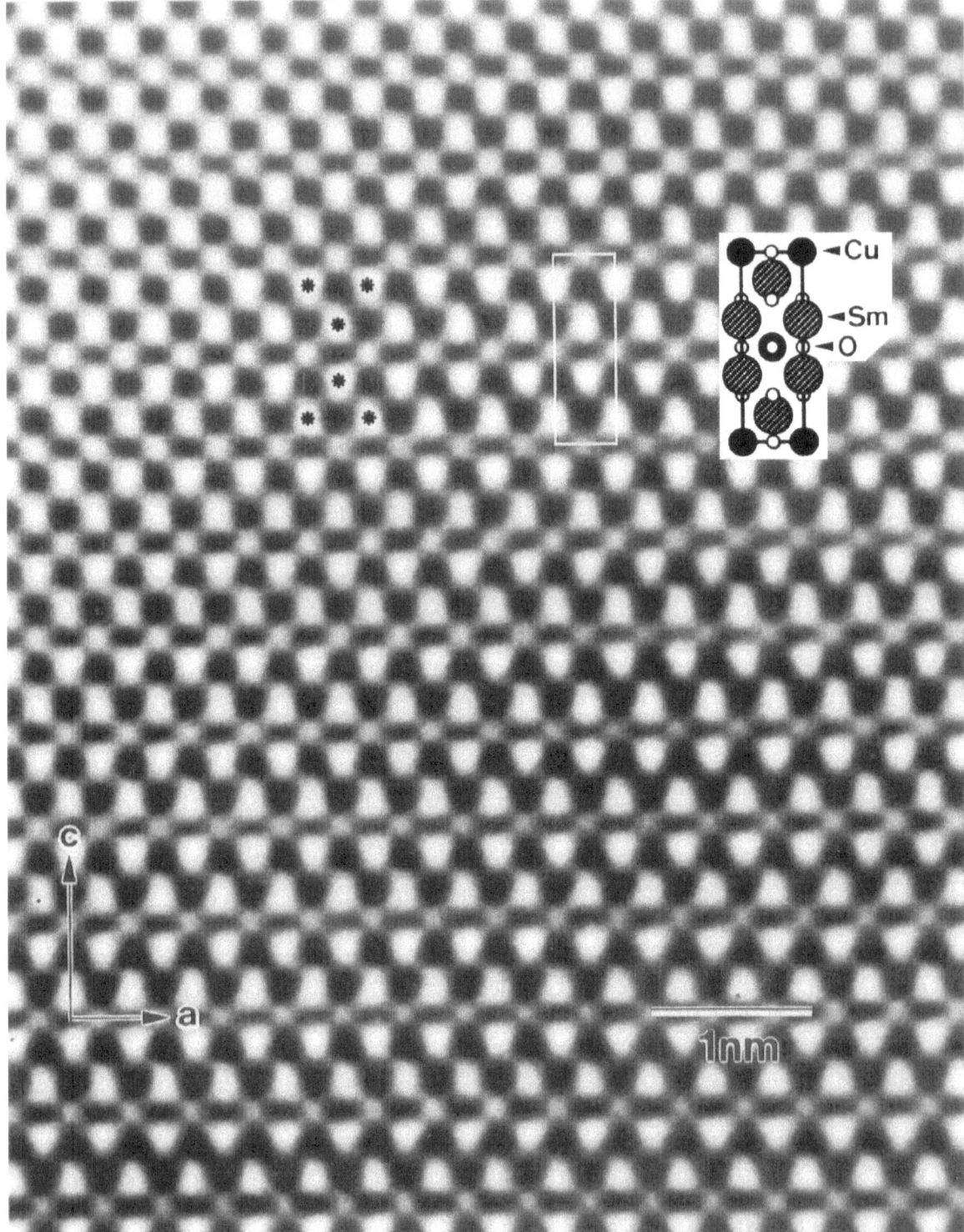

Fig. 3.48. Structure image of a Sm_2CuO_4 superconductor.
Specimen: Sm_2CuO_4; **Preparation**: crushing; **Observation**: 400 kV EM, [010] incidence

3.2.3 Ordered Alloys

In some alloy systems, the constituent metallic elements occupy random lattice points at high temperatures, but with a decrease in temperature those elements tend to occupy specific regular positions. This structure change with temperature is called the order–disorder phase transformation, and alloys forming ordered atomic arrangements at low temperatures are called ordered alloys [56]. Figure 3.49 shows some typical ordered structures. Ordered alloys show various structure changes due to phase transformation and also to small compositional changes. To clarify these structure changes on the atomic scale, high-resolution electron microscopy is very useful. We now discuss characteristic dynamical diffraction effects in ordered alloys, and then the imaging process which produces so-called superstructure images (see Sect. 2.1.5). Through these superstructure images, various ordered atomic arrangements and their structure changes due to temperature and compositional changes will be presented.

3.2.3.1 Dynamical Diffraction Effect in Ordered Alloys

In an electron diffraction pattern of an ordered alloy composed of constituent elements A and B, superlattice reflections corresponding to the formation of the superlattice appear in addition to the fundamental reflections. While the structure factors for the fundamental reflections do not depend on the ordered state, those for the superlattice reflections, in general, strongly depend on the ordered state and are given by

$$F(\boldsymbol{u}) = cT(\boldsymbol{u})\big(f_A(\boldsymbol{u}) - f_B(\boldsymbol{u})\big)\sum_i \gamma_i \exp(-2\pi i \boldsymbol{u} \boldsymbol{r}_i)$$
(3.1)

where c and $T(\boldsymbol{u})$ indicate a constant and a Debye-Waller factor, respectively, and $f_A(\boldsymbol{u})$ and $f_B(\boldsymbol{u})$ are the scattering factors of atoms A and B. The parameter γ_i indicates the manner of occupation at the lattice point i, and is given by

$$\gamma_i = \begin{cases} m_B & (\text{for atom A at } \boldsymbol{r}_i) \\ -m_A & (\text{for atom B at } \boldsymbol{r}_i) \end{cases}$$
(3.2)

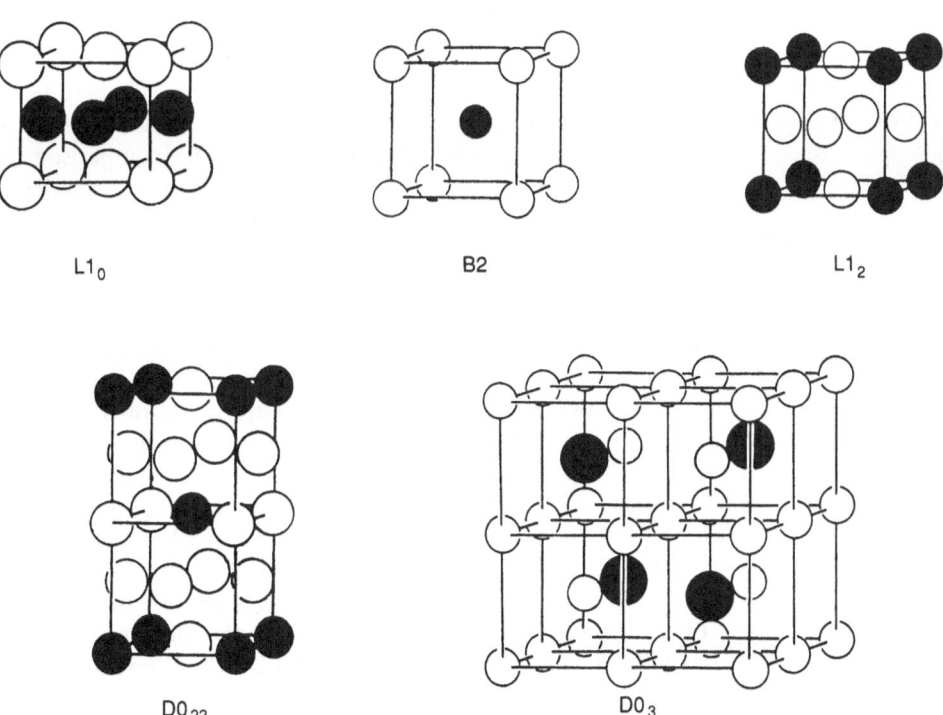

Fig. 3.49. Some typical ordered structures of alloys of the AB type (L1$_0$, B2) and A$_3$B type (L1$_2$, D0$_{22}$, D0$_3$). A and B atoms are shown as *open* and *solid circles*, respectively

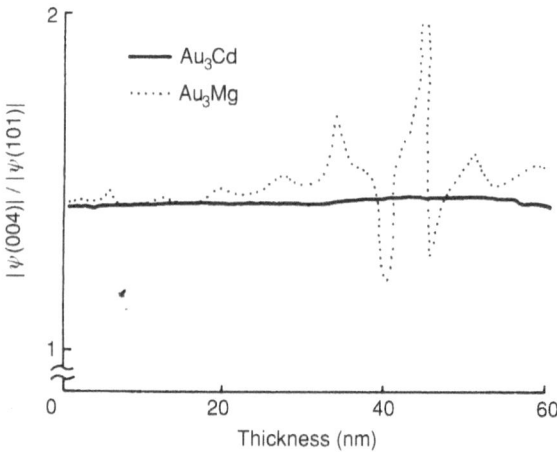

Fig. 3.50. Ratios of amplitudes for the superlattice reflections 004 and 101 for Au_3Cd and Au_3Mg with the $D0_{22}$-type structure, calculated as a function of crystal thickness

where m_A and m_B are the atomic fractions for the elements A and B ($m_A + m_B = 1$). In the structure factor for the superlattice reflections shown in Eq. 3.1, the scattering factor at the lattice point i is interpreted as the deviation from the average scattering factor ($f_A m_A + f_B m_B$), i.e., for atom A, the scattering factor is $f_A - (f_A m_A + f_B m_B) = (f_A - f_B) m_B$, while for atom B, $- (f_A - f_B) m_A$. Thus, from Eq. 3.1, the intensity of the superlattice reflection is generally smaller than that of the fundamental reflection, and in cases where the differences between the scattering factors of the constituent elements is small, the intensity of the superlattice reflection becomes weaker, and little dynamical diffraction effect on these reflections is expected.

In superlattice reflections of ordered alloys based on the fcc and bcc lattices, there is a characteristic dynamical diffraction effect [57, 58]. That is, the scattering amplitude $\Psi(u)$ of the superlattice reflection is given by the structure factor $F(u)$:

$$\Psi(u) = DF(u). \qquad (3.3)$$

In the above equation, D is the *dynamical factor*, and shows the dynamical diffraction effect on the superlattice reflection. The dynamical factor does not directly depend on each superstructure, but can be evaluated if the composition and crystal thickness are known. It is known that the absolute value of the dynamical factor tends to increase rapidly with an increase in crystal thickness, but it changes monotonically in reciprocal space. In Fig. 3.50, the calculated ratios of the amplitudes for superlattice reflections 004 and 101 of Au_3Cd with a $D0_{22}$-type structure are compared with those of Au_3Mg. A hypothetical Au_3Mg with a $D0_{22}$-type structure is used for the comparison. Note that owing to the large differences in the scattering factors of the constituent elements, the structure factor of the superlattice reflection in Au_3Mg is much larger than that in Au_3Cd. While the ratios of the amplitudes in Au_3Mg change drastically as a function of crystal thickness, the ratios in Au_3Cd tend to take a constant value, being equal to the ratio of the absolute values of structure factors. It is known that this kinematical relation between superlattice reflections only holds for thicker crystals where the differences in the scattering factors are small and the accelerating voltage of the electron microscope is high [58]. Thus the high-resolution image obtained with these superlattice reflections reflects the ordered atomic arrangement in a relatively thick crystal. In the case of ordered alloys of A_3B or A_4B, B-atom positions projected along the incident electron beam are observed as bright or dark spots in the high-resolution image (see Fig. 2.14) [57, 59–61]. A

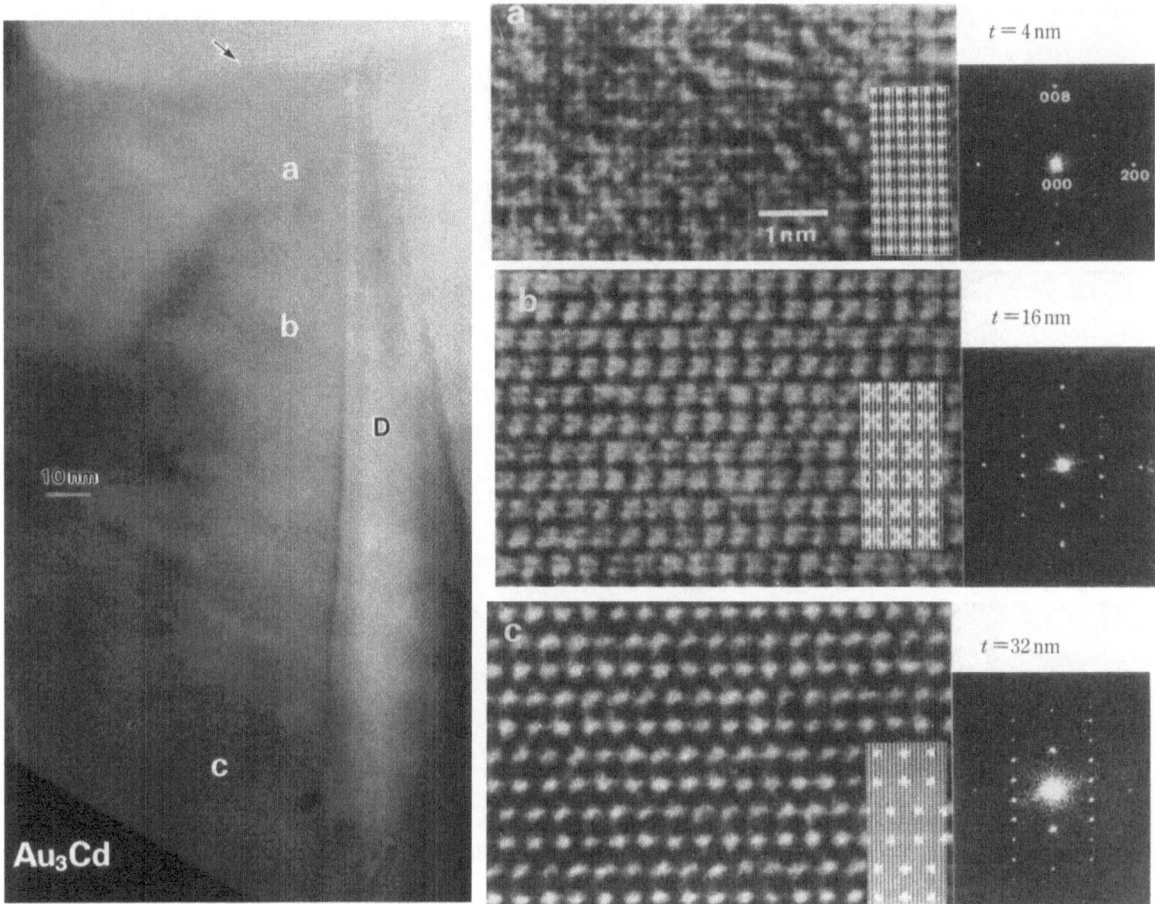

Fig. 3.51. *Left*, Bright-field image, *center*, high-resolution images of an Au$_3$Cd alloy with the D0$_{23}$-type structure, and *right*, their optical diffractograms.
Specimen: Au–24.0at%Cd; **Preparation**: jet electropolishing (CH$_3$COOH (133 ml) + H$_2$O (7 ml) + CrO$_3$ (25 g));
Observation: 1000 kV EM, [010] incidence.
Remarks: High-resolution images (*center*) with simulated images inserted show the changes of image contrast as a function of crystal thickness. From the optical diffractograms, it can be seen that superlattice reflections strongly contribute to the high-resolution image **c** obtained at a thick crystal region. *t* indicates crystal thickness

high-resolution image showing the potential difference of the A and B atoms in a superstructure is called a *superstructure image* [62].

Figure 3.51 shows high-resolution images of Au$_3$Cd with a D0$_{23}$-type structure. The high-resolution images in Fig. 3.51a–c correspond to the regions a, b, and c in the low-magnification image on the left-hand side, and crystal thickness increases from 3.51a to 3.51c [62]. The crystal thickness can be estimated from the width of the planar defects indicated by D. From the optical diffractograms on the right-hand side, it can be seen that the intensity of the superlattice reflections increases with an increase in crystal thick-

ness, and the contribution of the superlattice reflections is larger than that of the fundamental reflections in the thicker area. It should be noted that the kinematical relation among the superlattice reflections is retained in these crystal regions. In the high-resolution image observed at region c, Cd atom positions appear as bright spots. The contrast of the high-resolution image is consistent with simulated images inserted in the figure (see also Fig. 2.14).

Figure 3.52 shows a superstructure image of an Au–Mn alloy (Au–22.6at%Mn) [63]. Here, the bright spots in the image correspond to Mn atom positions projected along the incident electron

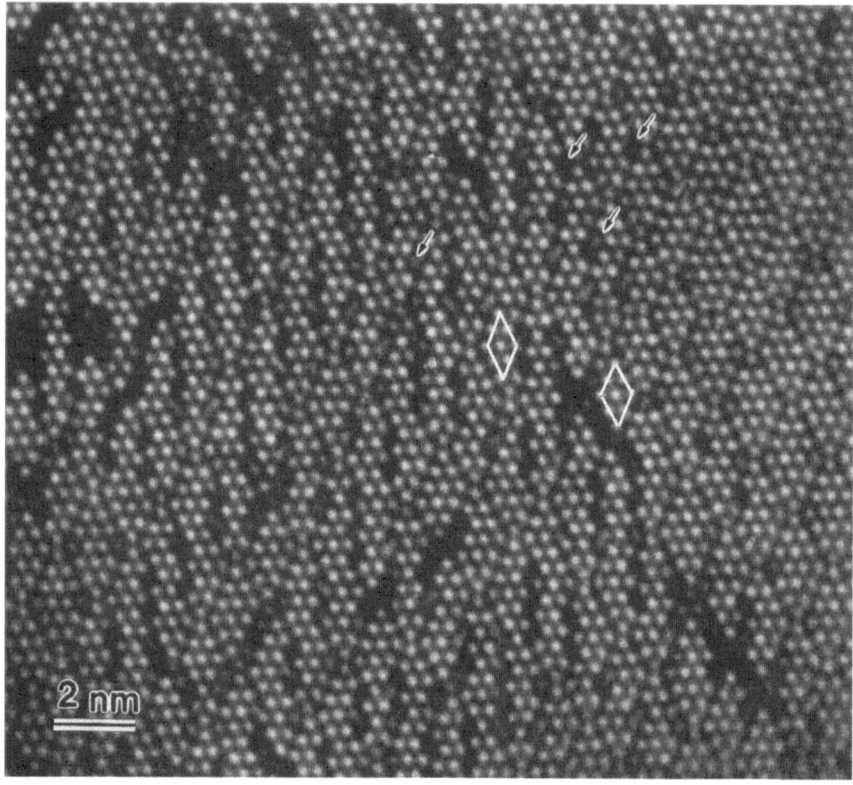

Fig. 3.52. Superstructure image of Au–22.6at%Mn.
Specimen: Au–22.6at%Mn; **Preparation**: jet electropolishing ($HClO_4$: H_2SO_4 : H_2O = 3 : 3 : 50);
Observation: 1000 kV EM, [001] incidence.
Remarks: The brightness of the white spots indicated by *arrows* is less than that of other white spots. It is considered that the Mn atom positions in those atomic columns are partially occupied by Au atoms

beam direction. The structure of this alloy basically consists of the superstructure indicated by the lozenges in Fig. 3.52, and it contains local disordering. For example, the brightness of the spots indicated by the arrows is less than that of the others, and the regions corresponding to these spots are interpreted as the atomic columns where some of the Mn atoms are replaced by Au atoms [64, 65]. The above interpretation can be confirmed in the image simulation shown in Fig. 3.53. On the basis of the correspondence between the contrast of superstructure images and the atomic arrangement, the probability of finding the atom pair can be evaluated by applying the auto-correlation function to the superstructure images (see Sect. 4.1.2). It should be noted that the brightness of the spots in the superstructure image is not proportional to the content of Au in the atomic columns, and the relation between the brightness and the atomic content is affected by the dynamical factor D in Eq.3.3. Characteristic structure changes of some ordered alloys with temperature and with composition are now considered using superstructure images.

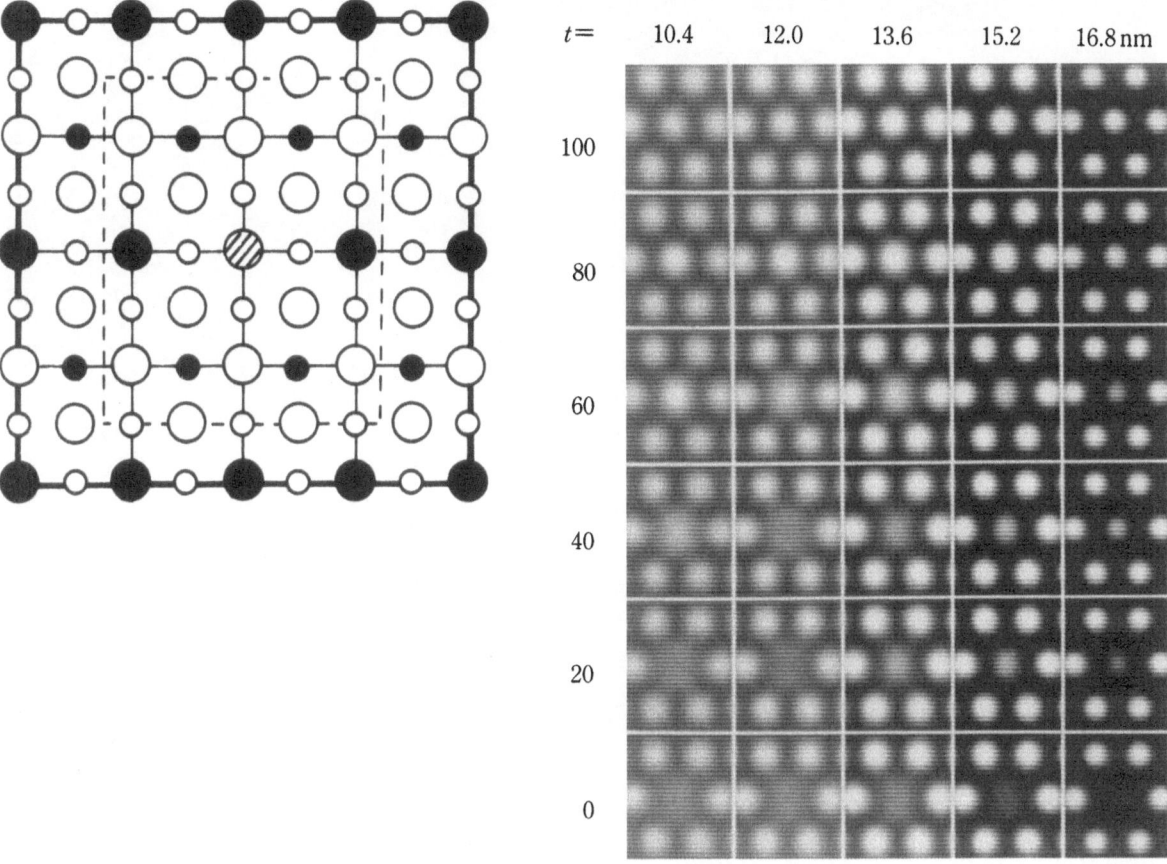

| $t=$ | 10.4 | 12.0 | 13.6 | 15.2 | 16.8 nm |

at % Mn

Fig. 3.53. Computer simulation on high-resolution images of Au–Mn alloys where some of the Mn atoms at the center of the model are replaced by Au atoms. High-resolution images are calculated as functions of crystal thickness (t) and the Mn atom content at the column in the center of the model

3.2.3.2 High-Resolution Images in Various Ordered Alloys

One-Dimensional Long-Period Superstructure in Cu–Au Alloys. It is well known that in the Cu–Au alloys, a superstructure of the $L1_0$-type (the so-called CuAu I) is formed at low temperature, while at high temperatures between 385°C and 410°C, a *one-dimensional long-period superstructure* (the so-called CuAu II) is formed [56]. As shown in Fig. 3.54, the structure of CuAu II is based on the $L1_0$-type structure with an atomic displacement of $a/2[011]$ at intervals of $5a[100]$. In a one-dimensional long-period superstructure, the

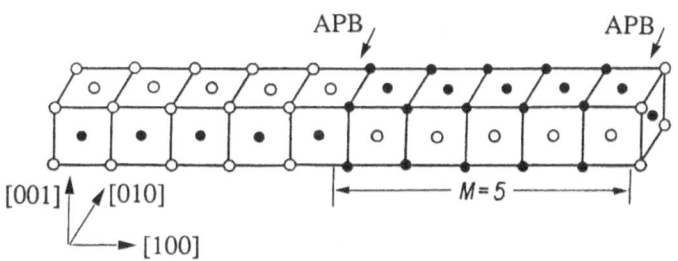

Fig. 3.54. One-dimensional long-period superstructure of CuAu II. *Open* and *solid circles* indicate Cu and Au atoms, respectively

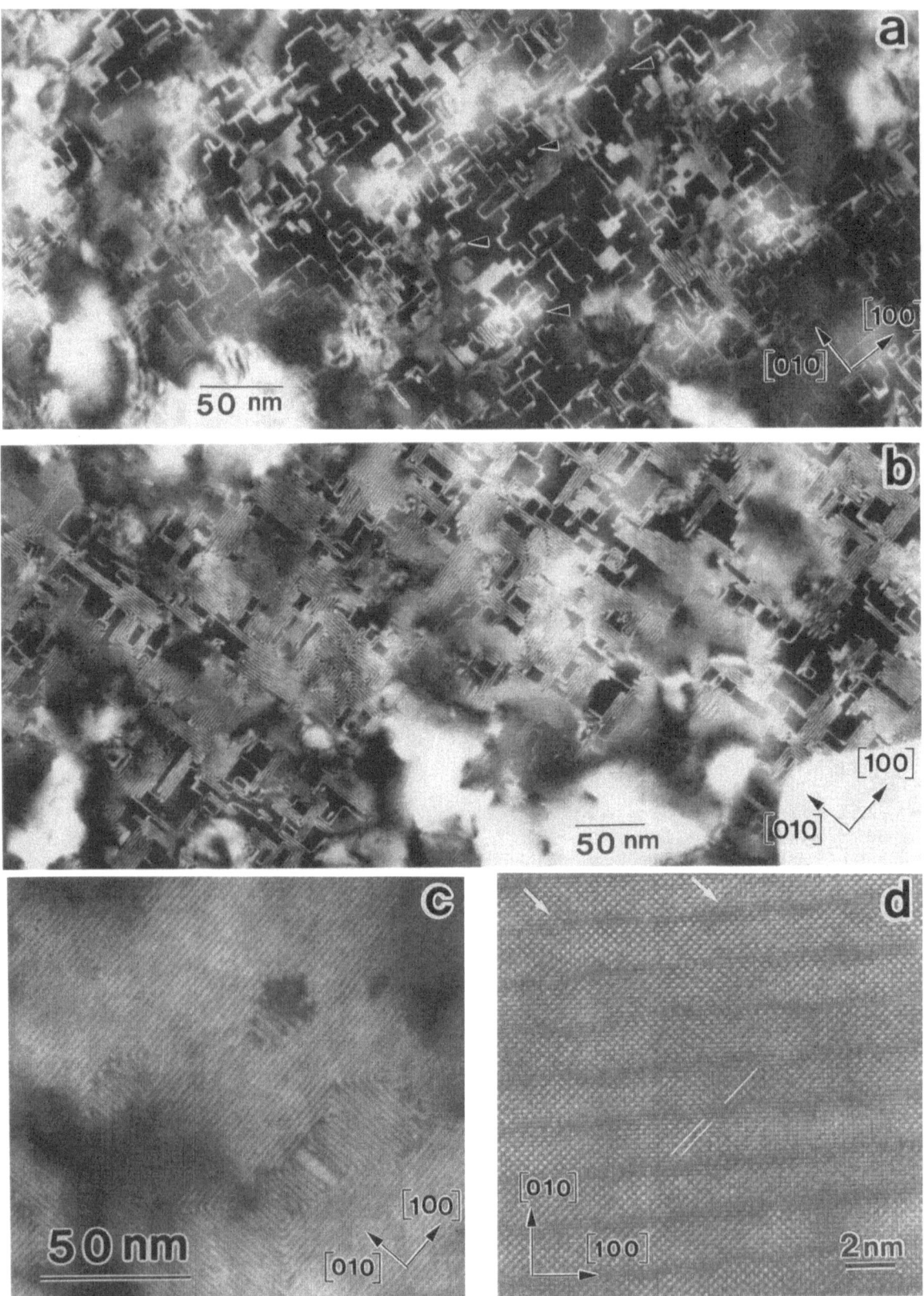

Fig. 3.55. **a–c** Bright-field images, and **d** a superstructure image of Cu–Au alloys.
Specimen: **a–c** Cu–50.0at%Au and **d** Cu–41.0at%Au; **Preparation**: jet electropolishing;
Observation: 200 kV EM (bright-field images) and 400 kV EM (superstructure image of **d**).
Remarks: The heat treatment was as follows; **a** 400°C 10 min; **b** 400°C 20 min; **c** 400°C 1 h; **d** 310°C 349 h.
Superstructure image **d** was kindly provided by Prof. D. Watanabe, Iwakimeisei University, Dr. T. Ohsuna, Tohoku University, and Dr. T. Kimoto, National Institute for Metals

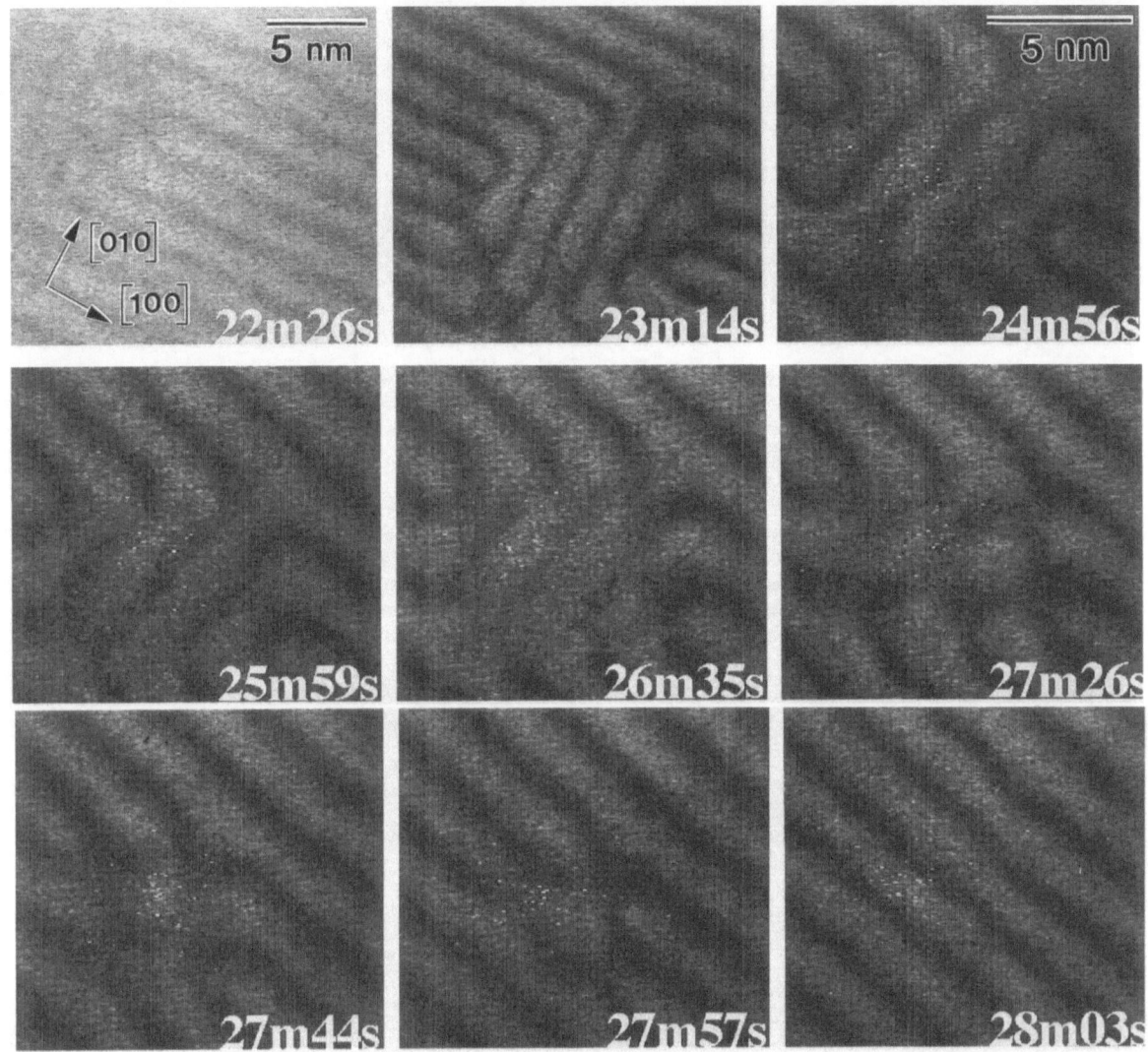

Fig. 3.56. In situ observation of a Cu–Au alloy with a one-dimensional long-period structure using a heating stage. **Specimen**: Cu–50.0at%Au; **Preparation**: jet electropolishing; **Observation**: 200kV EM, [001] incidence.
Remarks: *Black bands* correspond to APBs. APBs parallel to the [010] direction in the center of the images gradually rearrange to become parallel to the [100] direction. The passage of time is indicated on the images. Direct magnification is increased after 24 min 56 s

phase of the atomic arrangement is reversed at the (100) plane (indicated by the arrows in Fig. 3.54). Thus, the plane is called the *anti-phase boundary*, and the region on either side of the boundary is called the *anti-phase domain*. The length of the anti-phase domain is evaluated in the units of the lattice constant of the fundamental lattice, and the atomic displacement at the boundary is called the *displacement vector*. At 50at%Au, the length of the anti-phase domain in CuAu II was measured as about 5.1. Superstructure with a period that is a non-integral value is called an *incommensurate structure*.

In Fig. 3.55a–c, bright-field images of a Cu–50at%Au alloy are shown. From these images, the growth process of the anti-phase boundaries in the CuAu I structure can be seen. In the image of the alloy annealed at 400°C for 10 min, anti-phase boundaries are seen as short white lines parallel to the [100] and [010] directions. The small white circles, indicated by arrowheads, correspond to the anti-phase boundaries being generated. In the alloy annealed for 20 min, as shown in Fig. 3.55b, the number of anti-phase boundaries increases, and these boundaries are formed regularly with constant spacing. Furthermore, after annealing

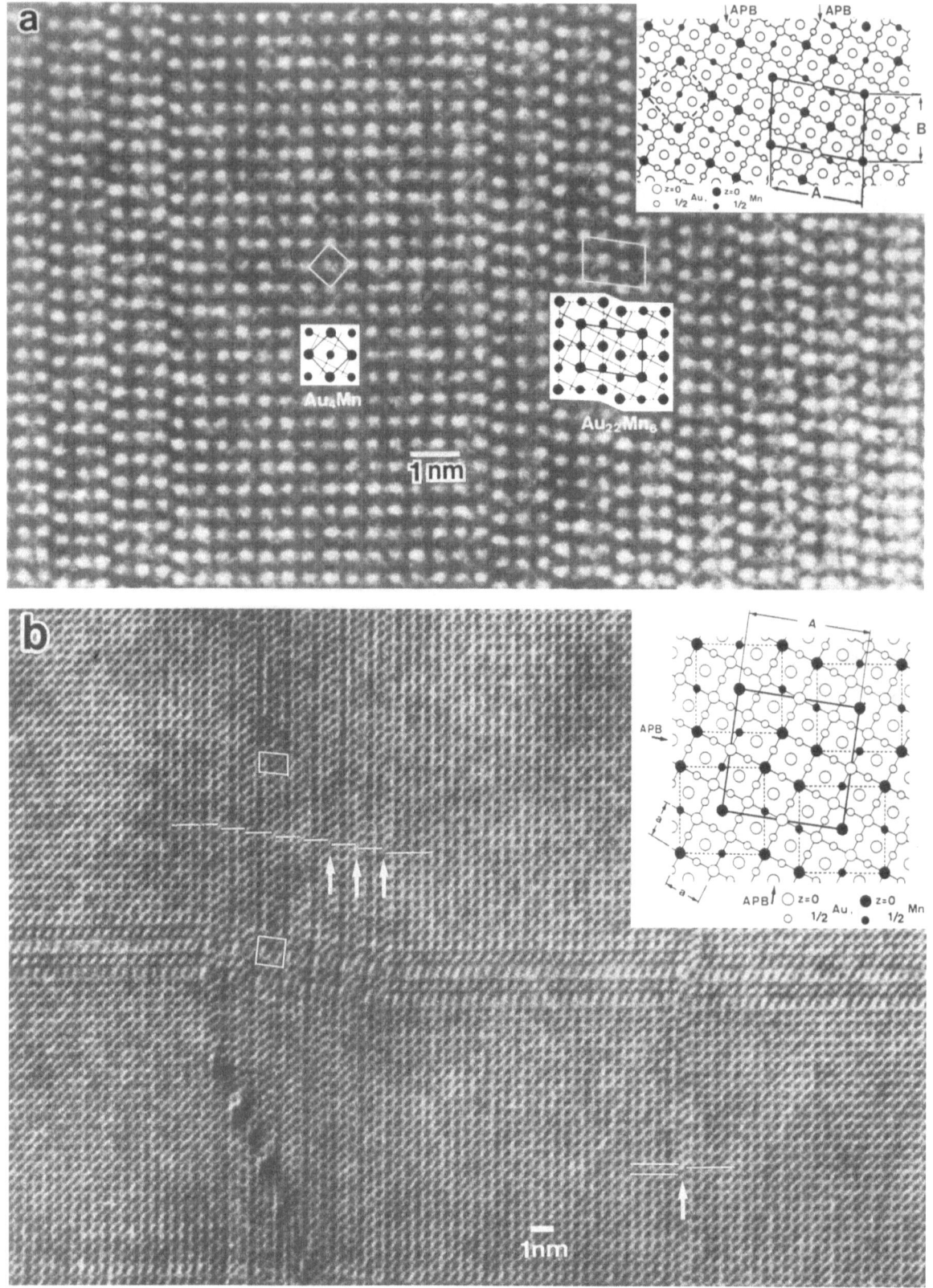

Fig. 3.57. Superstructure image of Au–20.7at%Mn ordered alloys.
Specimen: Au–20.7at%Mn; **Preparation**: jet electropolishing ($HClO_4 : H_2SO_4 : H_2O = 3 : 3 : 50$);
Observation: 1000 kV EM, [001] incidence.
Remarks: In image **a**, the Mn atom positions appear as *white spots* showing the formation of the Au_4Mn and $Au_{22}Mn_6$ structures. The structure models in the *insert* in **a** show unit cells of the Au_4Mn (*left*) and $Au_{22}Mn_6$ (*right*) structures. In image **b**, the Mn atom positions appear as *black spots*, and the periodic arrangement of one-dimensional APBs results in the formation of the $Au_{22}Mn_6$ and $Au_{31}Mn_9$ structures (see the structure model *upper right*)

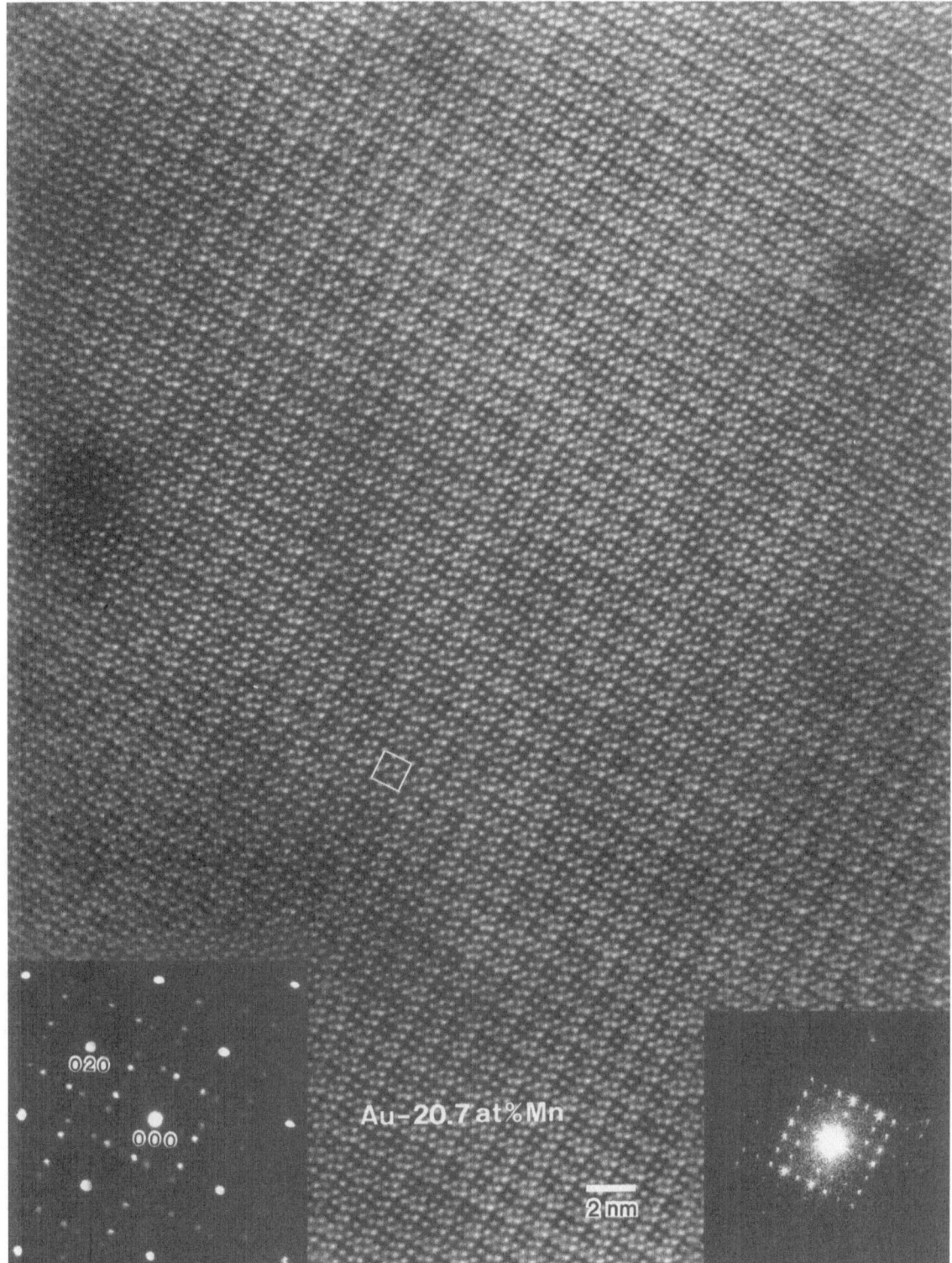

Fig. 3.58. Superstructure image of the $Au_{31}Mn_9$ structure.
Specimen: Au–20.7at%Mn heat-treated at 400°C;
Preparation: jet electropolishing ($HClO_4 : H_2SO_4 : H_2O = 3 : 3 : 50$); **Observation**: 1000 kV EM, [001] incidence.
Remarks: *White spots* correspond to Mn atom positions. The formation of two-dimensional APB in the Au_4Mn structure results in the $Au_{31}Mn_9$ structure (see structure model in Fig. 3.57b)

for 1 h, anti-phase boundaries form in the whole area and eventually form the CuAu II structure. Figure 3.55d shows the superstructure image of the CuAu II structure observed in a Cu–41at% Au, where the small white dots correspond to Au atoms projected along the incident beam direction parallel to the [001] direction, and the black horizontal bands are the anti-phase boundaries [66]. The boundaries are seen to be tilted slightly away from the [100] direction due to the step in the boundaries indicated by the white arrows.

In Fig. 3.55c, there are two types of anti-phase boundaries parallel to the (100) and (010) planes. The domains with different anti-phase boundaries are changed to a single domain after annealing for a long time. Figure 3.56 shows such a process of the single domain formation as observed by high-resolution electron microscopy. The specimen is kept at about 400°C by a heating stage in a transmission electron microscope. The anti-phase boundaries parallel to the (100) plane disappear, and the single domain with anti-phase boundaries parallel to the (010) plane is clearly observed.

Superstructures and Anti-Phase Boundaries in Au–Mn Alloys. In Au–Mn alloys, structure is sensitive to compositional variation and heat treatment, and several superstructures have been observed to date. Figure 3.57 shows one example of the superstructure images observed in Au–20.7at% Mn. In Fig. 3.57a, the bright dots correspond to the Mn atoms projected along the direction of the incident electron beam. As shown in the models inserted, there are the so-called Au_4Mn structure ($D1_a$-type structure), and also the $Au_{22}Mn_6$ structure [67]. As shown in the model inserted at the upper right, the $Au_{22}Mn_6$ structure

is formed from the Au_4Mn structure with the periodic arrangement of anti-phase boundaries being parallel to the vertical direction viz., on the (210) plane. In another superstructure image observed in the specimen of Au–20.7at% Mn, the Mn atom positions appear as dark spots, as shown in Fig. 3.57b. In addition to the Au_4Mn and $Au_{22}Mn_6$ structures, in the central part of the image there is a local $Au_{31}Mn_9$ structure. This is introduced from the Au_4Mn structure with the periodic arrangement of anti-phase boundaries being parallel to both the vertical and horizontal directions, which correspond to the (210) and ($\bar{1}$20) planes, respectively (see the model inserted at the upper right). As shown in Fig. 3.58, the $Au_{31}Mn_9$ structure forms homogeneously in the wide area when the specimen is annealed at 400°C [68]. At the bottom of Fig. 3.58, the electron diffraction pattern and the optical diffractogram are shown on the left and right, respectively. As in the case of Fig. 3.51, the intensity distributions of superlattice reflections in the electron diffraction pattern and the optical diffractogram are in good agreement.

Ordering Process of Ni_4Mo Alloys. Figure 3.59 shows electron diffraction patterns of an Ni_4Mo alloy which was quenched from 1000°C and subsequently annealed at 650°C [69]. The structure of this alloy is based on the fcc lattice, and in the electron diffraction pattern obtained from the quenched specimen without annealing, characteristic weak diffuse scattering is observed around the 1 1/2 0 reciprocal point. This diffuse scattering indicates the so-called short-range ordered state, where ordering in the atomic arrangement extends in the short range [70]. The microstructure of the short-range ordered state has been extensively studied by conventional electron mi-

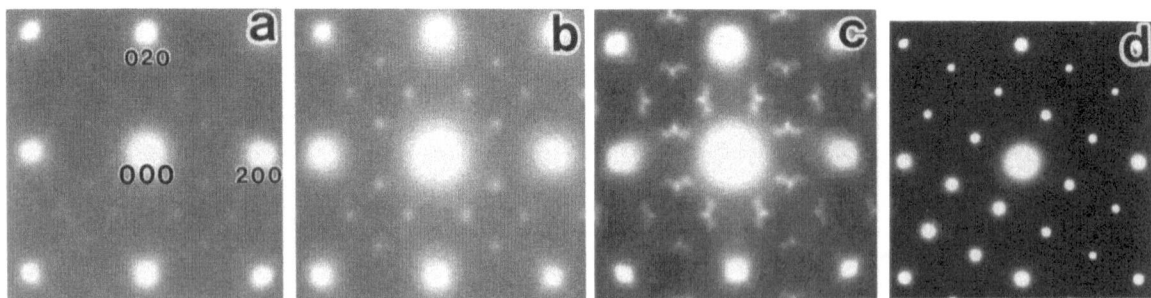

Fig. 3.59. Electron diffraction patterns of Ni_4Mo alloys annealed at about 650°C after quenching at 1000°C. **a** As quenched; **b** 5 min annealing; **c** 15 min annealing; **d** 336 h annealing

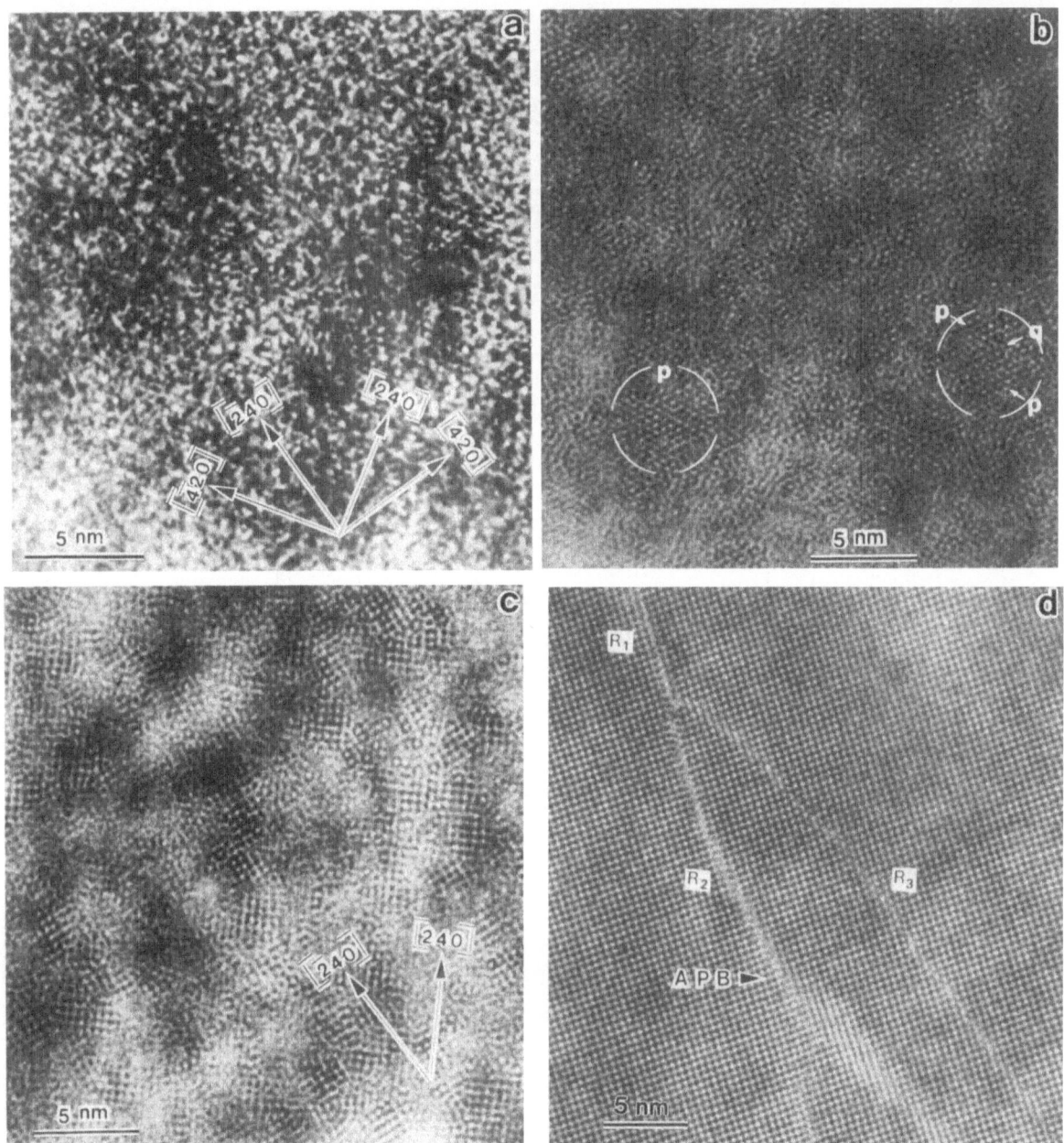

Fig. 3.60. High-resolution images of Ni$_4$Mo alloys annealed at about 650°C after quenching at 1000°C.
Specimen: Ni–20.1at%Mo; **Preparation**: jet electropolishing (H$_2$SO$_4$:CH$_3$OH = 1:4);
Observation: 200kV EM, [001] incidence.
Remarks: **a** As quenched; **b** 5 min annealing; **c** 15 min annealing; **d** 336h annealing. In the quenched state, no clear white or black spots are seen, indicating the so-called short-range ordered state. With an increase in annealing time, ordered domains become larger

croscopy as well as high-resolution electron microscopy [69, 71, 72]. In Fig. 3.59, it can be seen that the intensity of the diffuse scattering becomes stronger and sharper after 5 min annealing, and with an increase in annealing time, the diffuse scattering of the electron diffraction pattern changes to the characteristic shape shown in Fig. 3.59c, and finally forms the superlattice reflections corresponding to a superstructure of the D1$_a$-type. Figure 3.60 shows high-resolution images

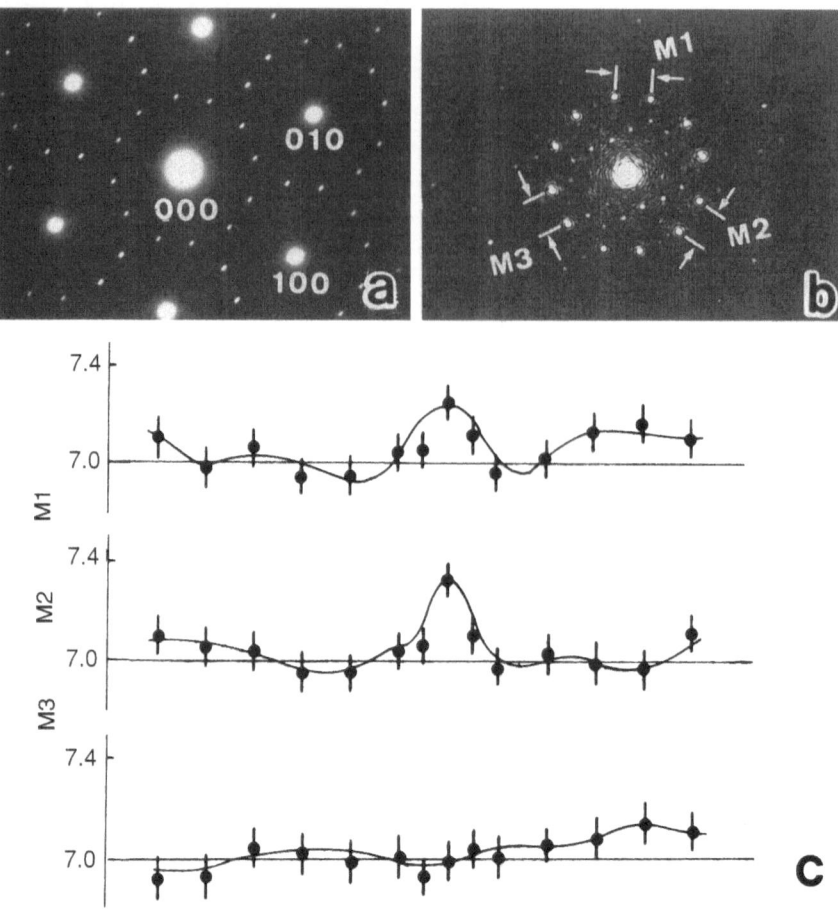

Fig. 3.61. **a** Electron diffraction pattern of Au–32.0at%Cd. **b** Optical diffractogram of the high-resolution image in Fig. 3.62b. **c** Change of periods in three directions evaluated from the optical diffractogram in **b**

which correspond to the electron diffraction patterns in Fig. 3.59. In the high-resolution image of the alloy quenched from 1000°C, no sharp bright or dark spots can be seen, indicating no regularly arranged Mo and Ni atoms projected along the incident electron beam parallel to the [001] direction. After annealing for 5 min, regularly arranged bright spots, corresponding to the atomic columns consisting of Mo atoms, are seen in the regions indicated by p and q. From the arrangement of the bright spots, it can be seen that superstructures of the $D0_{22}$-type and $D1_a$-type are formed at p and q, respectively. After annealing for 15 min, larger domains consisting of a superstructure of the $D1_a$-type can be seen (Fig. 3.59c). In the alloy annealed for 336 h, a superstructure of the $D1_a$-type forms in the whole area of Fig. 3.59d. Note that antiphase boundaries exist parallel to the [001] direction in the regions indicated by R_1, R_2, and R_3.

Two-Dimensional Long-Period Superstructures with Incommensurate Periods in Au–Cd Alloys. Figure 3.61a shows an electron diffraction pattern of an Au–32.0at%Cd alloy with a hexagonal structure. The incident electron beam is parallel to the [001] direction. As indicated by the superlattice reflections in the electron diffraction pattern, this alloy forms a two-dimensional long-period structure with incommensurate periods ($M = 7$–8) on the (001) plane. The incommensurate period increases continuously with a decrease in Cd content. Figure 3.62a and b show high-resolution images of Au–30.5at%Cd ($M = 7.5$) and Au–32.0at%Cd ($M = 7.1$), respectively [73]. The dark spots in Fig. 3.62a and the bright spots in Fig. 3.62b correspond to regions consisting of Au atoms, and the domains outlined by white lines correspond to a commensurate superstructure with period 7. At the boundaries between the domains, there is another superstructure with period 9. The incommensurate period can be investigated with

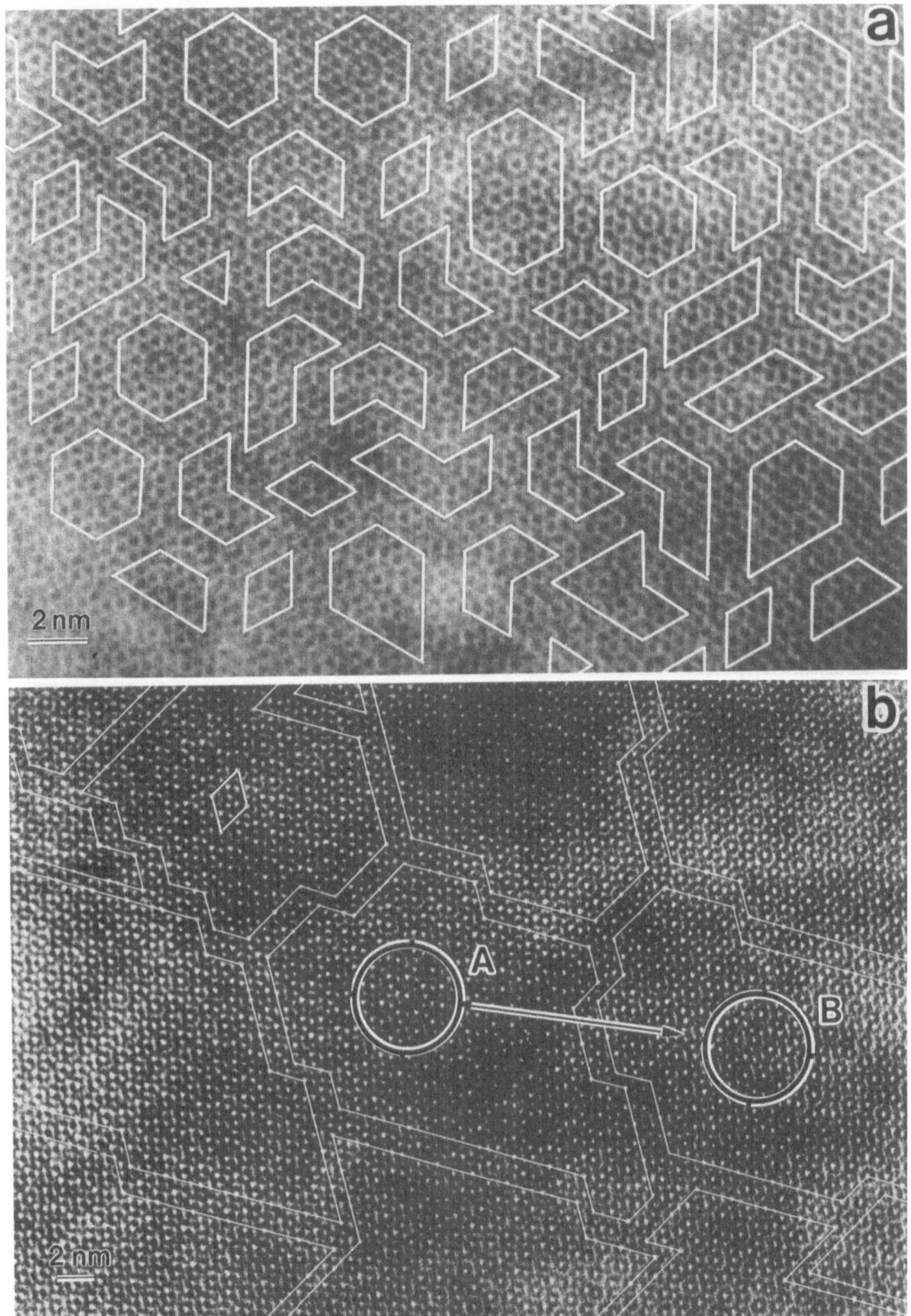

Fig. 3.62. High-resolution images of Au–Cd alloys with two-dimensional incommensurate structures. **a** M (period of incommensurate structure) = 7.5; **b** M = 7.1.
Specimen: **a** Au–30.5at%Cd and **b** Au–32.0at%Cd; **Preparation**: jet electropolishing (CH_3COOH (133 ml) + H_2O (7 ml) + CrO_3 (25 g)); **Observation**: 1000 kV EM, [001] incidence.
Remarks: *White lines* indicate the domains of a two-dimensional commensurate structure with M = 7. The *circles A and B* in **b** indicate the starting and ending areas from which the optical diffractograms are taken. The image contrasts of **a** and **b** are reversed

111

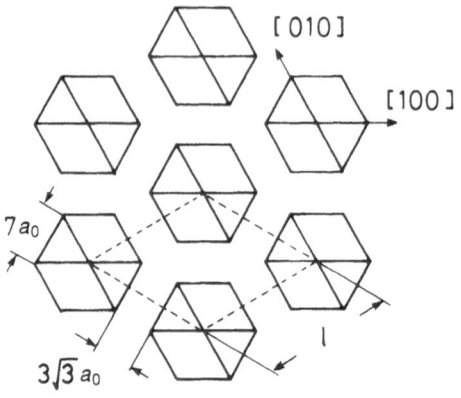

a

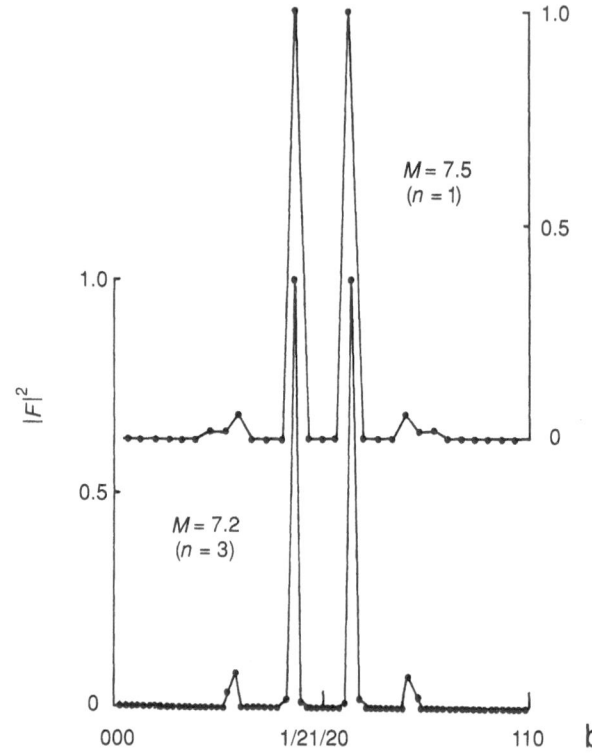

b

Fig. 3.63. a Model of a two-dimensional long-period structure with incommensurate periods. **b** Intensity of the calculated *hh0* reflection based on the model in **a**. With an increase in the domain size (*n*), the period (*M*) becomes smaller

an optical diffractogram such as shown in Fig. 3.61b. Figure 3.61c shows the change of the incommensurate period from the region A to the region B through the boundary in Fig. 3.62b. The periods *M1* and *M2* measured along the two directions which are not parallel to the boundary deviate largely from period 7 at the boundary.

A model of an incommensurate structure obtained from the high-resolution images is shown in Fig. 3.63a. In general, the length of the edges of the hexagonal domains of the commensurate structure ($M = 7$) is given by $7na_0$, where n is an integer. The model shown in Fig. 3.63a corresponds to the case of $n = 1$. In general, the period M is determined using the volume ratio of the two commensurate structures. As shown in Fig. 3.63a, the unit cell of the supercell is given by

$$l = \left(7n + 3\right)\sqrt{3}a_0 \qquad (3.4)$$

assuming that the superstructure has hexagonal symmetry. Then the paired superlattice reflections h_1 and h_2 ($h_2 > h_1$) are given by

$$h_1 = \frac{1}{3\left(7n + 3\right)}\left[\frac{9}{7}\left(7n + 3\right) + 1\right] \qquad (3.5)$$

$$h_2 = \frac{1}{3\left(7n + 3\right)}\left[\frac{12}{7}\left(7n + 3\right)\right] \qquad (3.6)$$

where $[R]$ is the Gauss' notation, indicating the largest integer which is smaller than a real number R. Thus, the incommensurate periodicity M is given by

$$M = \frac{1}{h_2 - h_1}$$

$$= \frac{3\left(7n + 3\right)}{\left[\frac{12}{7}\left(7n + 3\right)\right] - \left[\frac{9}{7}\left(7n + 3\right) + 1\right]}. \qquad (3.7)$$

Figure 3.63b shows the calculated intensity distribution along the *hh0* direction for the models of $n = 1$ and 3. Only superlattice reflections of the first order have strong intensity. The periods and the intensity profiles are consistent with observations. It should be noted that the position of superlattice reflections corresponding to the incommensurate structure changes with the number of valence electrons per atom (*e/a*), and the phase stability can be explained by the theory proposed by Sato and Toth [74].

3.2.4 Quasicrystals

The 1984 discovery by Shechtman et al. [75] of an electron diffraction pattern with five-fold rotational symmetry in a rapidly solidified Al–14at%Mn alloy had a strong impact on solid state physics. For a long time, it had been thought that solids were divided into two structural classes: crystalline structures with periodic atomic arrangements, and amorphous structures with random atomic arrangements. It had also been recognized that only crystals with periodic structures produced sharp diffraction peaks, and that five-fold rotational symmetry was not allowed in crystals with periodic arrangements. The discovery by Shechtman et al. brought about a drastic change in thinking concerning the structure of solids. Materials showing diffraction patterns with non-crystallographic symmetries, but nonetheless consisting of sharp peaks, were referred to as *"quasicrystals"* [76]. Quasicrystals have new aperiodic structures producing sharp diffraction peaks, and they also have *quasiperiodicity* with new ordering, instead of the periodicity of crystals.

Since the discovery by Shechtman et al., a variety of quasicrystals have been found as metastable or stable phases in many alloys. Quasicrystals are classified into two main groups: *icosahedral quasicrystals* and *decagonal quasicrystals*. The former are three-dimensional quasicrystals with icosahedral symmetry having six five-fold symmetry axes, ten three-fold symmetry axes, and 15 two-fold symmetry axes (called an icosahedral phase). The latter is a two-dimensional quasicrystal with two-dimensional quasiperiodic planes and one-dimensional periodicity perpendicular to the planes (called a decagonal phase).

3.2.4.1 Electron Diffraction Patterns of Quasicrystals

Figure 3.64 shows the electron diffraction patterns of a stable Al–Cu–Fe icosahedral phase, taken with the incident beams parallel to the two-fold, three-fold, and five-fold symmetry axes [77]. They were taken at two different camera lengths to observe diffraction spots on the zeroth Laue zone as well as those on the higher Laue zones, and Kikuchi patterns. In Fig. 3.64e and f, patterns taken with different exposure times are inserted to give a simultaneous view of the diffraction spots in higher Laue zones as well as the Kikuchi patterns. The patterns show a number of sharp diffraction spots, which are located at strict icosahedral symmetry positions indexed with the golden ratio τ ($\tau = (1 + \sqrt{5})/2$). In Fig. 3.64, the diffraction patterns in the zeroth Laue zone and the Kikuchi bands indicate two-fold, six-fold, and ten-fold rotational symmetries corresponding to the projection symmetry of the icosahedral phase, whereas the diffraction spots in the higher Laue zones and the Kikuchi lines show two-fold, three-fold, and five-fold symmetries corresponding to three-dimensional symmetry. For example, in Fig. 3.64e the three-fold rotational symmetry can clearly be seen in the intensity distribution of diffraction spots on higher Laue zones and in the Kikuchi pattern formed with bright and dark lines. On the other hand, pentagons shown by bright and dark Kikuchi lines can be seen in Fig. 3.64f. Hence, the diffraction patterns show icosahedral symmetry.

Figure 3.65 shows electron diffraction patterns of an Al–Ni–Co decagonal quasicrystal [78]. In the pattern in Fig. 3.65a, which was taken with the incident beam parallel to the ten-fold rotational symmetry axis, i.e., the periodic axis, there are a number of diffraction spots located at the positions (indexed with the golden ratio) with ten-fold symmetry. However, Fig. 3.65b, which was taken with the incident beam perpendicular to the ten-fold symmetry axis, shows a section of the parallel planes consisting of Bragg spots with the same interval, which indicates the periodicity of the structure in the vertical direction. This decagonal quasicrystal has 0.4 nm periodicity, and some polytypes with 0.8 nm, 1.2 nm, or 1.6 nm periodicity have been found in other alloys.

Figures 3.64 and 3.65 were taken from highly ordered quasicrystals with long-range correlation and show a number of weak diffraction spots, but in most quasicrystals formed as metastable phases, the weak spots disappear, as can be seen in Fig. 3.68b which shows a lack of long-range correlation.

3.2.4.2 Theoretical Interpretation of the Structure of Quasicrystals

In order to understand a structure with quasiperiodicity, an elegant theoretical model has been proposed. Figure 3.66 shows how to construct one-dimensional quasiperiodicity. A one-dimensional quasiperiodic lattice can be formed by the projection of a two-dimensional square lattice on a straight line with an irrational slope ($1/\tau$ in Fig. 3.66). The straight line passes through the original

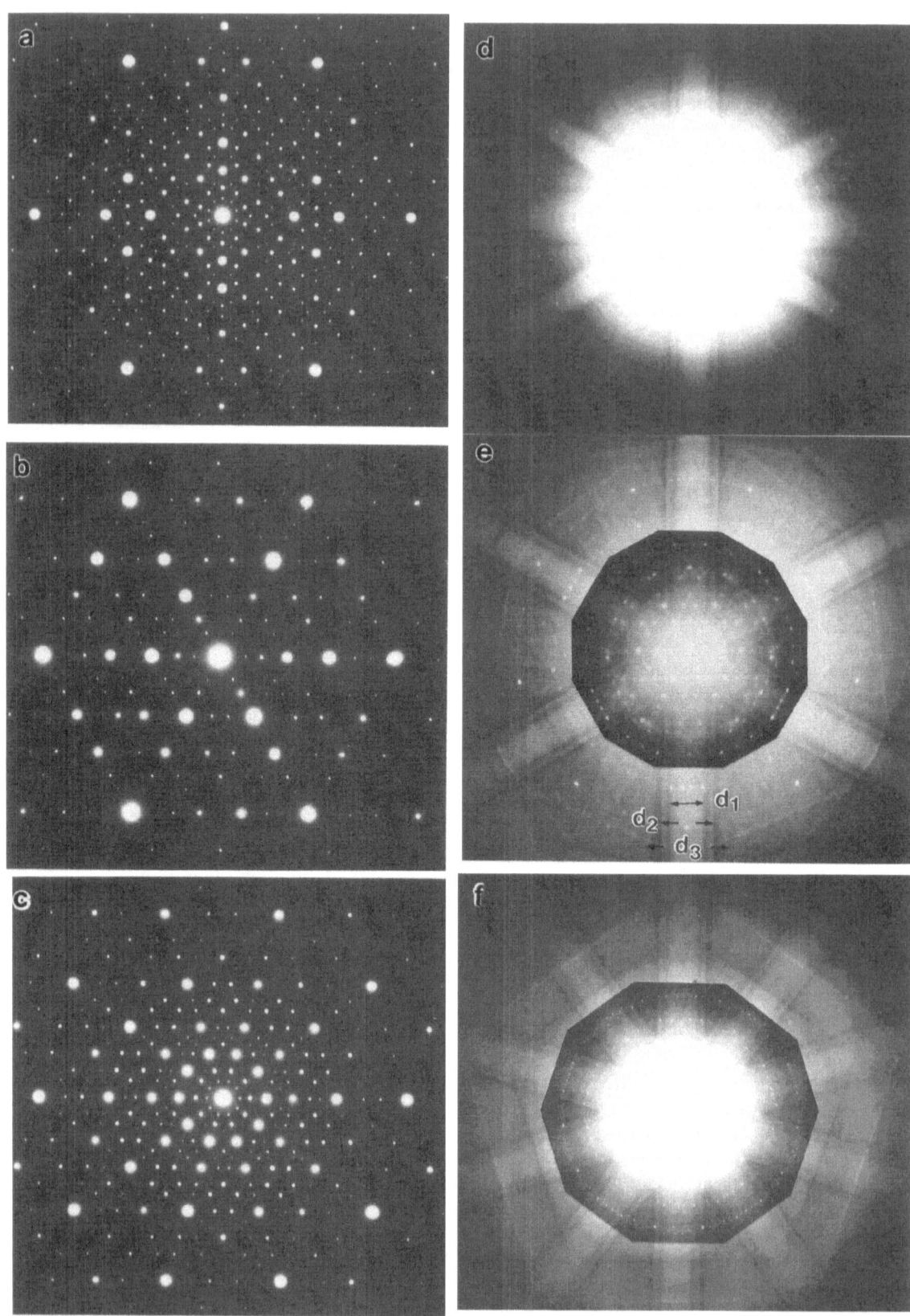

Fig. 3.64. Electron diffraction patterns and Kikuchi patterns of a stable Al–Cu-Fe icosahedral phase.
Specimen: $Al_{65}Fe_{15}Cu_{20}$; **Preparation**: crushing; **Observation**: 200 kV EM, incident beams parallel to the (**a,d**) two-fold, (**b,e**) three-fold, and (**c,f**) five-fold symmetry axes

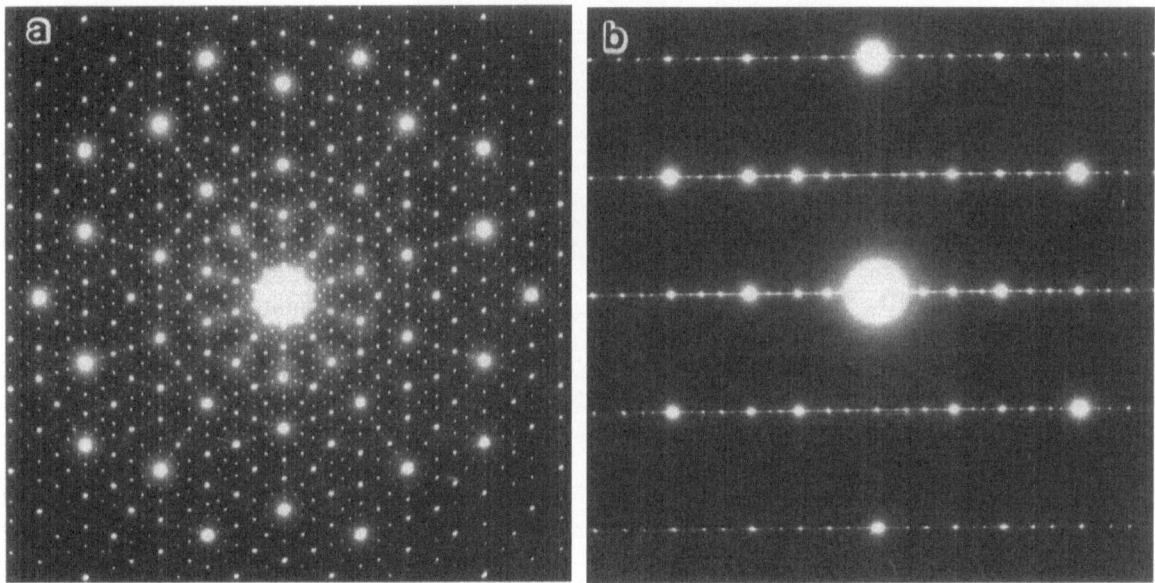

Fig. 3.65. Electron diffraction patterns of an Al–Ni–Co decagonal quasicrystal.
Specimen: $Al_{70}Ni_{15}Co_{15}$; **Preparation**: crushing;
Observation: 400 kV EM, incident beams **a** parallel and **b** perpendicular to the ten-fold symmetry axis

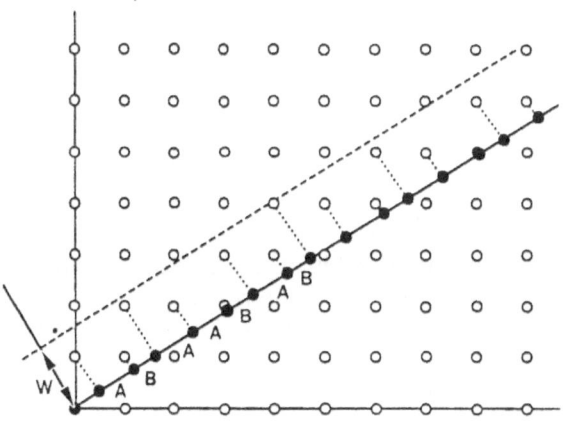

Fig. 3.66. Illustration showing how to construct one-dimensional quasiperiodicity. A one-dimensional quasiperiodic lattice can be formed by the projection of a two-dimensional square lattice on a straight line with an irrational slope $(1/\tau)$

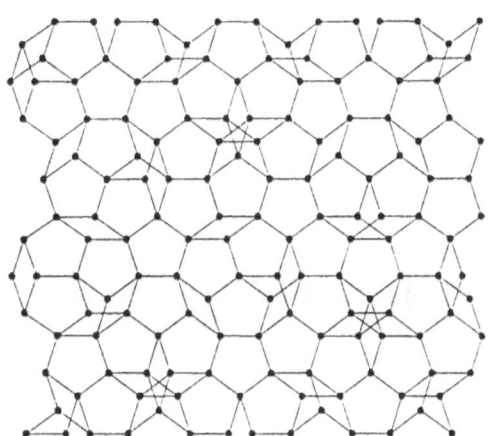

Fig. 3.67. Example of pentagonal tiling made by the projection method

lattice point, but never passes through another lattice point owing to the irrational slope. Therefore, by projecting the lattice points in the zone between two parallel lines (the solid and dotted lines in the figure) onto the solid line, one can obtain an aperiodic arrangement of two types of interval ($A = \cos\theta$, $B = \sin\theta$) ($\theta = \tan^{-1}(1/\tau)$, $A = B \cdot \tau$). This arrangement is a *quasiperiodic arrangement* (also called a *Fibonacci sequence*) of

two units, A and B. It was found mathematically that the Fourier transform of an infinite quasiperiodic arrangement made sharp diffraction peaks, described as a delta function.

A two-dimensional quasiperiodic lattice can be made by the projection of a five-dimensional hypercubic lattice on a two-dimensional plane. Figure 3.67 shows an example of pentagonal tiling made by the projection method. Lattice points in the tiling can be indexed by five unit vectors parallel to the angles $2n\pi/5$ ($n = 0, 1, 2, 3, 4$).

The diffraction pattern obtained by putting scattering subjects at lattice points in the tiling is known to show a ten-fold symmetry pattern formed with sharp diffraction spots. By viewing obliquely along the ten-fold directions of $2n\pi/10$, one can see straight lines of lattice points and a quasiperiodic array of two intervals of the line of lattice points, i.e., the Fibonacci sequence of two intervals. This type of ordering produces a diffraction pattern with sharp peaks. Does this strange type of tiling occur in the quasicrystals of alloys? How are atoms arranged in the quasicrystal alloys in relation to this tiling? These two questions are of great interest to research workers. High-resolution electron microscopy is an indispensable tool for studying the real structures of quasicrystals.

3.2.4.3 High-Resolution Images of Icosahedral Quasicrystals

Diffraction patterns with five-fold symmetry have been observed from multiple-twinning particles, and so a multiple twinning model was proposed in an attempt to understand the diffraction patterns with five-fold symmetry in the early stages of quasicrystal studies [79]. However, a high-resolution image of an Al–Mn icosahedral phase in a rapidly solidified alloy brought a conclusion (Fig. 3.68 [80]). This micrograph, which was taken from the Al–Mn quasicrystal at its earliest stage, ruled out the multiple-twinning model and clearly showed the existence of five-fold symmetry (in an average structure) in real space.

At present, we cannot decide whether the bright dots in the image in Fig. 3.68 correspond to atom positions or channels between atoms, but a contrast distribution in the image undoubtedly reflects a topological feature of the quasicrystal structure. Hereafter, this type of image is called a "lattice image" because it includes little information about atomic arrangements. Structure images, taken with a 400 kV electron microscopy with a higher resolution, will be presented later. The bright dots in the image in Fig. 3.68 are apparently arrayed in a well-regulated manner like those of crystals, but no periodicity can be found in this bright dot array. It can be seen from the image that the bright dots are arranged in straight lines parallel to the five-fold symmetry directions indicated by arrows. That is, the bright dot array has no five-fold rota-

tional symmetry, strictly speaking, but it has five-fold symmetry in an average structure forming a diffraction pattern. It should be also noted that a pentagonal tiling pattern can be formed in the image by connecting the bright dots by lines [81], as shown in Fig. 3.68d. This tiling resembles the pentagonal tiling in Fig. 3.67, and this feature produces a diffraction pattern, formed with sharp spots, with five-fold symmetry.

A close oblique examination of Fig. 3.68a reveals frequent displacements of the bright dot arrays. This defect is called *linear phason strain*, and the frequency of the displacements depends on the direction of the bright dot arrays. The appearance of linear phason strain results in shifts of the diffraction spots from the positions of five-fold symmetry [82]. This type of defect appears in most quasicrystals, and results from their directional growth. In order to remove the linear phason strain, annealing at high temperatures is required, but it is impossible to remove it in the Al–Mn quasicrystal, because this quasicrystal is transformed to a crystalline phase by annealing at high temperatures. In the parts of quasicrystals formed as stable phases, the linear phason strain can be removed completely by annealing at high temperatures [83–85].

Dislocations can exist in quasicrystals as in crystals. The high-resolution electron micrographs in Fig. 3.69 show the existence of dislocations in the Al–Mn–Si icosahedral phase for the first time [86]. Since the strain fields of the dislocations are extended, it is hard to determine the dislocation cores clearly. However, the existence of edge-type dislocations from extra half-planes can be seen by counting numbers of lattice planes (with d_l and d_s ($d_l = \tau \cdot d_s$)). For example, if one counts number of lattice planes (indicated by black lines) parallel to the A direction in the image of Fig. 3.69a, from S to U and from S to T, the difference in the number of lattice planes, Δ_A, is $10d_s - 6d_l = (10 - 6\tau)d_s$. In this way, the differences for the five directions are

$$\Delta_A = \left(10 - 6\tau\right)d_s$$

$$\Delta_B = \Delta_E = \left(2\tau - 3\right)d_s$$

$$\Delta_C = \Delta_D = \left(5\tau - 8\right)d_s.$$

For the Y dislocation, they are

$$\Delta_A = \Delta_B = \left(5 - 3\tau\right)d_s$$

$$\Delta_C = \Delta_E = \left(5\tau - 8\right)d_s$$

$$\Delta_D = 0$$

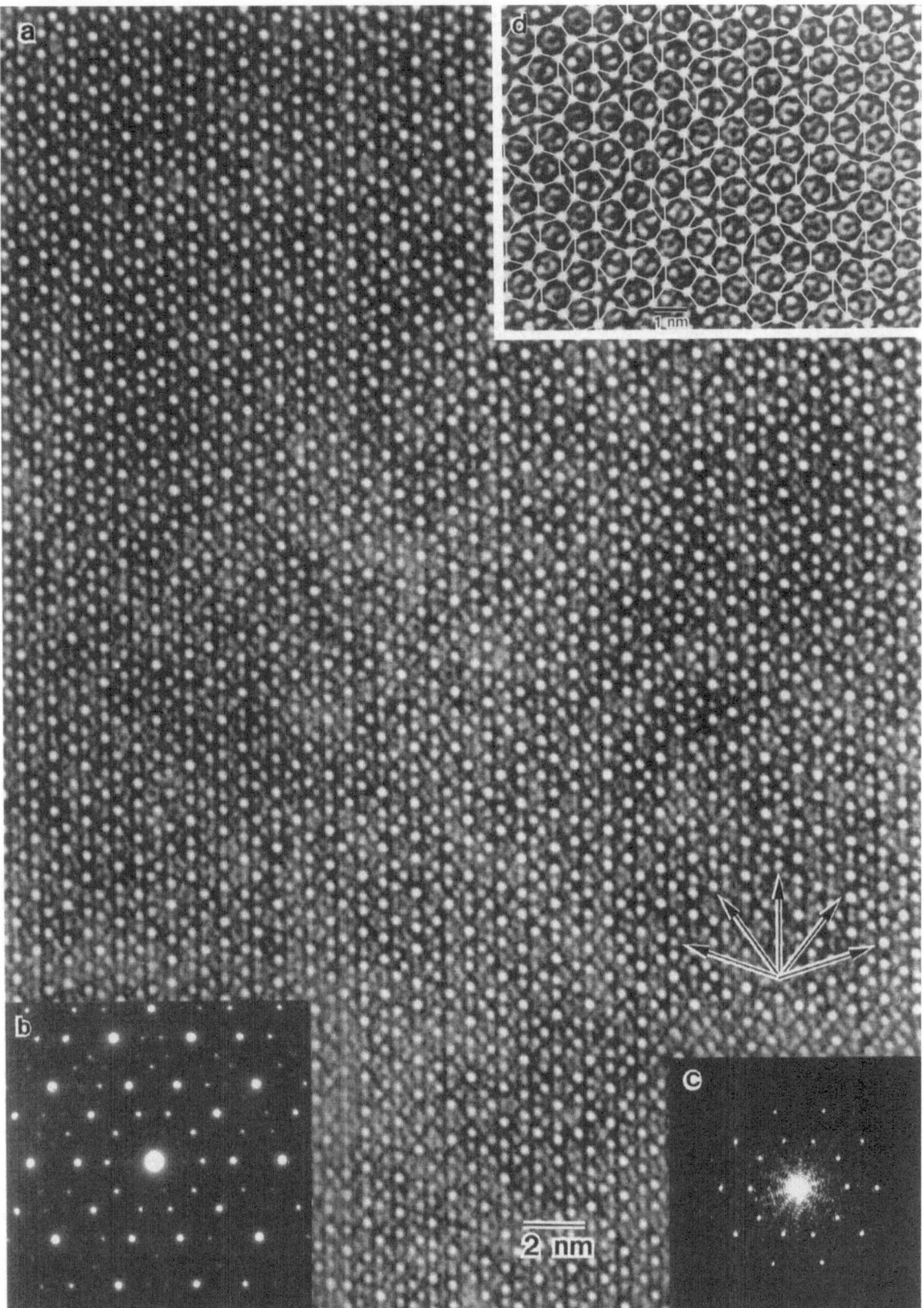

Fig. 3.68. a Lattice image and **b** electron diffraction pattern of an Al–Mn icosahedral quasicrystal. **c** Optical diffractogram of the lattice image. **d** "Tiling" which was constructed by connecting bright dots in the image with lines.

Specimen: $Al_{80}Mn_{20}$ rapidly cooled from the liquid state; **Preparation**: electropolishing ($HClO_4 : CH_3OH = 1:9$); **Observation**: 200 kV EM, incidence of five-fold symmetry axis

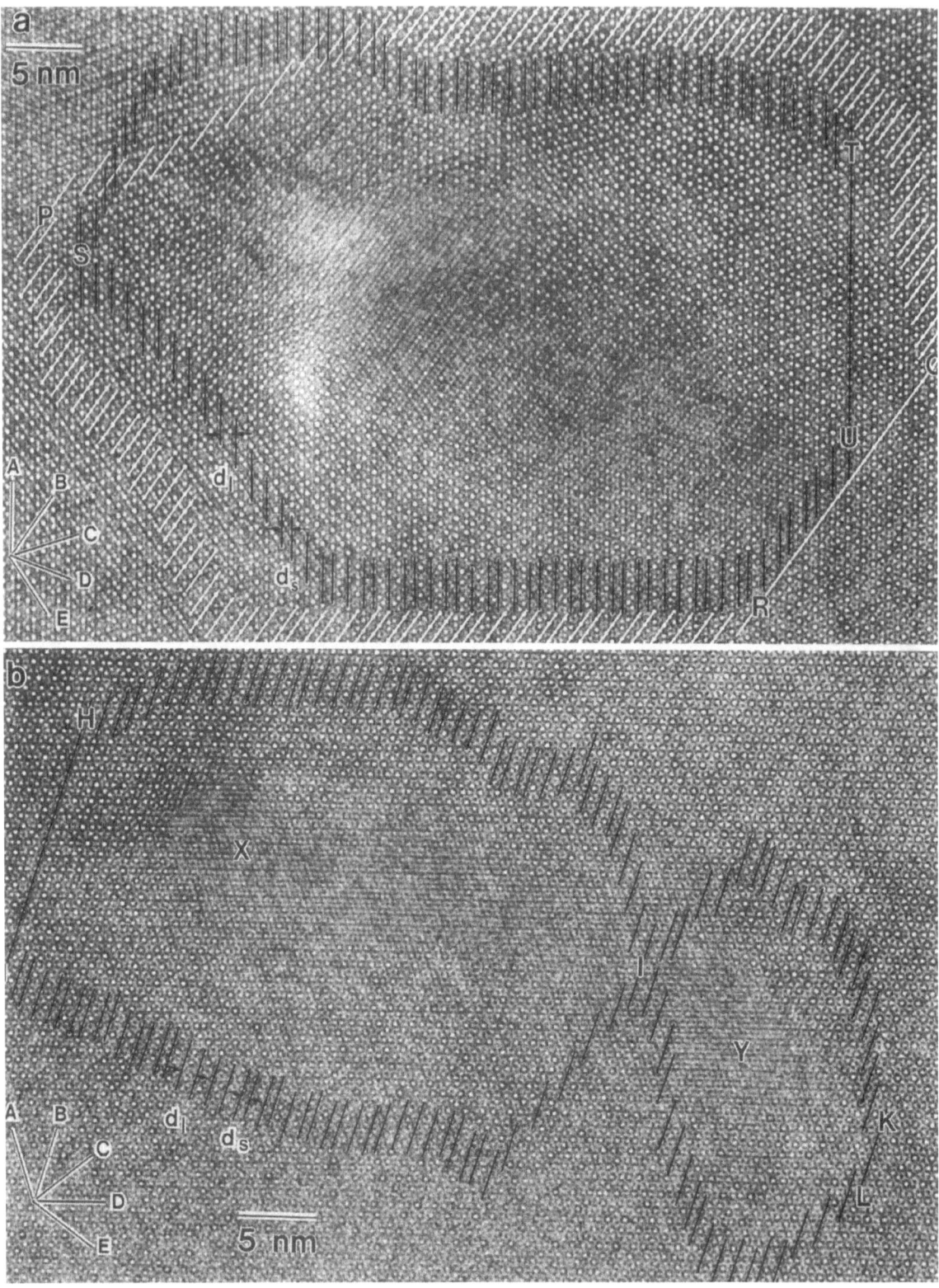

Fig. 3.69. Lattice images of dislocations in an Al–Mn–Si icosahedral quasicrystal.
Specimen: $Al_{74}Mn_{20}Si_6$ rapidly cooled from the liquid state; **Preparation**: electropolishing (ice-cold $HClO_4$: CH_3OH = 1 : 9); **Observation**: 200 kV EM, incidence of five-fold symmetry axis.
Remark: The black and white lines are inserted to count the numbers of d_s and d_l lattice spacings along the paths around the dislocations

and consequently the existence of extra half-planes showing an edge-type dislocation can be seen.

A close examination of Fig. 3.69 in this manner shows that the X and Y dislocations in Fig. 3.69b have Burgers vectors which are positive and negative in relation to each other, and that the dislocation in Fig. 3.69a has a Burgers vector with a double value, which can be interpreted as being the existence of two dislocations in the same direction.

The lattice images also provide information about topological features in the structure of quasicrystals and defects, but it is necessary to study structure images to obtain information about atomic arrangements. For quasicrystals, structure images from thin specimens have been observed. It is relatively easy to observe the structure images of oxides or inorganic materials composed of light elements, while observing the structure images of alloys is very difficult because they contain relatively heavy elements and have a high density. Success in observing and studying the structure images of some quasicrystals has been achieved because

1. the quasicrystals are mainly formed in alloys based on a relatively light element of aluminum;
2. there are many diffraction spots around the transmitted beam, as can be seen in Figs. 3.64 and 3.65, and consequently the intensity of the strongest spot is not too strongly enhanced by the dynamical diffraction effect, and so structure images can be observed in relatively thick specimens (see Sect. 2.1);
3. these quasicrystals are so brittle that thin crystals can easily be obtained by a crushing method.

It is worth mentioning the difference between a lattice image and a structure image of quasicrystals [87, 88]. Figure 3.70 shows a lattice image and a structure image of icosahedral phases, taken with the incident beam parallel to the five-fold axis. They were taken with 200 kV and 400 kV electron microscopes with resolutions of 0.23 nm and 0.17 nm, respectively. The left-hand image is a typical lattice image of the Al–Mn–Si icosahedral phase, and the right-hand image is a structure image of an Al–Pd–Mn icosahedral phase. The lattice image was obtained from a relatively thick region of a few tens of nanometers, whereas the structure image was taken from a thin region of less than 5 nm. As can be see from

pentagons of the same size in the micrographs, the structure image has finer contrast than the lattice image.

Figure 3.70b and c are optical and digital diffractograms taken from the images, together with an electron diffraction pattern in Fig. 3.70a. The diffraction spots indicated by the arrows in the electron diffraction pattern have strong intensity in a kinematical approximation such as X-ray diffraction, as indicated in the X-ray diffraction pattern at the bottom of the figure. The strong reflections, which result from nearest-neighbor atom pairs, have lattice spacings of about 0.2 nm. The optical diffractogram (Fig. 3.70b) taken from the lattice image is formed by reflections, which are enhanced by multiple scattering in the relatively thick specimen, corresponding to lattice spacings larger than 0.2 nm. On the other hand, the diffractogram in Fig. 3.70c, which is taken from the structure image, shows that the strong reflections coming from the nearest-neighbor atom pairs contribute to form the structure image. That is, the lattice and structure images are distinguished by whether the reflections from the nearest-neighbor atom pairs contribute to the images or not. The lattice image contains information about quasilattices, but little information about atomic arrangements, whereas the structure image contains information about atomic arrangements. It is clear that the structure images taken under strict conditions from thin specimens faithfully reflect the atomic arrangements projected along the five-fold symmetry axis, and the dark and bright regions in the observed images correspond to the high and low potential regions, respectively (see Sect. 2.1). The detailed interpretation of lattice and structure images is discussed later.

Structure images of quasicrystals can be observed in very thin specimens, and they change to lattice images with increasing specimen thickness because of the excitation of weak reflections near the transmitted beam. The change from structure image to lattice image can be seen in the upper right-hand side of Fig. 3.70. The structure image in the thin region at the lower left gradually changes to the lattice image in the thick region at the upper right [87].

Figure 3.71 is a structure image of an Al–Li–Cu icosahedral quasicrystal [89, 90]. The alloy includes a light element Li, so that the structure image can be obtained in a relatively thick region and also in a wide region. In this image, the bright areas correspond to regions of low potential

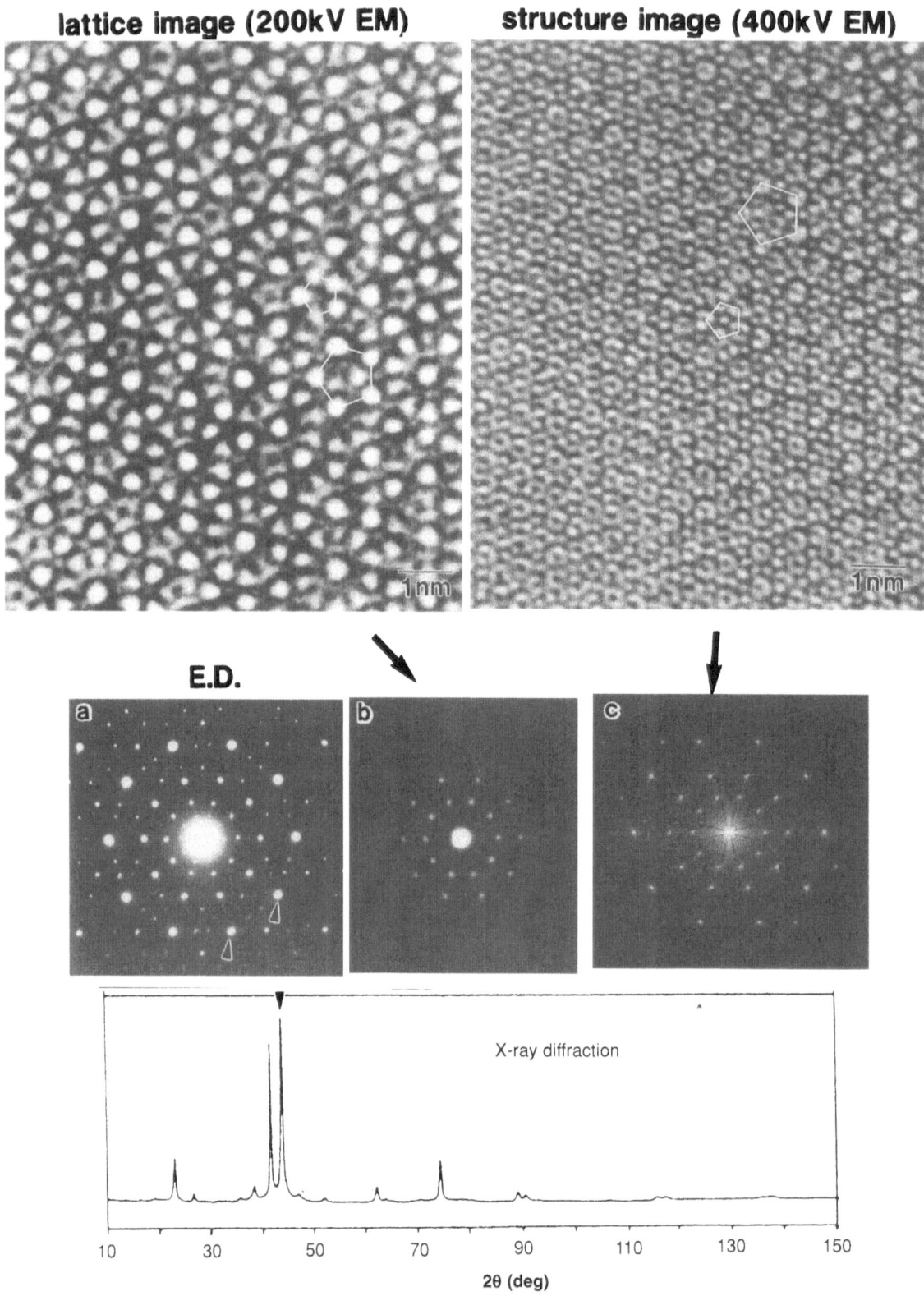

Fig. 3.70. Comparison between a *lattice image* (upper left) and a *structure image* (upper right) of icosahedral quasicrystals, electron diffraction pattern **a**, optical diffractogram of the lattice image **b**, digital diffractogram of the structure image **c**, and X-ray powder diffraction pattern (bottom).
Specimen: Rapidly solidified $Al_{74}Mn_{20}Si_6$ (lattice image), conventionally solidified $Al_{70}Pd_{20}Mn_{10}$ (structure image); **Preparation**: electropolishing (ice-cold solution of $HClO_4$: CH_3OH = 1 : 9) for $Al_{74}Mn_{20}Si_6$, crushing for $Al_{70}Pd_{20}Mn_{10}$; **Observation**: 200 kV EM (lattice image), 400 kV (structure image), incident beams parallel to the five-fold symmetry axis

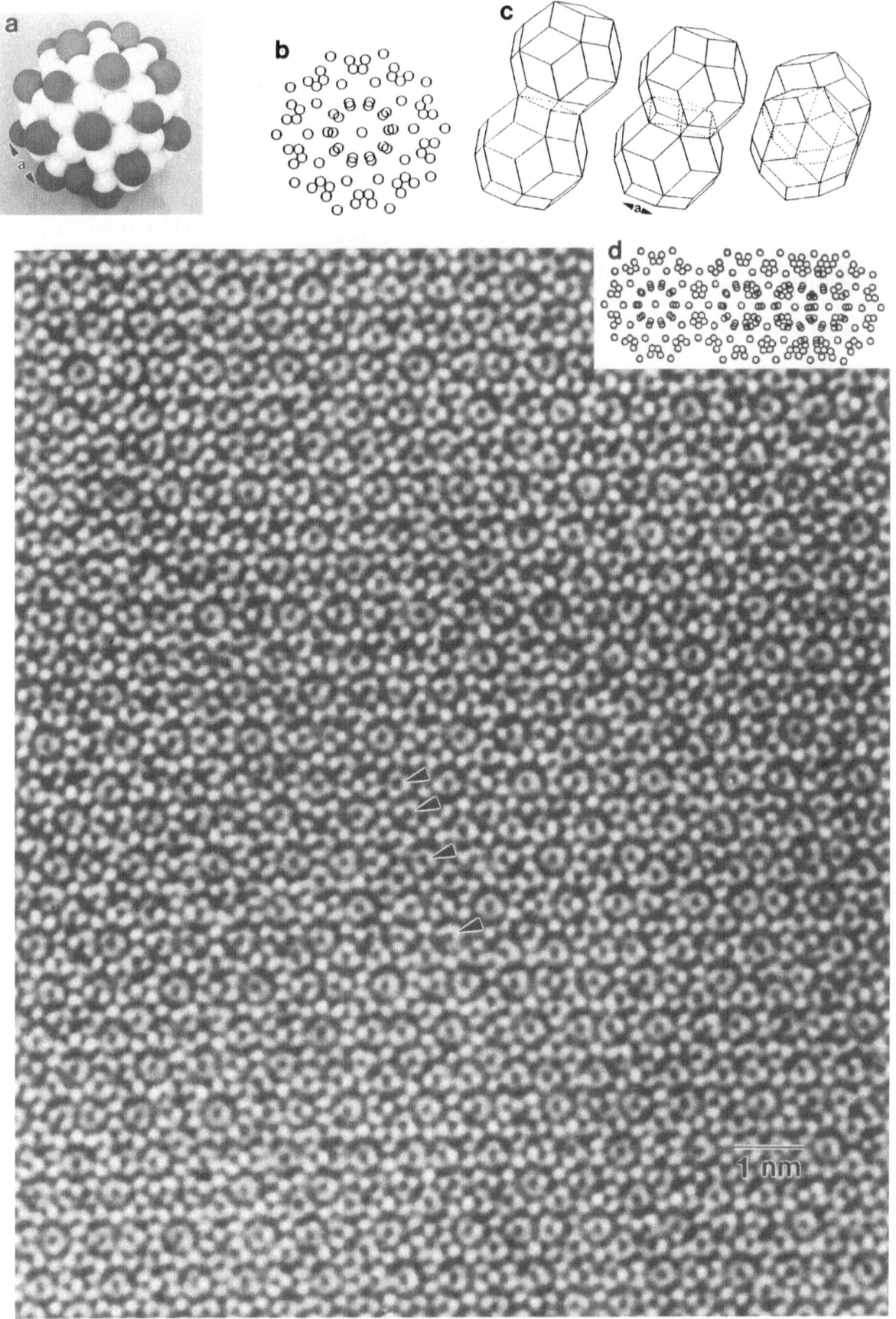

Fig. 3.71. Structure image of an Al–Li–Cu icosahedral quasicrystal and atomic clusters.
Specimen: Al–Li–Cu; **Preparation**: crushing; **Observation**: 400 kV EM, incidence of five-fold symmetry axis

(channels in atomic arrangements), and the dark areas to regions of high potential (atom positions). The image shows the characteristic distribution of an image unit consisting of a bright ring with a central dark dot and ten bright dots surrounding the ring. The image units are arrayed at intervals associated with the golden ratio τ, as indicated by the arrowheads. The image contrast of the bright ring and bright dots can be understood by an atomic arrangement (Fig. 3.71b) of a *triacontahedral atom cluster* (Fig. 3.71a), projected along the five-fold symmetry axis. From a comparison between the atomic arrangement (Fig. 3.71b) and image contrast, it can be seen that the central atom and the double decagonal rings correspond to a central dark dot and a dark ring, respectively, while the ring channel between the central atom and the double rings corresponds to a bright ring, and the ten channels outside correspond to ten bright dots.

The triacontahedral atom clusters are considered to be connected with each other by three types of linkage (Fig. 3.71c), and Fig. 3.71d shows an arrangement of the atom clusters with intervals associated with τ, projected along the five-fold symmetry axis. From the atomic arrangement in Fig. 3.71d, it is easy to understand the characteristic distribution of the image unit indicated by the arrowheads, but unfortunately it is impossible to see the three-dimensional aperiodic arrangement of the atom clusters from the structure image corresponding to the projected structure along the incidence beam.

3.2.4.4 High-Resolution Images of Decagonal Atom Clusters

In contrast to icosahedral quasicrystals with three-dimensional aperiodic structures, two-dimensional aperiodic arrangements in decagonal quasicrystals can be determined from high-resolution electron microscopy. For example, structure images of a decagonal quasicrystal, taken with the incident beam parallel to the periodic axis (ten-fold symmetry axis), make it possible to determine aperiodic structures directly from the observed images. However, in general, it is impossible to determine accurate atomic arrangements from the high-resolution images observed, except for materials with simple structures, such as high-Tc superconducting oxides. Hence, structure analysis must be done with the aid of X-ray diffraction.

Some crystalline phases, referred to as *crystalline approximants*, have been found to be formed in alloys with similar compositions to those of the quasicrystalline phases. The structures of the crystalline phases are closely related to those of the quasicrystalline phases, and are considered to be formed with structural units similar to those in the quasicrystals. Therefore, if the structure units in the approximants are determined in detail by X-ray diffraction, we can determine the aperiodic arrangements of the structure units in the quasicrystals by high-resolution electron microscopy. That is, by taking advantage of high-resolution observations (from which it is hard to determine accurate atomic arrangements) and X-ray diffraction (from which it is impossible to determine aperiodic structures), the structures of decagonal quasicrystals have been revealed.

An Al_3Mn crystalline approximant is an important phase in attempts to understand the structures of decagonal quasicrystals with a period of 1.2 nm. X-ray diffraction analysis shows that the Al_3Mn structure is formed with a definite linkage of two types of atom column, which have a two-fold screw relationship, as shown in Fig. 3.72a and b [91]. The atom columns are composed of a pentagonal atom column, which is formed with stacking pentagonal arrangements of atoms and central atoms along the columnar axis, and decagonal atom rings surrounding the pentagonal atom column. The atom columns are connected by edge-sharing of pentagons with an edge length of 0.46 nm, as shown in the projected atomic arrangements at the bottom of Fig. 3.72a and b, and form two-dimensional arrangements. Figure 3.73 shows some structure units, which are important in attempting to understand the structures of the decagonal and crystalline phases, formed by the edge-sharing of the pentagons of the atom columns. The hexagonal (H-), star-shaped pentagonal (P-), and decagonal (D-) units, with an edge length of 0.65 nm, are units forming aperiodic or periodic arrangements of the pentagonal atom columns in the decagonal and crystalline phases.

Figure 3.74a shows a structure image of the Al_3Mn crystalline phase, taken with the incident beam parallel to the columnar axis of the pentagonal atom columns. In the image, there are ring contrast distributions consisting of a dark ring surrounding a bright ring with a central dark dot. The contrast distribution can easily be understood from the projected atomic arrangement at the

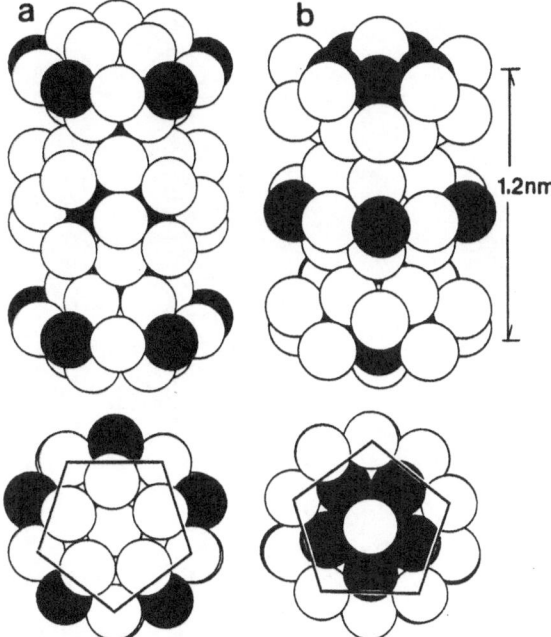

Fig. 3.72. Two types of atom column which are the bases of structures in crystalline approximants and decagonal quasicrystals. Projected atomic arrangements are below. *White* and *black circles* are Al and Mn atoms, respectively

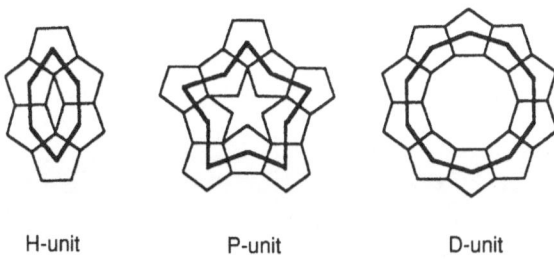

H-unit   P-unit   D-unit

Fig. 3.73. Three structure units formed by the edge-sharing of the pentagons of the atom columns in Fig. 3.72

bottom left; that is, a decagonal atom ring becomes a dark ring, a ring channel between the atom ring and a central atom becomes a bright ring, and a central atom corresponds to a dark dot. Therefore the image in Fig. 3.74a shows that the structure of the Al$_3$Mn phase is constructed by a periodic tiling of the H-units formed by the pentagonal atom columns in Fig. 3.72.

Figure 3.74b is a structure image of another crystalline phase with a structure formed with the H- and P-units in Fig. 3.73 [92]. The image shows a periodic tiling of the H- and P-units of the pentagonal atom columns. This phase was found by

electron diffraction, and its structure was determined by high-resolution electron microscopy.

What is the structure of decagonal quasicrystals? A structure image of an Al–Pd–Mn decagonal quasicrystal, taken with the incident beam parallel to the ten-fold symmetry axis, is shown in Fig. 3.75 [93]. The image shows an arrangement of ring contrasts forming D-units consisting of a decagonal arrangement of ten ring contrasts. The D-units are arrayed with a definite linkage by sharing two ring contrasts, and so all the D-units remain in the same direction. Gaps in the arrangement of the D-units are completely filled by H- and P-units, as shown schematically at the upper right-hand side. That is, the structure of the decagonal quasicrystal is characterized as an aperiodic arrangement of the H-, P-, and D-units of the pentagonal atom columns [94]. Hence, if the atomic arrangements in the three units are known, the atomic arrangement of the quasicrystal can be determined.

Atomic arrangements in the H- and P-units are determined unequivocally by placing the pentagonal atom columns in Fig. 3.72 inside pentagons in the units, and an atomic arrangement in an uncertain area inside the D-unit was derived from the contrast distribution in the image [93]. A simulated image calculated from the atomic model proposed is inserted in the image in Fig. 3.75. Good correspondence with the observed image contrast can be seen.

Local structures of quasicrystals can be determined by structure images, whereas the arrangements of D-units in wide regions can be seen in lattice images. Figure 3.76 is a lattice image of the Al–Pd–Mn decagonal quasicrystal taken from a relatively thick specimen. From the change in image contrast with increasing specimen thickness, the bright ring contrast in the image was found to correspond to a decagon, which is enhanced by the dynamical diffraction effect, inside the D-unit. Hence, from Fig. 3.76, one can determine the arrangement of D-units in a wide region of about 100 nm × 100 nm. This quasicrystal includes linear phason strain, so there are some steps on the arrays of bright ring contrasts in the p direction, but there are also straight arrays of ring contrasts longer than 100 nm in the r direction. This result shows the existence of a long-range correlation in the arrangement of D-units.

Many decagonal quasicrystals have been found in a variety of alloys. Their structures can be interpreted by definite linkages of decagonal atom

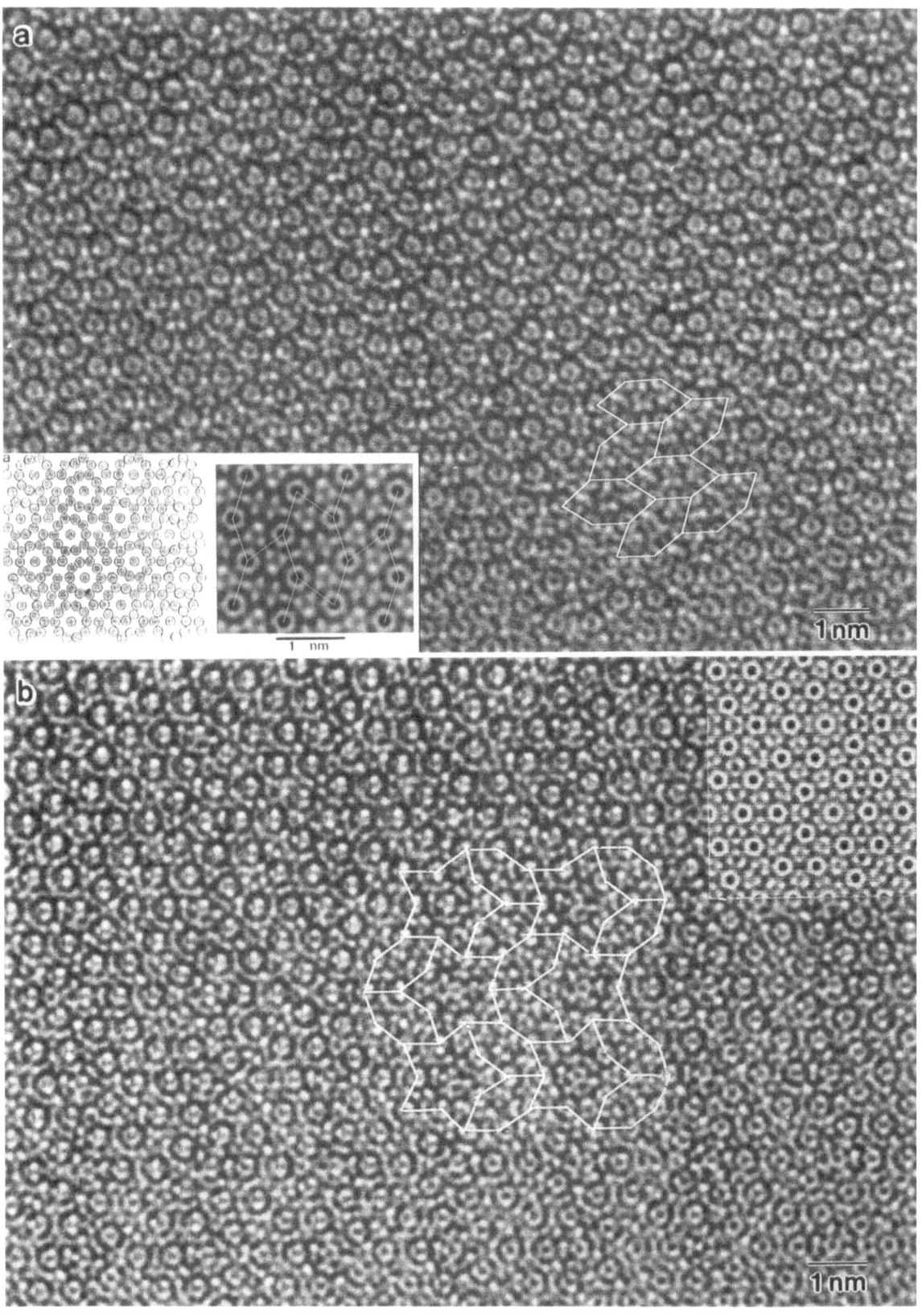

Fig. 3.74. Structure images of two types of crystalline approximant related to decagonal quasicrystals. **Specimen**: **a** Al_3Mn and **b** $Al_{72}Pb_{18}Cr_{10}$; **Preparation**: crushing; **Observation**: 400 kV EM

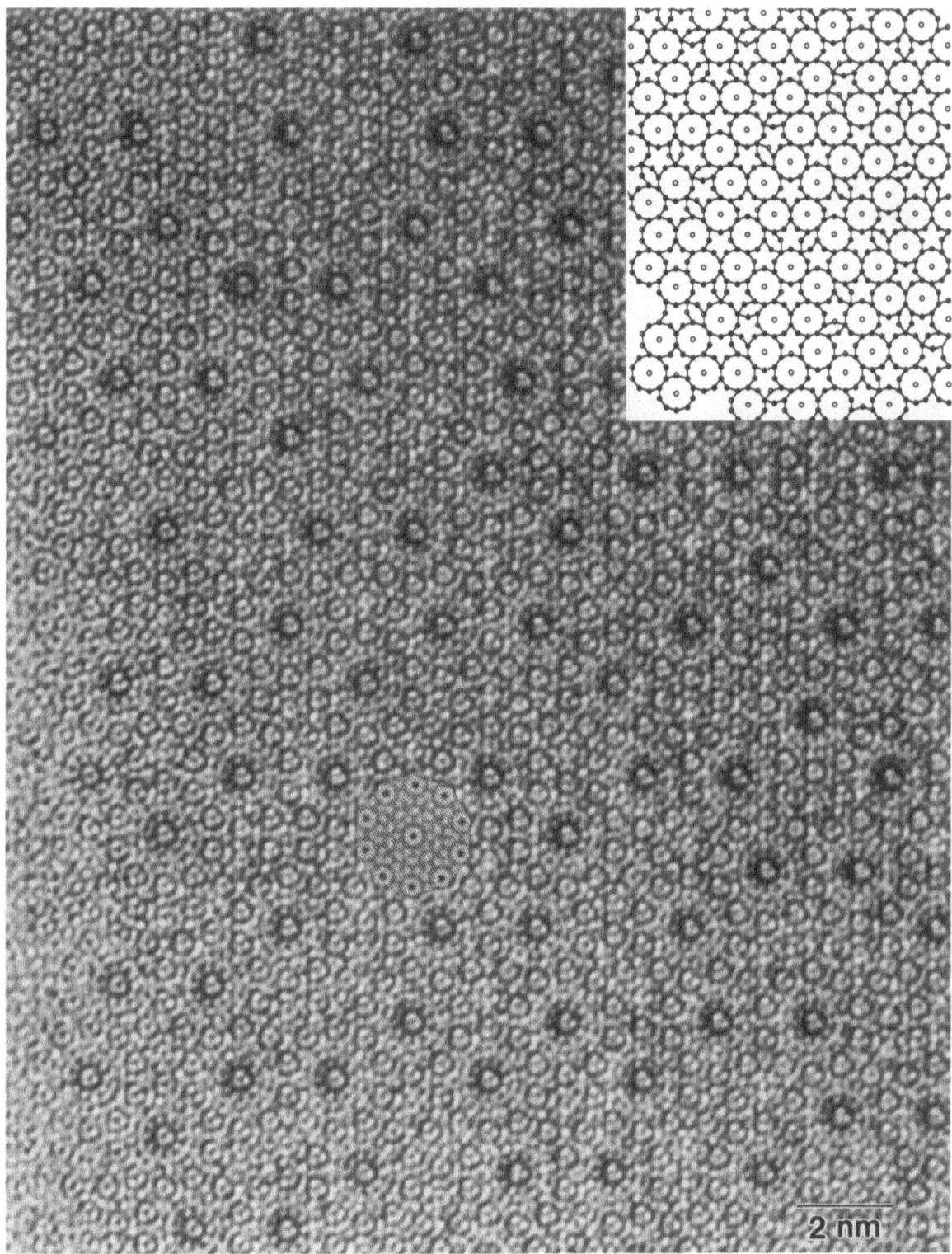

Fig. 3.75. Structure image of an Al–Pd–Mn decagonal quasicrystal.
Specimen: $Al_{70}Pb_{13}Mn_{17}$; **Preparation**: crushing; **Observation**: 400 kV EM, incidence of ten-fold symmetry axis.
Remarks: The *insert* is a schematic illustration of atom columns (*small circles*) and decagonal atom clusters (*large circles*) obtained from the image. A simulated image of a decagonal atom cluster from a structure model is inserted in the main figure

Fig. 3.76. Lattice image of an Al–Pd–Mn decagonal quasicrystal.
Specimen: $Al_{70}Pb_{13}Mn_{17}$; **Preparation**: crushing; **Observation**: 400 kV EM, incidence of ten-fold symmetry axis

columns, which have a variety of atomic arrangements depending on quasicrystals [95–99]. Those structures, particularly the arrangements of decagonal atom columns have been determined by high-resolution electron microscopy. Thus, high-resolution electron microscopy has become an indispensable tool for understanding the real structures of decagonal quasicrystals [100].

References

1. For example, Hirth JP, Lothe J (1982) Theory of dislocations. Wiley, New York
2. Bollmann W (1956) Phys Rev 103:1588
3. Hirsch PB, Horne RW, Whelan MJ (1956) Philos Mag 1:677
4. Shindo D, Yoo MH, Hanada S, Hiraga K (1991) Philos Mag A64:1281
5. Sato M, Hiraga K, Sumino K (1980) Jpn J Appl Phys 19:L155
6. Hiraga K, Hirabayashi M, Sato M, Sumino K (1982) Crystallogr Res Tech 17:189
7. Hirabayashi M, Hiraga K, Shindo D (1982) Ultramicroscopy, 9:197
8. Kawabata T, Shindo D, Hiraga K (1992) Mater Trans JIM 33:565
9. Shindo D, Yoshida M, Lee BT, Takasugi T, Hiraga K (1995) Intermetallics 3:167
10. Mills MJ, Miracle DB (1993) Acta Metall Mater 41:85
11. Crimp MA, Tonn SC, Zhang Y (1993) Mater Sci Eng A170:95
12. Takasugi T, Hanada S, Yoshida M, Shindo D (1995) Philos Mag A71:347
13. Hiraga K, Oku T, Shindo D, Hirabayashi M (1989) J Electron Microsc Technol 12:228
14. Alexander H, Spence JCH, Shindo D, Gottschalk H, Long N (1986) Philos Mag A53:627
15. Shindo D, Spence JCH, Alexander H, Long N Vanderschaeve G (1986) Proceedings of the XIth International Congress on Electron Microscopy, The Japanese Society of Electron Microscopy, Tokyo, Japan, p 785
16. Kolar HR, Spence JCH, Alexander H (1996) Phys Rev Lett 77:4031
17. Crawford RC, Ray ILF, Cockayne DJH (1973) Philos Mag 27:1
18. Lee BT, Pezzotti G, Hiraga K (1994) Mater Sci Eng A177:151
19. Pezzotti G, Lee BT, Hiraga K, Nishida T (1994) J Mater Sci 29:1786
20. Hiraga K (1984) Sci Rep RITU A32:1
21. Hiraga K, Hirabayashi M, Niihara K, Hirai T (1984) Proceedings of the IXth International Conference on Chemical Vapor Deposition, The Electrochemical Society, Pennington, NJ, p 575
22. Pope DP, Ezz SS (1984) Int Metall Rev 29:136
23. Aoki K, Izumi O (1979) J Jpn Inst Metals 43:1190
24. Emori H, Takasugi T, Hiraga K (1997) Philos Mag A75:1403
25. Takasugi T, Emori H, Hiraga K (1997) Philos Mag A75:1417
26. Ohnishi N, unpublished results (1992)
27. Hiraga K, Hirabayashi M, Ishigaki N (1986) J Micros 142:201
28. Lee BT, Chun BS, Hiraga K (1994) J Mater Res 9:2519
29. Hayashi S, Hirai T, Hiraga K, Hirabayashi M (1982) J Mater Sci 17:3336
30. Hiraga K, Hirabayashi M, Hayashi S, Hirai T (1983) J Am Ceram Soc 66:539
31. Higashi K (1993) Mater Sci Eng A166:109
32. Jeong Ha-Guk, Hiraga K, Mabuchi M, Higashi K (1996) Philos Mag Lett 74:73
33. Takayanagi K, Tanishiro Y, Takahashi S, Takahashi M (1985) Surface Sci 164:367
34. Suzuki T, Tanishiro Y, Minoda H, Yagi K (1993) Surface Sci 298:473
35. Ichihashi T, Iijima S (1994) Proceedings of the 13th International Congress on Electron Microscopy, Paris, vol 2B, Les Editions de Physique, Les Ulis, France, p 1013
36. Oku T, Hiraga K, Shindo D, Nakajima S, Tokiwa A, Kikuchi M, Syono Y (1996) In: Narliker A (ed) Studies of high temperture superconductors. Noba Science, Commack, NY
37. Sugimoto T, Muramatsu A, Sakata K, Shindo D (1993) J Colloid Interface Sci 158:420
38. Shindo D, Park GS, Waseda Y, Sugimoto T (1994) J Colloid Interface Sci 168:478
39. Terasaki O, Ohsuna T (1995) Catal Today 23:201
40. Hyde BG, Andersson S (1988) Inorganic crystal structures. Wiley-Interscience, New York
41. Kikuchi M, Kusaba K, Bannai E, Fukuoka K, Syono Y, Hiraga K (1985) Jpn J Appl Phys 24:1600
42. Barbier J, Hiraga K, Otero-Diaz LC, White TJ, Williams TB, Hyde BG (1985) Ultramicroscopy 18:211
43. Shindo D, Hiraga K, Oku T, Oikawa T (1991) Ultramicroscopy 39:50
44. Lee BT, Hiraga K (1993) Mater Trans JIM 34:930
45. Lee BT, Hiraga K, Shindo D, Nishiyama A (1994) J Mater Sci 29:959
46. Lee BT, Hiraga K (1994) J Mater Res 9:1199
47. Lee BT, Hiraga K unpublished results (1993)
48. Bednorz JG, Müler KA (1986) Z Phys B64:189
49. Wu MK, Ashburn JR, Torng CJ, Hor PH, Meng RL, Huang ZJ, Wang YQ, Chu CW (1987) Phys Rev Lett 58:908

50. Hiraga K, Shindo D, Hirabayashi M, Kikuchi M, Oh-ishi K, Syono Y (1987) Jpn J Appl Phys 26:L1071

51. Hiraga K, Shindo D, Hirabayashi M, Kikuchi M, Syono Y (1987) J Electron Microsc 36:261

52. Oku T, Hiraga K, Shindo D, Kikuchi M, Nakajima S, Syono Y (1991) Proceedings of the 3rd International Symposium on Superconductivity (ISS '90), Springer, Berlin Heidelberg Tokyo, Japan, p 367

53. Hiraga K, Shindo D, Hirabayashi M, Kikuchi M, Kobayashi N, Syono Y (1988) Jpn J Appl Phys 27:L1848

54. Hiraga K, Shindo D, Kikuchi M, Nakajima S (1988) JEOL News 26E:28

55. Oku T, Kajitani T, Hiraga K, Hosoya S, Shindo D (1991) Physica C 185–189:547

56. Ogawa S, Hirabayashi M, Watanabe D, Iwasaki H (1997) Long-period ordered alloys. Agne Gijyutsu Center, Tokyo, Japan

57. Shindo D (1982) Acta Crystallogr A38:310

58. Shindo D, Hirabayashi M (1988) Acta Crystallogr A44:954

59. Kuwano N, Toki M, Tanaka N, Eguchi T (1980) 7th European Congress on Electron Microscopy Foundation, Antwerp. Electron Microsc 4:166

60. Van Dyck D, Van Tendeloo G, Amerinckx S (1982) Ultramicroscopy 10:263

61. Mäki J (1986) Phys Stat Solidi 95a:51

62. Hiraga K, Shindo D, Hirabayashi M (1981) J Appl Crystallogr 14:185

63. Terasaki O, Watanabe D, Hiraga K, Shindo D, Hirabayashi M (1981) J Appl Crystallogr 14:392

64. Shindo D, Hiraga K, Hirabayashi M (1984) Sci Rep RITU A32:32

65. Tanaka N, Cowley JM (1987) Acta Crystallogr A43:337

66. Watanabe D, Ohsuna T, Kimoto T (1993) Ultramicroscopy 52:465

67. Hiraga K, Hirabayashi M, Terasaki O, Watanabe D (1982) Acta Crystallogr A38:269

68. Hiraga K, Shindo D, Hirabayashi M, Terasaki O, Watanabe D (1980) Acta Crystallogr B36:2550

69. Lee KH, Hiraga K, Shindo D, Hirabayashi M (1988) Acta Metall 36:641

70. Spruiell JE, Stanbury EE (1965) J Phys Chem Solids 26:811

71. De Ridder R, Van Tendeloo G, Amerinckx S (1976) Acta Crystallogr A32:216

72. Hata S, Fujita H, Matsumura S, Kuwano N, Oki K, Shindo D (1995) Abstract of the 117th Meeting of the Japan Institute for Metals in Hawaii. The Japan Institute of Metals, Sendai, Japan, p 96

73. Hirabayashi M, Hiraga K, Shindo D (1981) J Appl Crystallogr 14:169

74. Sato H, Toth RS (1961) Phys Rev 124:1833

75. Shechtman D, Blech I, Gratias D, Cahn JW (1984) Phys Rev Lett 53:1951

76. Levine D, Steinhardt PJ (1984) Phys Rev Lett 53:2477

77. Hiraga K, Bo-ping Zhang, Hirabayashi M, Inoue A, Masumoto T (1988) Jpn J Appl Phys 27:L951

78. Hiraga K, Lincoln FJ, Sun W (1991) Mater Trans JIM 32:308

79. Field RD, Fraser HL (1984–85) Mat Sci Eng 68:L17

80. Hiraga K, Hirabayashi M, Inoue A, Masumoto T (1985) Sci Rep RITU A32:309

81. Hiraga K, Hirabayashi M, Inoue A, Masumoto T (1987) J Microsc 146:245

82. Hiraga K, Hirabayashi M (1987) J Electron Microsc 36:353

83. Hiraga K (1989) Mater Res Soc Symp Proc 139:125

84. Hiraga K, Lee KH, Hirabayashi M, Tsai AP, Inoue A, Masumoto T (1989) Jpn J Appl Phys 28:L1624

85. Guryan CA, Goldman AI, Stephens PW, Hiraga K, Tsai AP, Inoue A, Masumoto T (1989) Phys Rev Lett 62:2409

86. Hiraga K, Hirabayashi M (1987) Jpn J Appl Phys 26:L155

87. Hiraga K (1991) Quasicrystals: The state of the art. Directions in: DiVincenzo DP, Steinhardt PJ (eds) Condensed Matter Physics 11:95

88. Hiraga K (1992) Electron Microscopy, EUREM 92. Secretariado de Publicaciones de la Universidad de Granada, vol 2, p 485

89. Hiraga K (1991) J Electron Microsc 40:81

90. Hiraga K (1990) Springer Ser Solid-State Sci 93:68

91. Hiraga K, Kaneko M, Matsuo Y, Hashimoto S (1993) Philos Mag B67:193

92. Sun W, Yubuta K, Hiraga K (1995) Philos Mag B71:71

93. Hiraga K, Sun W (1993) Philos Mag Lett 67:117

94. Hiraga K, Sun W (1993) J Phys Soc Jpn 62:1833

95. Hiraga K, Abe E, Matsuo Y (1993) Philos Mag Lett 70:163

96. Matsuo Y, Hiraga K (1994) Philos Mag Lett 70:155

97. Hiraga K (1991) Sci Rep RITU A36:115

98. Hiraga K, Sun W, Yamamoto A (1994) Mater Trans JIM, 35:657

99. Hiraga K (1995) In: Chapuis G, Paciorek W (eds) Proceedings of the International Conference on Aperiodic Crystals (Aperiodic '94), World Scientific Publishing, Singapore, p 341

100. Hiraga K (1997) In: Hawkes PW (ed) Advances in imaging and electron physics. Academic Press, New York, p 37

4. Peripheral Instruments and Techniques for High-Resolution Electron Microscopy

4.1 Image Processing

In order to extract structure information from high-resolution images, various image processing techniques have been used. For example, noise on a high-resolution image of a crystal can be removed by over-printing with translating the image in a dark room. With an optical bench consisting of a laser beam and optical lenses, optical diffractograms of electron microscope images have been obtained, and by inserting a phase filter and a mask, image reconstruction has been carried out by modulating the phases and the amplitudes of the scattered beams.

With the recent rapid improvements in computer technology, various hardware systems have been developed to digitize and output electron microscope images. By connecting the system to a personal computer or an engineering work station (EWS), it is easy to handle the digital data of electron microscope images. This section considers image processing of high-resolution images which have been digitized from EM film or printing paper. Quantitative image analysis based on the digital data obtained by the *imaging plate* and the *slow-scan CCD* (charge-coupled device) camera will be discussed in Sect. 4.2.

4.1.1 Input and Output of High-Resolution Images

4.1.1.1 Input of High-Resolution Images

One of the most popular devices for inputting or digitizing a high-resolution image on film or printing paper is a so-called TV camera such as a vidicon camera or a CCD camera. The main advantage of a TV camera over other devices is the ease of operation. Nowadays, various image scanners are also available, where film or printing paper set in a frame or on a rolling drum are scanned while they are read. Although these scanners generally tend to take longer to input an image than TV cameras, they have a higher resolution and larger pixel numbers. The capability of TV cameras and image scanners can be determined by the *resolution*, the *pixel number*, and the *gray level*. The resolution is usually indicated in μm or dpi.[1]

Since the resolution of conventional EM film is about $10\,\mu$m, the resolution of TV cameras and image scanners should also be about $10\,\mu$m. However, for the actual data input process, the resolution of the high-resolution image and its magnification in printing should be taken into account. It should also be noted that *sampling* with the smallest pixel size may result in using a large amount of computer memory. The pixel size needed for inputting data can be determined in a similar way to sampling in the image simulation considered in Sect. 1.4. In image simulation of high-resolution images, the sampling interval should be small enough to represent the distribution of crystal potential, or in reciprocal space, higher-order reflections whose amplitudes are not neglected should be taken into account. In the input of high-resolution images, the sampling interval should be small enough to represent their intensity distribution accurately. For example, the sampling interval may be 10–$30\,\mu$m (2500 dpi–850 dpi) for EM film, while it may be 0.1–0.5 mm (250 dpi–50 dpi) for printing paper, and the number of sampling points is generally $1024\ (=2^{10}) \times 1024$ or $2048\ (=2^{11}) \times 2048$.

In conventional image processing systems, the number of gray levels for the input of images is $256\ (=2^8)$, but for special cases needing higher precision, $512\ (=2^9)$ or 1024 gray levels are used. In order to check the gray levels used for the inputted image, a so-called *histogram* of gray levels should be consulted. As shown in Fig. 4.1, the histogram indicates the number of pixels as a function of the gray level. In Fig. 4.1a and b, the

---

[1] dpi means "dots per inch" and indicates the number of pixels in one inch ($=25.4$ mm), and 2500 dpi corresponds to about $10\,\mu$m.

gray levels of the inputed image shift to lower and higher levels, respectively, while the image is inputted with appropriate gray levels in Fig. 4.1c. The gray levels of the image can be adjusted by changing the incident electron intensity and the input condition of the image processing system, e.g., the light intensity on film or printing paper. It should be noted that the darkness of conventional EM film is not proportional to the electron intensity, and the relation between the darkness of the film and the electron intensity is given by the sensitivity curve. Furthermore, the printing of an image also changes the intensity distribution on the printing paper so that it becomes even further removed from the original electron intensity distribution. In order to obtain an accurate reproduc-

tion of the intensity of electron microscope images, it is necessary to use image recording systems such as the imaging plate or the slow-scan CCD camera.

4.1.1.2 Output of High-Resolution Images

Processed images are usually outputted as half-tone images with 256 gray levels, pseudo-color images, or contour maps. Being similar to devices for inputting images, the performance of printers is determined by resolution, the pixel number, the gray level, and also the sharpness of the color. Various new printers have recently been developed, and some of them output fine images whose quality is comparable to that of printing paper. In general, printers and output paper for producing high-quality images are expensive. Like the input of images, the image contrast is adjusted by refering to the gray-level histogram for the output. In order to enhance the contrast, various non-linear functions relating the input image intensity to the output image intensity are used. A histogram such as that shown in Fig. 4.2 is normally considered to be appropriate for outputting high-resolution images.

For the output of simulated images, keyboard characters used to be used, but recently half-tone images with about 256 gray levels have replaced them (see Fig. 1.11).

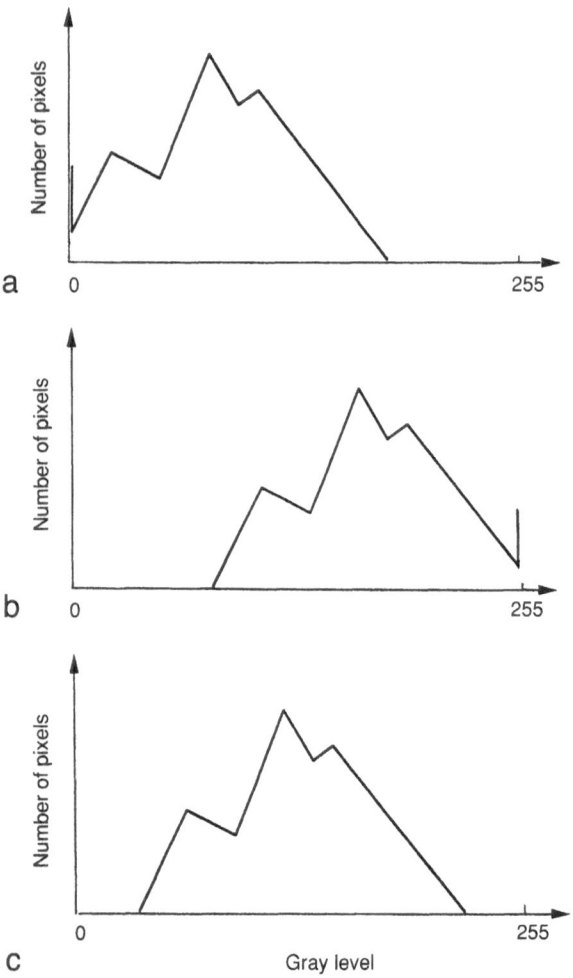

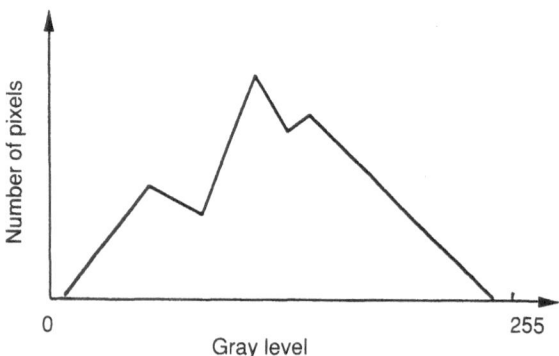

Fig. 4.1. The number of pixels as a function of the gray level. Gray levels of the input image shift to **a** lower and **b** higher levels, while the image is inputted with appropriate gray levels **c**

Fig. 4.2. The number of pixels as a function of the gray level which is appropriate for the output of a high-resolution image

4.1.2 Practice in Processing High-Resolution Images

Five kinds of image processing techniques which are widely used for the analysis of high-resolution images are now presented.

4.1.2.1 Processing with Spatial Filters

Various spatial filters have been used for the purpose of smoothing and enhancing the intensity distribution of an image. The process is carried out by the operation of a spatial filter $W(r)$ on the image data $I(r)$, i.e.,

$$F^{SM}(r) = \frac{1}{N} \sum I(r+q)W(q) \qquad (4.1)$$

where N is a constant corresponding to the sampling number in the summation (Σ) or the pixel number in the spatial filter.

One of the most popular spatial filters applied to high-resolution images is the filter for smoothing, and this is used to remove noise on high-resolution images. The main noise on high-resolution images is caused by silver halide particles in the photographic emulsion, contamination of a specimen, or quantum noises due to low electron intensity. Two examples of spatial filters for smoothing are shown in Fig. 4.3.

4.1.2.2 Averaging by a Translation Process

In general, the noise on high-resolution images does not indicate any periodicity. Thus, for crys-

1	1	1
1	1	1
1	1	1

1	1	1
1	2	1
1	1	1

Fig. 4.3. Two examples of spatial filters for smoothing

talline specimens, that noise can be effectively removed by a translation process, which is also used in printing using plural exposures in a dark room. For example, a translation process with a vector a is given by

$$F^{AV}(r) = \frac{1}{3}\left[I(r) + I(r+a) + I(r-a)\right]. \qquad (4.2)$$

This translation process can only be applied to a region in a high-resolution image which has a periodicity, and should not be applied to regions where a lattice defect exists or where the crystal thickness changes drastically. Figure 4.4 demonstrates the translation process performed on a high-resolution image of α-Fe_2O_3 with the translation vectors a_1 and a_2, i.e.,

$$F^{AV}(r) = \frac{1}{5}\left[I(r) + I(r+a_1) + I(r-a_1)\right.$$
$$\left. + I(r+a_2) + I(r-a_2)\right]. \qquad (4.3)$$

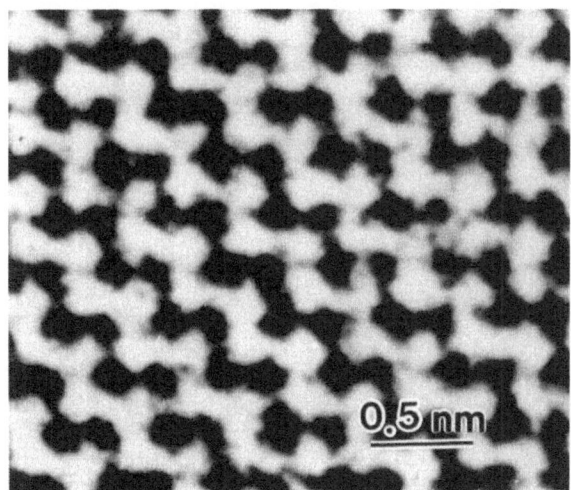

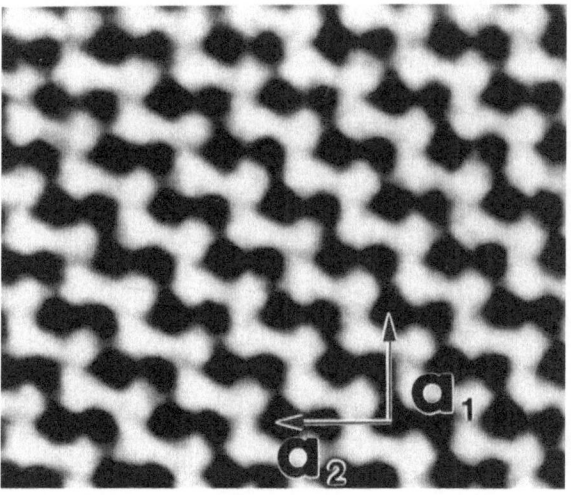

a   b

Fig. 4.4. a High-resolution image of α-Fe_2O_3. **b** Processed image obtained by translating **a** with the translation vectors a_1 and a_2

4.1.2.3 Filtering with the Fourier Transform

In a digital diffractogram of a high-resolution image, the noise on the image forms a monotonic background with no sharp peaks. Thus, by selecting diffraction spots on a digital diffractogram and carrying out the inverse Fourier transform, this noise may be removed. However, special care should be given to any region containing lattice defects or modulated structures, as pointed out above, because information about deviations from the periodic structure can be lost with this process. For example, in the processed images obtained with the Fourier transform in Figs. 3.2 and 3.7, one

can clearly identify the dislocation cores, but the evaluation of atomic displacements around the cores becomes more difficult.

We now present an example of image processing with the Fourier transform performed to extract the weak signal on a high-resolution image. Figure 4.5a shows a high-resolution image of an alloy semiconductor $Ga_{0.5}In_{0.5}P$ [1], which is attracting much attention as a laser material. In the electron diffraction pattern of this material in Fig. 4.6, characteristic diffuse scattering between the fundamental reflections can be seen. The intensity distribution of the diffuse scattering is very sensitive to the growth conditions in the

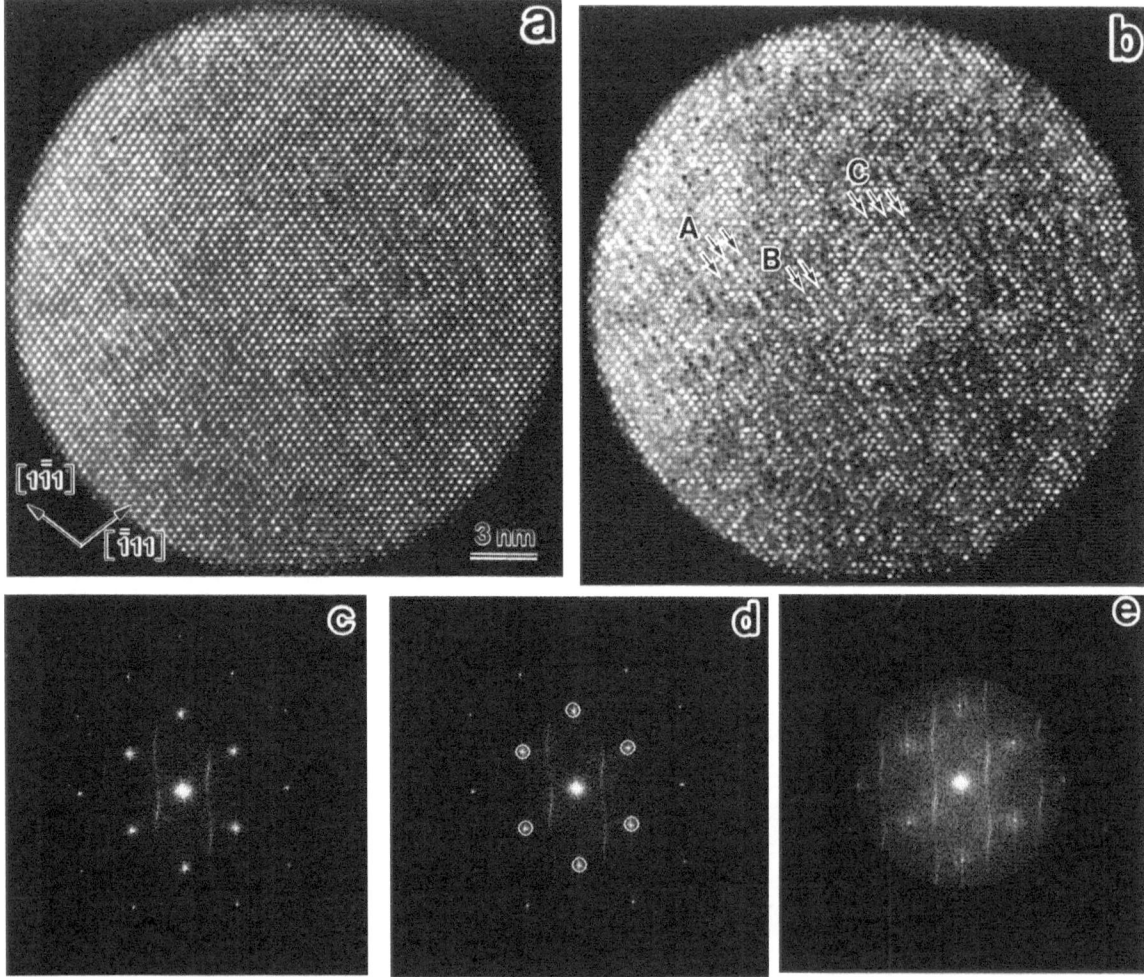

Fig. 4.5. a High-resolution image of $Ga_{0.5}In_{0.5}P$ with [110] incidence. **b** Processed image obtained by decreasing the amplitudes of the fundamental reflections *circled* in **d. c** Digital diffractogram of **a. d** Digital diffractogram with the fundamental reflections *circled*. **e** Digital diffractogram of **b**

Fig. 4.6. Electron diffraction pattern of $Ga_{0.5}In_{0.5}P$ with [110] incidence

clearly seen, and the diffuse scattering is considered to be formed by the arrangement of these white dots. In the digital diffractogram of the processed image in Fig. 4.5e, the diffuse scattering is strongly enhanced over the fundamental reflections. In order to further enhance the contribution of the diffuse scattering to the high-resolution image, another processed image can be obtained with a fundamental reflection and the diffuse scattering in a digital diffractogram, as shown in Fig. 4.7. In the processed image, wavy white bands, indicated by A, B, and C, can be seen. These wavy bands correspond to the regions consisting of regular arrangements of white dots indicated by A, B, and C in Fig. 4.5b. A structure model of this material is discussed in Sect. 4.1.2.6.

4.1.2.4 Use of Image Correlation

Autocorrelation Function. In the intensity distribution of an electron microscope image $I(r)$, the correlation of the two points connected by vector r is given by the autocorrelation function $F^{AC}(r)$

$$F^{AC}(r) = \frac{1}{N}\sum I_1(q - r)I_1(q)$$
$$= \frac{1}{N}\sum I_1(q)I_1(q + r) \qquad (4.4)$$

where N indicates the number of sampling points in the summation corresponding to the shaded region in Fig. 4.8. The autocorrelation function of Eq. 4.4 can also be obtained by the Fourier transform of the intensity of a digital diffractogram. The autocorrelation function is especially useful to obtain the probability of finding specific atom pairs connected by a vector r on a high-resolution image.

Figure 4.9 shows an autocorrelation function obtained from a high-resolution image of an Au–Mn alloy [3]. In Fig. 4.9a and b, the brightness of the white dots reflects the Mn atom content in the atomic column parallel to the incident electron beam (see Fig. 3.53). Thus, the bright regions in the center in Fig. 4.9c and d, specified by the vector r_1 in the autocorrelation function, indicate a high probability of finding Mn atom pairs connected by the vector r_1 projected along the direction of the incident electron beam in Fig. 4.9a and b. Owing to the slight compositional change, the probability of finding Mn atom pairs connected by the vector c is drastically altered. Note that Eq. 4.4

metal organic chemical vapor deposition (MOCVD) method, such as the temperature of the substrate, and this is strongly correlated to the characteristics of the laser [2]. Weak diffuse scattering can be seen in the digital diffractogram in Fig. 4.5c, but its intensity is very weak compared with those of the fundamental reflections. Thus, the effect of the diffuse scattering on the high-resolution image is very weak, and only slight fluctuations in the brightness of the white dots are observable. In order to enhance the contribution of the diffuse scattering to the high-resolution image, the processed image in Fig. 4.5b was obtained by decreasing the amplitudes of the fundamental reflections circled in Fig. 4.5d while keeping the phases of these reflections unchanged. In order to avoid complicated interference among multiple reflections, higher-order reflections are removed with an artificial aperture (Fig. 4.5e). In the processed image, as indicated by the arrows, a couple of one-dimensional arrays, consisting of several white dots arranged with a spacing of twice the fundamental lattice, are

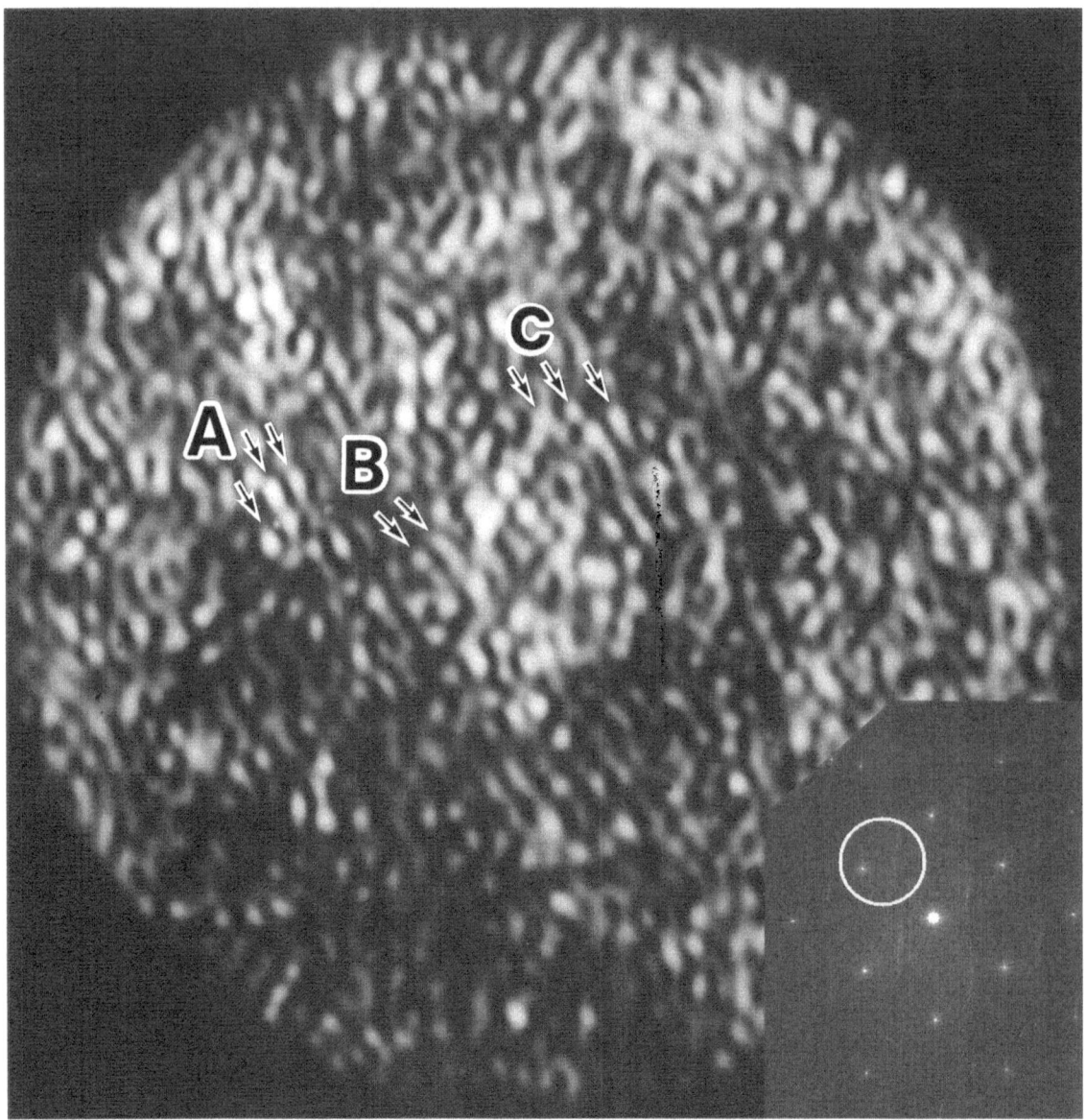

Fig. 4.7. Processed image of Fig. 4.5a obtained with a fundamental reflection and the diffuse scattering indicated in the digital diffractogram in the *inset*

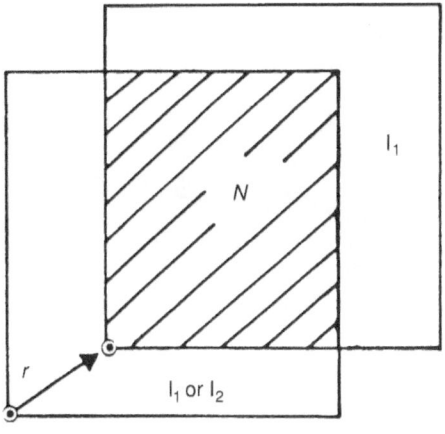

corresponds to the so-called *Patterson function* if $I_1(r)$ is considered to be the electron charge distribution.

Cross-Correlation Function. In contrast with the autocorrelation function, the cross-correlation function shows a correlation between the two points on the two images $I_1(r)$ and $I_2(r)$ connected by a vector r, and is given by

$$F^{CC}(r) = \frac{1}{N} \sum I_1(q - r) I_2(q)$$
$$= \frac{1}{N} \sum I_1(q) I_2(q + r). \quad (4.5)$$

Fig. 4.8. Sampling points, N, in the summation used to obtain the correlation functions

The shift of one image to the other due to drift, for example, can be accurately determined from a sharp peak in the cross-correlation function [4].

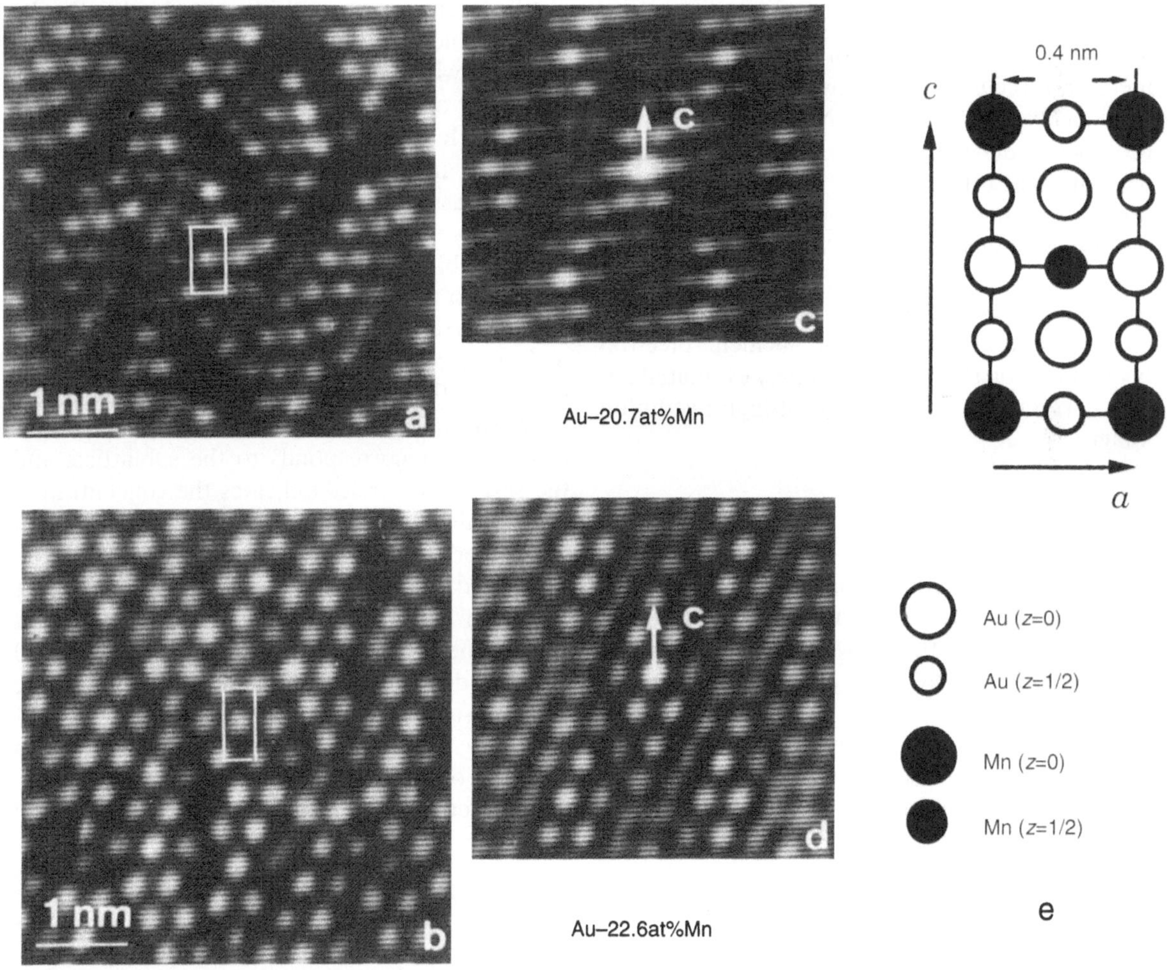

Au–20.7at%Mn

Au–22.6at%Mn

0.4 nm

c

a

○ Au ($z=0$)

○ Au ($z=1/2$)

● Mn ($z=0$)

● Mn ($z=1/2$)

e

Fig. 4.9. a,b High-resolution images of Au–20.7at%Mn and Au–22.6at%Mn, respectively. **c,d** Autocorrelation functions obtained from **a** and **b**, respectively. **e** Atomic arrangement of the DO_{22}-type structure on which the structures in **a** and **b** are based

4.1.2.5 Retrieval of High-Resolution Images

The resolution of a transmission electron microscope is limited by such factors as the spherical aberration and chromatic aberration. Thus, in order to improve the resolution of observed high-resolution images, image retrieval has to be carried out. For example, under the weak-phase object approximation, a so-called *Wiener filter* [5] can be applied to correct the effect of the contrast transfer function, i.e.,

$$F^{WF}(r) = \mathscr{F}^{-1}\left[\frac{\mathscr{F}[I(r)][B(u)]}{[B(u)]^2 + C}\right] \qquad (4.6)$$

where $[B(u)]$ indicates the imaginary part of the contrast transfer function in Eq. 4.30. By applying the Wiener filter to a high-resolution image of a quasicrystal, a processed image which more accurately reflects the projected atomic arrangement has been obtained [6]. In the Wiener filter method, a reconstruction image can be obtained from an observed high-resolution image to correct the aberration of the contrast transfer function, while in some other methods a processed image is obtained from a series of high-resolution images observed by changing the defocus value or the direction of the incident electron beam [7, 8]. In the latter case, the experimental conditions of the focus setting and the incident electron beam direction should be accurately evaluated before carrying out the retrieval of high-resolution images.

4.1.2.6 Construction of a Structure Model

When a high-resolution image is obtained from a very thin specimen, or using superlattice reflections or diffuse scattering which are not strongly modulated by the dynamic diffraction effect, the image reflects the ordered atomic arrangement projected along the incident electron beam. For example, in a digital diffractogram of $Ga_{0.5}In_{0.5}P$, weak diffuse scattering corresponding to that of the electron diffraction pattern and also of X-ray

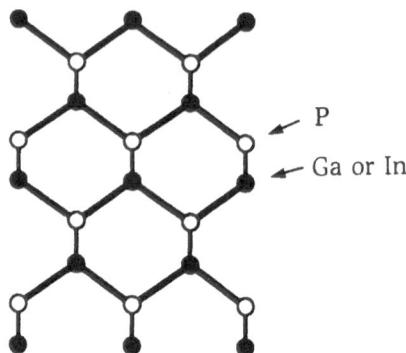

Fig. 4.10. Zinc-blende-type structure in $Ga_{0.5}In_{0.5}P$

diffraction patterns [9] is well reproduced, and thus it was found that the high-resolution image contained information corresponding to the diffuse scattering. We now discuss the construction of a structure model of $Ga_{0.5}In_{0.5}P$.

Alloy semiconductors such as $Ga_{0.5}In_{0.5}P$ take the zinc-blende (ZnS)-type structure shown in Fig. 4.10, where column-III atoms such as Ga and In occupy one of the sublattices, while column-V atoms such as P occupy the other sublattice. Furthermore, from the image simulation [1], the image contrast, or the brightness of the white dots in a high-resolution image, changes depending on the probability of occupation by Ga or In at each position in the sublattice. Based on these results, and by correlating the brightness of the white dots with occupancy of the constituent element in the sublattice, the structure model can be constructed, as shown in Fig. 4.11. In the model, the positions of the circles corresponds to the sublattice, and the size of the circles indicates the concentration of In atoms in the column parallel to the direction of the electron beam, i.e., the [110] direction. The calculated amplitudes based on the model are shown in Fig. 4.12. The amplitudes corresponds well to the intensity distribution of the observed electron diffraction pattern in Fig. 4.6, indicating that the structure model is basically correct. Furthermore, by comparing the calculated intensities with the observed intensities of electron diffraction patterns, refinement of the structure model can be carried out.

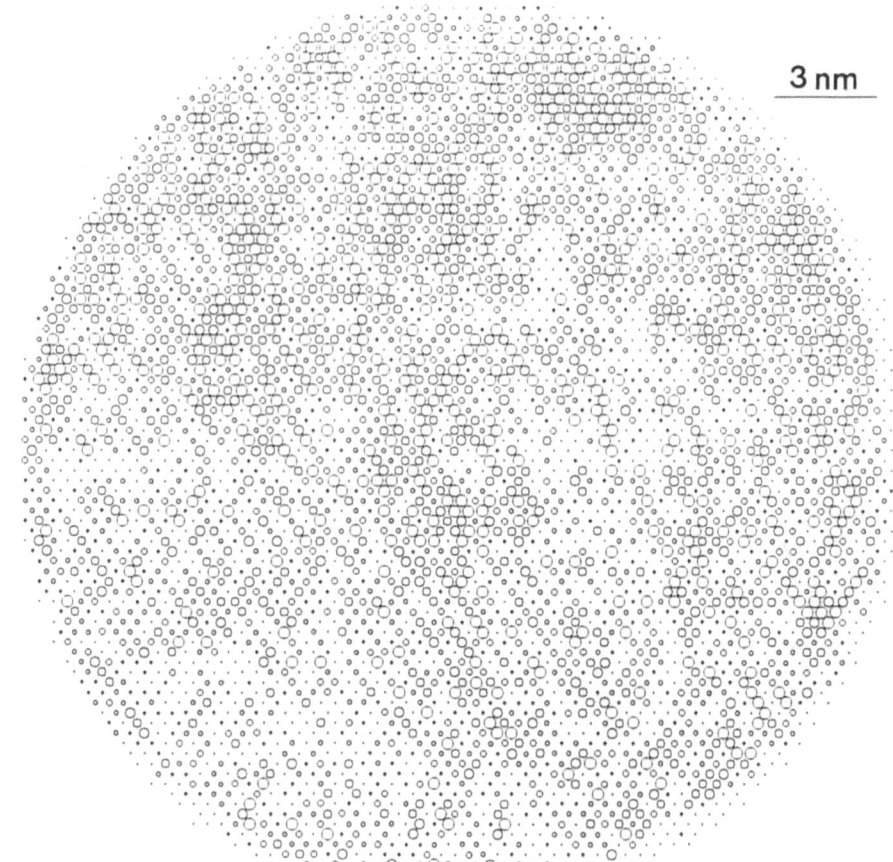

Fig. 4.11. Structure model obtained from the high-resolution image in Fig. 4.5a. The positions and size of the *circles* correspond to the sublattice and the concentration of In atoms in the column parallel to the [110] direction

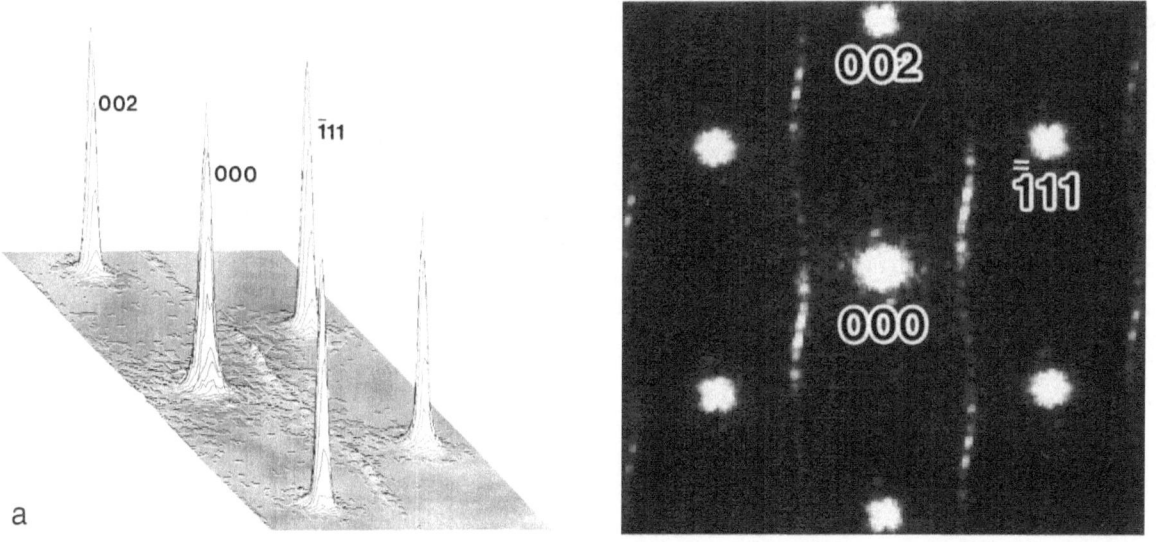

Fig. 4.12. Absolute values of the scattering amplitudes calculated on the basis of the model in Fig. 4.11. The amplitudes are shown as a bird's-eye view **a**, and a halftone image **b**

4.2 Quantitative Analysis

In the preceding section, we showed that digital data obtained from high-resolution images can be processed to extract structural information. Recently, new recording systems such as the imaging plate [10–12] and the slow-scan CCD camera [13] have been developed, and digital data of high-resolution images and electron diffraction patterns can be obtained directly from computers. Thus, with these new recording systems, extensive quantitative analyses can be carried out. In this section, we discuss the principles and properties of these recording systems, and then present a quantitative analysis of a high-resolution image with a residual index.

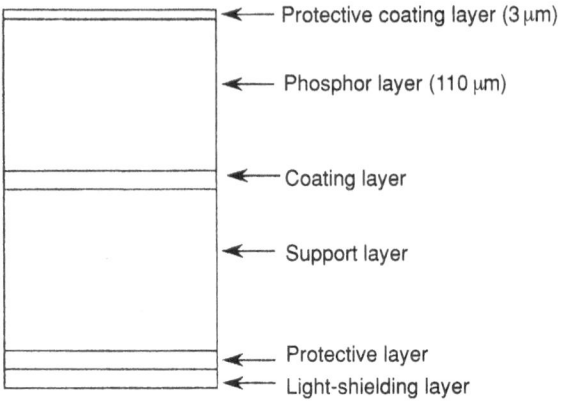

Fig. 4.13. Cross section of the imaging plate (FDL–UR–V type) showing its constituent parts

4.2.1 Principles of New Recording Systems

4.2.1.1 Imaging Plate

The imaging plate was originally developed as a high-sensitivity recording material for X-rays [10]. The imaging plate developed for electron microscopes [11, 12] is basically the same as that for X-rays and is made of phosphor, which consists of Ba, F, and halogen X (X = Cl, Br, I) doped with Eu^{2+}. The size of the imaging plate (FDL–UR–V type: 99.6 mm × 80.9 mm) is smaller than that for X-rays. The digital data on the imaging plate consist of 3700 × 3000 pixels. Since the size of one pixel in the imaging plate data is 25 μm, it is usually called the 25 μm imaging plate.[2] A cross section of an imaging plate is shown in Fig. 4.13. The front surface is colored blue because blue pigment is added in the phosphor. When the electron beam impinges on the phosphor layer, pairs consisting of an electron and a hole are formed. The electrons in these pairs are trapped at defects in the phosphor, i.e., vacancies in negative ions, while the holes are trapped at the Eu^{2+}. If the phosphor layer is irradiated with a laser beam in an imaging plate reader, the electrons combine with the holes, resulting in the emission of light. The emitted light is converted into an electrical signal through the photomultiplier. The blue pig-

ment is added to the phosphor in order to reduce the diffusion length of the laser beam in the phosphor. After the imaging plate is read, the electrons and holes left in the phosphor are completely removed by irradiating it with strong visible light, and it can be reused.

4.2.1.2 Slow-Scan CCD Camera

The elements of the slow-scan CCD camera used for electron microscopes are shown in Fig. 4.14. The incident electron beam is converted to light through the so-called YAG (yttrium aluminum garnet: $3Y_2O_3 \cdot 5Al_2O_3 : Ce^{3+}$). The light produced reaches the CCD through the fiber optic plate. The light is converted into an electrical charge which is temporarily stored for each channel at the electrode of the semiconductor on the CCD surface. The stored electrical charge is sequentially transferred to the next channel, and is then outputted as the electrical signal. Thus, by scanning slowly and storing the electrical charge for some time, a slow scan CCD camera has a higher sensitivity and a wider *dynamic range* than a conventional CCD. Also, if the CCD is cooled, the dark current which causes noise can be reduced.

---

[2] The size of one pixel on the imaging plate data is not limited by the imaging plate itself, but it is determined by the sampling interval of the imaging plate reader.

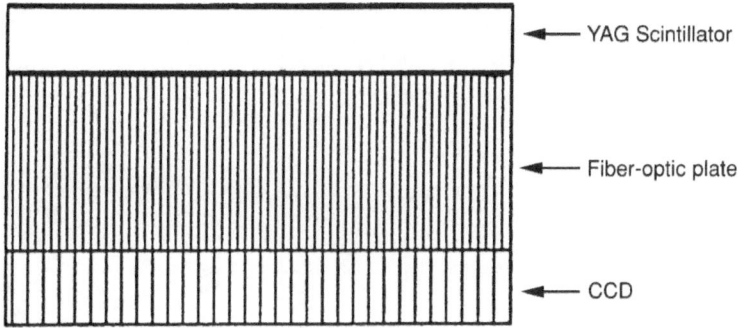

Fig. 4.14. Elements of the slow-scan CCD camera

4.2.2 Characteristics of New Recording Systems

We now discuss the characteristics of the imaging plate (FDL–UR–V type) and briefly compare them with those of the slow-scan CCD camera. The main characteristics of the imaging plate and the slow-scan CCD camera in some typical systems which are currently available are compared in Table 4.1. Figure 4.15 shows the signal intensity of the imaging plate measured at various electron intensities at 100 kV and 200 kV, under two gain conditions of the imaging plate reader. It can be seen that there is good linearity between the incident electron intensity and the output signal. The figure also shows that imaging plates have a wide dynamic range of about five orders, although the effective dynamic range under each gain condition is restricted to about four orders. Figure 4.16 shows the signal to noise ratio $(S/N)^2$ of the imaging plate as a function of the incident

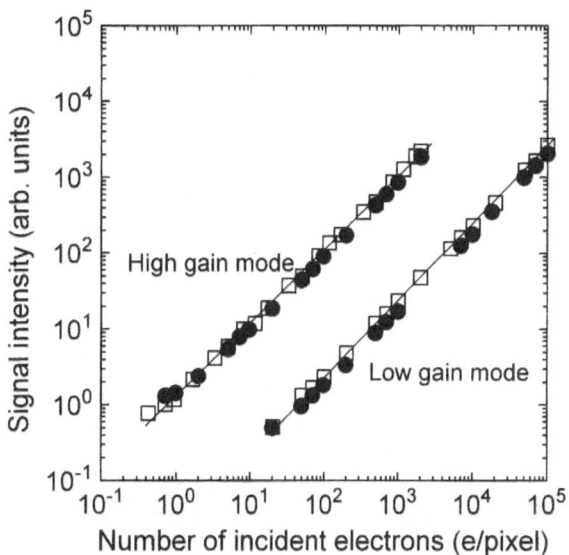

Fig. 4.15. Signal intensity of the imaging plate as a function of the electron intensity at 100 kV (□) and 200 kV (●)

Table 4.1. Comparison of charateristics of imaging plates and slow-scan CCD cameras

	Pixel size	Number of pixels	Dynamic range	Advantage	Notes
Imaging plates	$25\,\mu m \times 25\,\mu m$ ($50\,\mu m \times 50\,\mu m$)	3000×3760 (2048×1536)	4–5 figures	Possible to insert in any microscope	Fading characteristics
SS–CCD	$24\,\mu m \times 24\,\mu m$	1024×1024 (2048×2048)	4 figures	Digitized image can be seen a few seconds after exposure	Artifacts due to too strong electron intensity

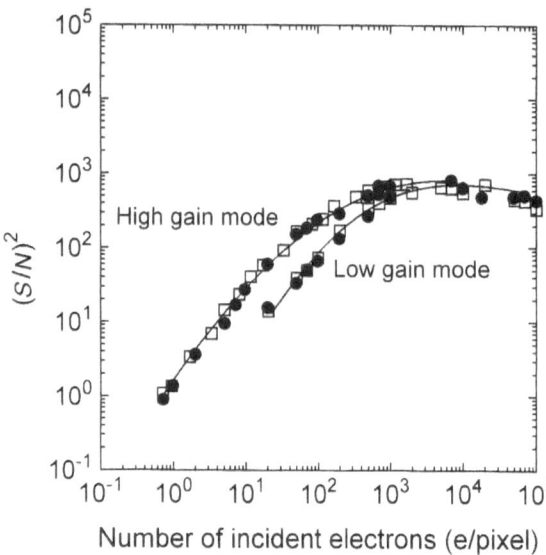

Fig. 4.16. Signal-to-noise ratio $(S/N)^2$ of the imaging plate as a function of the incident electron intensity at 100 kV ($\square$) and 200 kV ($\bullet$)

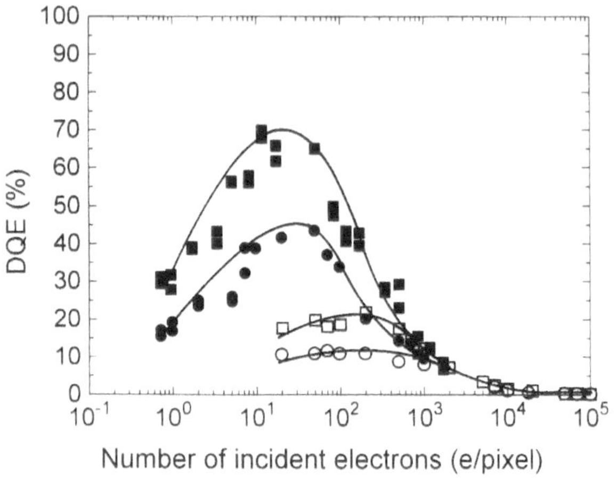

Fig. 4.17. Detective quantum efficiency (DQE) of the imaging plate as a function of the incident electron intensity at 100 kV ($\blacksquare$, high gain mode; $\square$, low gain mode) and 200 kV ($\bullet$, high gain mode; $\bigcirc$, low gain mode)

electron intensity. At conditions of both high gain and low gain, $(S/N)^2$ tends to saturate at a high intensity region, although good linearity between the incident electron intensity and the output signal is reserved in this region. In order to discuss the signal to noise ratio in detail, it is also useful to evaluate the *detective quantum efficiency* (DQE), which is defined using the input and output signal to noise ratio, i.e.,

$$\text{DQE} = \left(S/N\right)_{\text{out}}^{2} \Big/ \left(S/N\right)_{\text{in}}^{2}. \qquad (4.7)$$

Figure 4.17 shows the DQE of the imaging plate as a function of the incident electron intensity. It can be seen that the DQE of the imaging plate tends to have smaller values in both lower- and higher-intensity regions [14, 15]. This tendency is also noted for the slow-scan CCD camera [16, 17]. In Fig. 4.17, for 100 keV electrons, the DQE of the imaging plate has maximum values of about 70% at 2×10 e/pixel and 20% at 2×10^2 e/pixel under high gain mode and low gain mode, respectively. For 200 keV electrons, it has maximum values of about 45% (2×10 e/pixel) and 10% (2×10^2 e/pixel) under high gain mode and low gain mode, respectively. It is also noted that DQE has smaller values at higher accelerating voltages [14].

In general, it may be said that higher quality images can be obtained for higher S/N. Thus the highest quality image can be obtained in an inten-

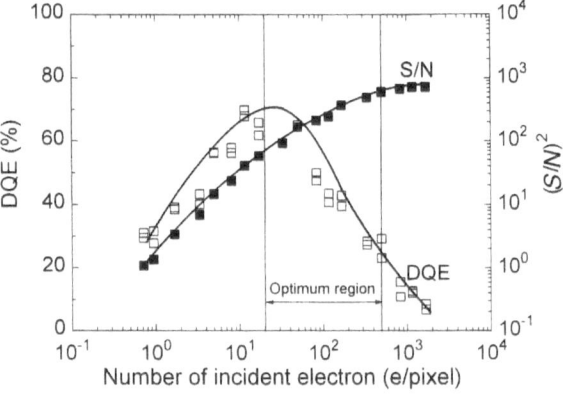

Fig. 4.18. Comparison of S/N and DQE as a function of the electron intensity for 100 keV electrons

sity region not less than the intensity of the S/N saturation. However, if radiation damage or contamination increases with an increase in electron exposure, not only S/N but also DQE should be taken into account. Figure 4.18 shows a comparison of S/N and DQE as a function of electron intensity for 100 keV electrons. In the region above 2×10 e/pixel, S/N increases with an increase in electron intensity, while DQE tends to decrease from its maximum value. Thus, in cases of radiation damage and contamination, optimum electron exposure may be around the intensity

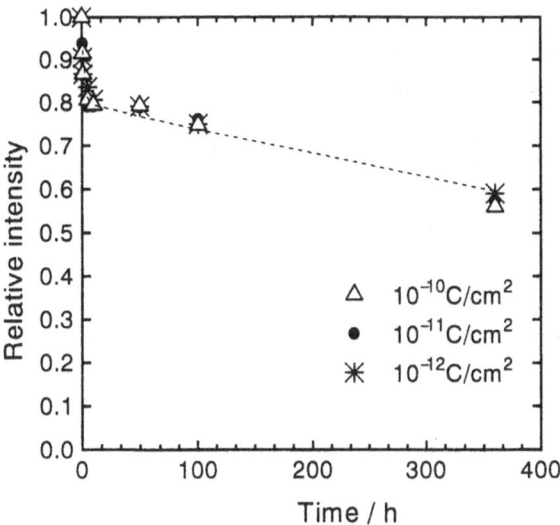

Fig. 4.19. Fading characteristic of the imaging plate

time it can take to read the imaging plate data. Also, c_1 and c_2 have the relation

$$c_1 + c_2 = 1 \qquad (4.9)$$

and T_1 and T_2 may correspond to the half-value periods of the well-known formula for the disintegration of radioactive elements. For an imaging plate (FDL–UR–V type), these constants are evaluated at 25°C as

$$c_1 = 0.2, \qquad c_2 = 0.8,$$
$$T_1 = 1.2\text{h}, \qquad T_2 = 850\text{h}. \qquad (4.10)$$

It has also been reported that the fading phenomenon is less pronounced at lower temperature such as 0°C [18]. As the fading phenomenon is related to the passage of time, t (Eq. 4.8), the intensity of the output signal is proportional to the incident electron intensity, and thus quantitative analysis can be carried out using the intensity of the output signal. Furthermore, by using a sensitivity curve such as that shown in Fig. 4.15, which gives the relation over time between the incident electron intensity and the output signal intensity, the absolute value of the incident electron intensity can also be obtained from the output signal intensity.

The most useful property of the slow-scan CCD camera compared with the imaging plate is the quick display of the output signal on a monitor. It takes only a few seconds to digitize and display the data. However, the intensity of the output signal is affected by some fluctuation in the gain from channel to channel, and thus the gain change for each channel must be corrected to obtain an accurate signal intensity. It has also been reported that a very strong electron intensity will have an effect on the output signal intensity around the region being irradiated, which is due to an oversaturation of electron charges in the CCD. This must be taken into account, especially for the region around the transmitted beam in electron diffraction patterns. These points are also noted in Table 4.1.

where DQE has its maximum value and where S/N also has a fairly high value. The optimum intensity regions for the imaging plate at various accelerating voltages were reported previously [14].

In addition to S/N and DQE, the fading phenomenon should be taken into account in the use of an imaging plate. The *fading phenomenon* indicates that the signal intensity gradually decreases over time. Figure 4.19 shows the fading characteristic of an imaging plate. The phenomenon of signal decrease is especially pronounced for the first 5 h after electron exposure. In general, the signal intensity, $I(t)$, of an imaging plate in $(t + 0.1)$ h after electron exposure is given by

$$I(t) = I_0\big[c_1 \cdot \exp(-0.638 \cdot t/T_1)$$
$$+ c_2 \cdot \exp(-0.638 \cdot t/T_2)\big] \qquad (4.8)$$

where I_0 is the signal intensity in 0.1 h (= 6 min) after electron exposure, which may be the shortest

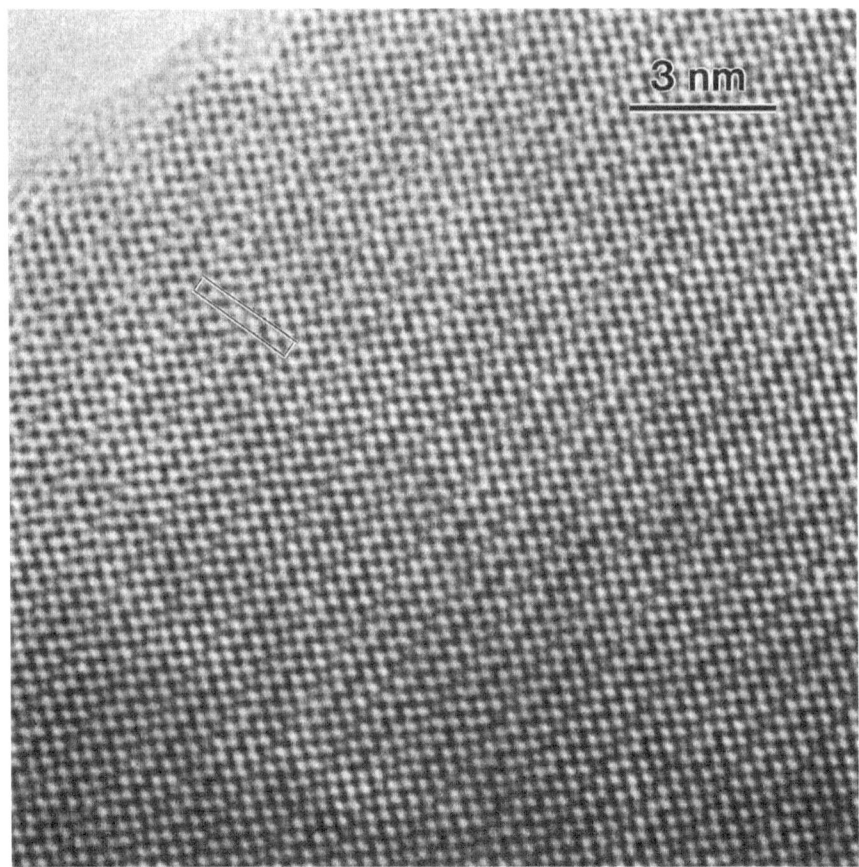

Fig. 4.20. High-resolution image of $Tl_2Ba_2CuO_6$

4.2.3 Quantitative High-Resolution Electron Microscopy

4.2.3.1 Quantitative Analysis of High-Resolution Images with a Residual Index

To observe high-resolution images accurately with the recording systems discussed above, not only the incident electron intensity but also the magnification should be appropriately adjusted. In general, the direct magnification of the observation can be adjusted by taking into account the resolution of the electron microscope image and the recording system, i.e.,

$$[\text{Resolution of a high-resolution image}]$$
$$\times [\text{Magnification}]$$
$$> [\text{Resolution of the recording system}]. \quad (4.11)$$

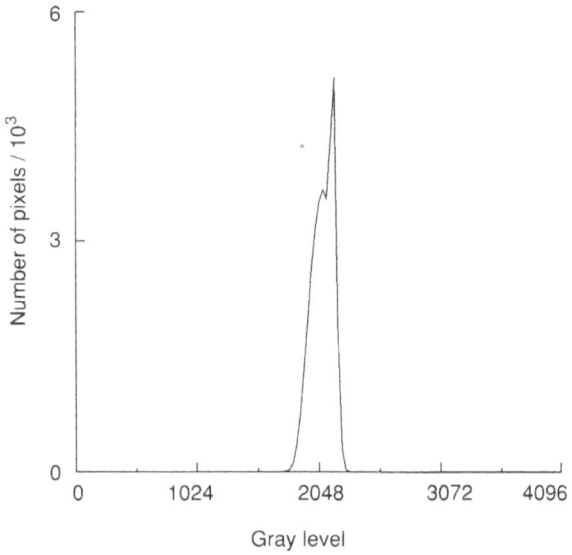

Fig. 4.21. Histogram for the high-resolution image in Fig. 4.20

For a quantitative comparison of an observed image and a calculated image, a *residual index*

$$R_{HREM} = \sum |I_{obs} - I_{cal}| / \sum I_{obs} \quad (4.12)$$

or the so-called *normalized Euclidean distance* (NED)

$$\text{NED} = \left[\sum \left(I_{obs} - I_{cal} \right)^2 / \left(\sum I_{obs}^2 \cdot \sum I_{cal}^2 \right)^{1/2} \right]^{1/2}$$

$$(4.13)$$

can be evaluated [19]. Basically, in both cases, agreement between observations and calculations is better as the residual index or NED becomes smaller.

We now discuss the quantitative analysis of a high-resolution image observed with an imaging plate using the residual index. A high-resolution image of $Tl_2Ba_2CuO_6$ is shown in Fig. 4.20 [20]. In the image, small dark spots show heavy atom positions projected along the incident electron beam. Figure 4.21 shows the histogram (see Sect. 4.1.1.1) for the high-resolution image in Fig. 4.20. Although the number of gray levels needed for recording high-resolution images seems to be relatively smaller than that for electron diffraction patterns, it can be seen that about 600 gray levels were used for recording the high-resolution image.

In general, in the quantitative analysis of high-resolution images, the first task is image simulation based on a basic structure model. By minimizing the residual index, experimental parameters such as crystal thickness and defocus value are determined. After determination of these parameters, structure refinement can be carried out using the experimental parameters obtained.

Figure 4.22a shows a model of the atomic arrangement of $Tl_2Ba_2CuO_6$. Figure 4.22b is a contour map showing the intensity distribution of a part of the image near the crystal edge. Figure 4.22c and d show high-intensity and low-intensity regions, which may correspond to low-potential and high-potential regions, respectively. The rectangles in the model in Fig. 4.22a and in the intensity distributions in Fig. 4.22b–d indicate unit cells of $Tl_2Ba_2CuO_6$, which has a tetragonal structure with the lattice constants $a = 0.3866\,nm$ and $c = 2.324\,nm$. In order to remove noise such as the quantum noise, the contour map was produced by smoothing the data with a spatial filter of 2×2 pixels (see Sect. 4.1.2.1) and averaging the intensity after displacing the image by $+a$ and $-a$ (see Sect. 4.1.2.2). The latter process may correspond to triple exposures in photographic printing. The observed intensity of the high-resolution image

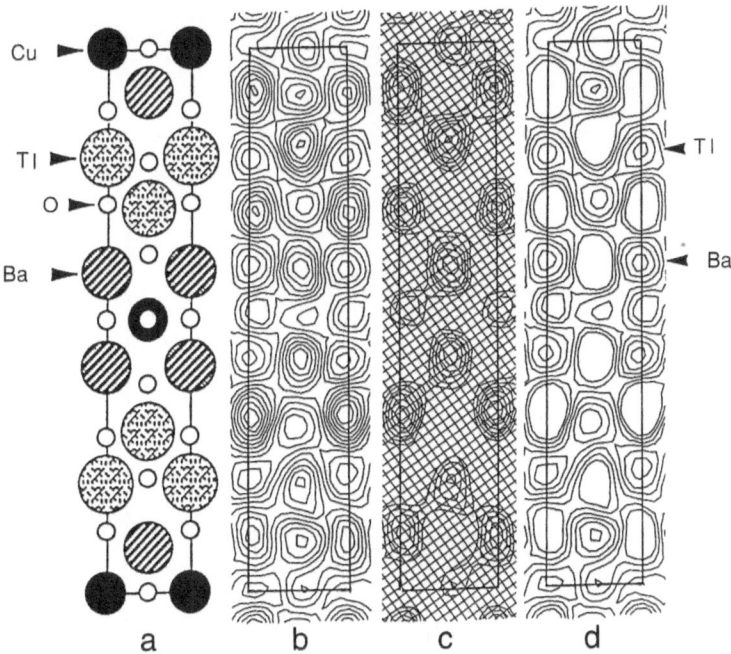

Fig. 4.22. **a** Model of the atomic arrangement of $Tl_2Ba_2CuO_6$. **b** Intensity distribution of a part of the image near the crystal edge shown as a contour map. **c** and **d** High intensity and low intensity regions, respectively, in **b**

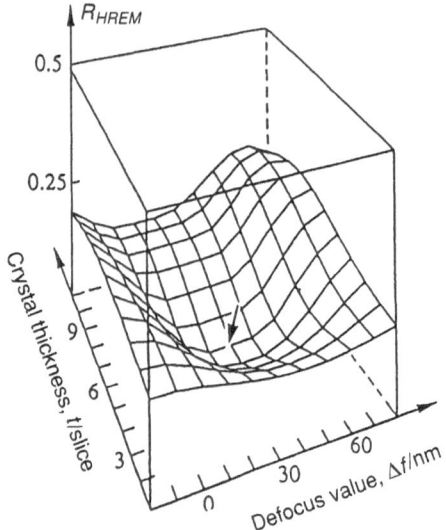

Fig. 4.23. Variation of R_{HREM} as a function of the crystal thickness t and defocus value Δf

Table 4.2. Parameters used for the calculation of the high-resolution images in Fig. 4.25

Wavelength	0.00164 nm
Spherical aberration constant	1.0 mm
Thickness of one slice	0.3866 nm
Number of beams	32 × 128
Defocus of objective lens*	23.0 nm
Defocus due to chromatic aberration*	24 nm
Crystal thickness*	5 slices (=1.93 nm)

*Variables in the calculation.

was divided by the intensity of the incident electron beam, which was measured in the vacuum region near the edge of the crystal. Thus we can directly compare the normalized observed intensity with the calculated intensity without any scaling factors. The grid in Fig. 4.22c indicates the positions where the observed intensity was measured with the imaging plate. The number of sampling points on the grid in the unit cell is 743.

The observed intensity at these sampling points was compared with calculated one. It can be seen that in the contour map in Fig. 4.22d, which shows low intensity, it is easy to distinguish the heavier atomic columns of Tl and Ba from those of Cu.

Based on the structure model in Fig. 4.22a, R_{HREM} can be evaluated by changing parameters such as the crystal thickness, defocus, and chromatic aberration. Figure 4.23 shows the variations of R_{HREM} as a function of the crystal thickness t and defocus value Δf. It can be seen that R_{HREM} changes gradually and has its minimum value at the point indicated by an arrow. The experimental parameters obtained for the minimum value $R_{HREM} = 0.0506$ are shown in Table 4.2. In order to

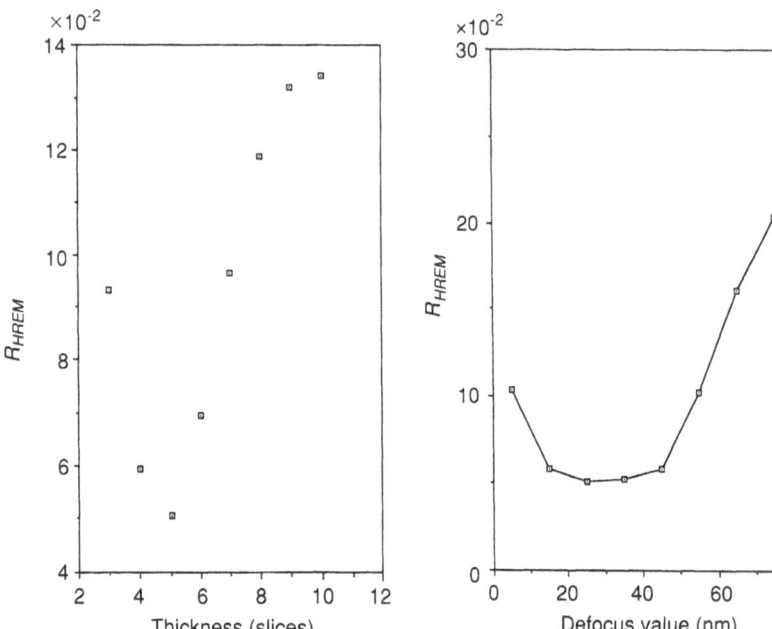

Fig. 4.24. Residual index R_{HREM} as a function of **a** the crystal thickness t and **b** the defocus value Δf

Fig. 4.25. Simulated images with $R_{HREM} = 0.0506$ shown by contour maps for **a** whole intensity, **b** high intensity, and **c** low intensity, compared with the observed images in Fig. 4.22b–d

see the detailed variations of R_{HREM}, cross sections of R_{HREM} through the minimum value are shown in Fig. 4.24a and b for t and Δf, respectively. In the calculation of Fig. 4.24a, all parameters except the crystal thickness were set equal to those in Table 4.2. It is noted that R_{HREM} is smaller than 0.07 in the crystal thickness range of four to six slices, and has a minimum value for five slices. Figure 4.24b shows the variations in R_{HREM} as a function of the defocus value Δf in the range 5–75 nm. It can be seen that R_{HREM} takes relatively small values in the range 20–50 nm, including the Scherzer focus (48.7 nm). Simulated images with $R_{HREM} = 0.0506$ are shown in Fig. 4.25, where three types of contour map, i.e., whole intensity, higher intensity, and lower intensity, are shown, corresponding to the intensity of the observed image in Fig. 4.22b–d. Based on the experimental parameters obtained, structure refinement can be carried out by further decreasing the residual index. Any deficiency of the constituent elements (such as Tl) could be clarified quantitatively with a residual index smaller than 5% [20].

4.2.3.2 Comparison of Quantitative High-Resolution Electron Microscopy with X-Ray Diffraction

In a diffraction study with X-rays and neutrons, the so-called R-factor, given by

$$R = \sum \left\| F_{obs} \right| - \left| F_{cal} \right\| / \sum \left| F_{obs} \right| \qquad (4.14)$$

is an estimate of the reliability of the analysis or proposed structure model, where $|F_{obs}|$ and $|F_{cal}|$ indicate the absolute values of observed and calculated structure factors, respectively. Thus, the R-factor is a residual index related to the absolute values of the structure factors. The summation is usually carried out in reciprocal space for as many reflections as possible. In this way, R, as defined in Eq. 4.14, directly shows the reliability of the model structure. R_{HREM}, as defined in Eq. 4.12 for the evaluation of high-resolution images, is basically different from R. R_{HREM} is a residual index obtained from the intensities of observed and calculated high-resolution images in real space. Since high-resolution images are generally obtained by a limited number of reflections around the origin of reciprocal space, the accuracy of high-resolution electron microscope analysis depends not only on R_{HREM}, but also on the resolution limit. In practice, no quantitative information about the atomic arrangement can be obtained, despite a small value of R_{HREM}, if the resolution is lower than 0.3 nm. It should also be noted that an increase in the number of sampling points in the computer simulation does not directly mean an increase in the accuracy of an analysis of high-resolution images.

4.2.3.3 Requirements for Quantitative High-Resolution Electron Microscopy

According to the above discussion, there are three important requirements for carrying out quantitative high-resolution electron microscopy, and these are summarized below [20, 21].

1. Use of a quantitative recording system. The recording system should have good linearity and a wide dynamic range of electron intensity. The characteristics of the recording system, such as the resolution, (S/N), and DQE, as discussed in Sect. 4.2.2, should be taken into account to optimize the experimental conditions. In addition to the intensity of the high-resolution image, the intensity of

the incident electrons should be measured accurately to normalize the image intensity.

2. High spatial resolution. This means a small spherical aberration, a reasonable objective aperture size, and a small chromatic aberration.

3. Small residual index. As demonstrated in the analysis of $Tl_2Ba_2CuO_6$, it is important that the assignment of experimental parameters as well as the structure refinement can be performed with a residual index smaller than 5% with electron microscopes of a resolution limit of about 0.17 nm. It should also be noted that the smallness of the residual index needed for structure refinement depends on the complexity of the structure.

In addition, quantitative high-resolution electron microscopy should be carried out with a thin crystal. This is because the effect of inelastic scattering and dynamical diffraction on a high-resolution image becomes stronger with an increase in crystal thickness, and it becomes more and more difficult to assign experimental parameters such as defocus value, crystal thickness, and chromatic aberration accurately as the crystal gets thicker. In order to remove the inelastic scattering effect due to plasmon excitation, energy filtering may be the most effective method for both quantitative high-resolution electron microscopy and electron diffraction studies.

Also, in the above analysis of $Tl_2Ba_2CuO_6$, a high-resolution image was investigated with the residual index. If the analysis is carried out on a series of high-resolution images taken with different defocus values (through-focus imaging), a more accurate structure refinement may be carried out.

4.3 Electron Diffraction

As mentioned in Sect. 1.1, the most significant characteristic of transmission electron microscopy is that it allows observation of both an image (information in real space) and an electron diffraction pattern (information in reciprocal space) from the same area. Hence, electron diffraction is important as a complementary tool of high-resolution electron microscopy for studying crystal structures. Some important mathematical equations for interpreting electron diffraction patterns are now presented, and then the interpretation of electron diffraction patterns is explained in detail.

4.3.1 Basis of Electron Diffraction

4.3.1.1 Basic Mathematical Equations for the Interpretation of Electron Diffraction Patterns

1. The scattering amplitudes of reflections in a diffraction pattern $Q(h, k, l)$ are given by the Fourier transform of a function $q(x, y, z)$ which corresponds to an object.

$$
\begin{aligned}
Q(h, k, l) &= \mathscr{F}\big[q(x, y, z)\big] \\
&= \iiint q(x, y, z) \\
&\quad \exp\big[-2\pi i(hx + ky + lz)\big] dx\, dy\, dz.
\end{aligned}
$$
(4.15)

2. Function $q(x, y, z)$ of a crystal is described by the convolution of two functions, viz., $f(x, y, z)$, showing an atomic arrangement in a unit cell of the crystal, and a delta function showing a periodic arrangement of the unit cell with the intervals of the lattice constants being a, b, and c of the unit cell.

$$
\begin{aligned}
q(x, y, z) &= f(x, y, z) \\
&\quad * \Bigg[\sum_{n_1, n_2, n_3} \delta(x - n_1 a, y - n_2 b, z - n_3 c)\Bigg].
\end{aligned}
$$
(4.16)

3. The Fourier transform of a periodic array of the delta functions arranged infinitely at regular intervals a, b, and c is equal to a periodic array of the delta functions arranged infinitely at regular intervals a^*, b^*, and c^*.

$$
\begin{aligned}
&\mathscr{F}\Bigg[\sum_{n_1, n_2, n_3} \delta(x - n_1 a, y - n_2 b, z - n_3 c)\Bigg] \\
&= \frac{1}{V} \sum_{n_1, n_2, n_3} \delta(h - n_1 a^*, k - n_2 b^*, l - n_3 c^*).
\end{aligned}
$$
(4.17)

V in Eq. 4.17 is the volume of a unit cell.

4. The Fourier transform of the convolution of two functions is the product of their Fourier transforms.

$$
\begin{aligned}
Q(h, k, l) &= \mathscr{F}\big[q(x, y, z)\big] \\
&= \frac{F(h, k, l)}{V}\Bigg[\sum_{n_1, n_2, n_3} \delta(h - n_1 a^*, \\
&\quad k - n_2 b^*, l - n_3 c^*)\Bigg]
\end{aligned}
$$
(4.18)

where

$$
F(h, k, l) = \mathscr{F}\big[f(x, y, z)\big].
$$
(4.19)

From these mathematical equations, the reciprocal lattice points defined by $\sum_{n_1, n_2, n_3} \delta(h - n_1 a^*, k - n_2 b^*, l - n_3 c^*)$ is determined by the real lattice points $\sum_{n_1, n_2, n_3} \delta(x - n_1 a, y - n_2 b, z - n_3 c)$. Also, $F(h, k, l)$ is the "structure factor" or the "kinematical scattering amplitude" at the hkl reciprocal lattice point, and is given by the atomic arrangement $f(x, y, z)$ in a unit cell.

4.3.1.2 Analytical Process of Electron Diffraction Patterns

Construction of a Reciprocal Lattice. The unit vectors, a^*, b^*, and c^* in the reciprocal lattice are given with the *fundamental translation vectors* a, b, and c in the real lattice as follows:

$$
a^* = \frac{b \times c}{a \cdot (b \times c)}, \qquad b^* = \frac{c \times a}{b \cdot (c \times a)},
$$
$$
c^* = \frac{a \times b}{c \cdot (a \times b)},
$$
(4.20)

where the directions of a^*, b^*, and c^* are perpendicular to the bc, ca, and ab planes, respectively. The denominators of the equations are equal to the unit cell volume.

In the cases of cubic, tetragonal and orthorhombic lattices, the vectors a, b, and c in the real lattice are perpendicular to each other, and $a*$, $b*$, and $c*$ in the reciprocal lattice are parallel to a, b, and c, respectively, Their lengths are

$$a* = 1/a, \qquad b* = 1/b, \qquad c* = 1/c. \quad (4.21)$$

Estimation of Intensity Distribution in a Reciprocal Lattice. Intensity at the reciprocal point *hkl* is determined by the atomic arrangement in a unit cell. The structure factor $F(h, k, l)$ (intensity is $|F(h, k, l)|^2$) is given by the coordinates and the kinds of atom in a unit cell as follows:

$$F(h, k, l) = \sum_n f_n \exp\left[-2\pi i\left(hx_n + ky_n + lz_n\right)\right]$$
$$(4.22)$$

where f_n is the scattering factor of the *n*th atom, and (x_n, y_n, z_n) is the coordinate of the *n*th atom. The structure factor is obtained by summing for all atoms in a unit cell. This gives a kinematical approximation of the scattering amplitudes, and in an analysis of the observed intensity of a diffraction pattern, the effect of dynamical diffraction should be taken into account (see Sect. 1.3.3).

4.3.1.3 Characteristics of Electron Diffraction

It is a significant characteristic of transmission electron microscopy that information concerning crystal structures can be obtained from electron diffraction while microstructures are being observed. However, electron diffraction is not often used for crystal structure analysis, being different from X-ray diffraction and neutron diffraction. This is because there is a strong interaction between the electrons and the material being analysed, the strong electron-scattering power of matter being about 10^4 times as large as the X-ray scattering power. This strong scattering power makes it possible to observe electron diffraction patterns from very small areas (a few nanometers in size), but it makes it difficult to apply electron diffraction to crystal structure analysis. That is, the intensity distribution of an X-ray diffraction pattern can be interpreted by the kinematical approximation, where the incident beam is scattered once by matter, and then the scattered beam will not be further scattered. However, electrons are scattered many times by the material being analysed (dynamical diffraction effect). The diffracted beam is scattered again and goes in the direction of either the transmitted beam or a different diffracted beam. This multiple scattering produces complicated changes in the amplitudes (and phases) of reflections (see Fig. 2.10), so it is very hard to perform a crystal structure analysis from the intensity distribution of observed electron diffraction patterns. However, if the electron diffraction patterns are interpreted by taking account of the effect of this multiple scattering of electrons by matter, it may be possible to obtain valuable information about crystal structures.

Most electron microscopes have an accelerating voltage higher than 100 kV, and so the electron beam used has a short wavelength (at 100 kV about 1/100 of that of X-rays, see Appendix A). Hence, the radius of the Ewald sphere, which is given by the inverse of the wavelength, is very long. Consequently, it can be said that electron diffraction patterns represent cross sections of the reciprocal lattice through the origin of reciprocal space. Therefore, if the intensity distribution in a reciprocal lattice is calculated from a known crystal structure, it is easy to interpret the various diffraction patterns observed by comparing the calculated intensity distributions in reciprocal planes through the origin.

4.3.2 Practice of Electron Diffraction

4.3.2.1 Measurement of a Camera Length

The interval (R) between the origin and a diffraction spot in an electron diffraction pattern and a lattice spacing (d) corresponding to that diffraction spot have approximately the following relationship:

$$Rd = L\lambda \qquad (4.23)$$

where λ is the wavelength of the electron and L is known as the *camera length*. Consequently, when lattice spacings are quantitatively estimated from diffraction spots, it is necessary to know the value of the camera length. Although the value of the camera length is given for an electron microscope, it is better to establish clear a relationship between a distance in a diffraction pattern and the lattice spacing by observing, e.g., Debye–Scherrer rings with a polycrystalline material having a known structure, such as vapor-deposited gold film. From the diffraction pattern formed by Debye–Scherrer rings, as shown in Fig. 4.26, the distortion of the diffraction pattern can also be quantitatively estimated, and this indicates different camera lengths depending on the orientation.

It should be noted that the camera length is sensitive to the specimen position at the objective lens, and a change of a few percent in the camera length can easily occur as the result of a position change. In order to avoid this type of mistake, a diffraction pattern should be taken after placing the specimen at the correct position using the Z-controller (see Sect. 2.2.2.4), with which the specimen can be moved along the microscope column axis. That is, the lens current of the objective lens is set at the optimum value, which is determined for each electron microscope, and then the specimen height is set so that the image of the specimen can be focused using the Z-controller. In this way it is always possible to obtain a diffraction pattern with the same camera length.

4.3.2.2 Determination of the Three-Dimensional Intensity Distribution

In order to analyze an unknown structure by the electron diffraction method, it is necessary to have a three-dimensional intensity distribution in reciprocal space. One can tilt a crystal through a large angle in a current transmission electron microscope. For instance, using a side-entry specimen holder with a double-tilt goniometer, a crystal can be tilted through an angle of 45° or 60°. Hence, a series of electron diffraction patterns from various angles can be obtained by tilting the crystal around one crystallographic axis. From these diffraction patterns, an intensity distribution in three-dimensional reciprocal space can be determined.

4.3.2.3 Points to Note in the Analysis of Electron Diffraction Patterns

Effect of Multiple Domain. When an intensity distribution is determined from electron diffraction patterns, it is necessary to ascertain that the diffraction patterns come from one domain with a specific crystallographic orientation. For instance, one domain with the same orientation for a fundamental structure is often divided into several domains with different crystallographic orientations for a superstructure. In this case domains should be examined with dark-field images, because in the bright-field method, the difference in contrast between domains cannot easily be seen. Figure 4.27 shows dark-field images taken with superlattice reflections, a and b, in the diffraction pattern. It can be seen that different regions are bright in Fig. 4.27a and b, and this is interpreted as meaning that the diffraction spots a and b come from different domains. That is, the diffraction pattern of one domain corresponds to only one of the reflections a, b, and c, and so the structure of

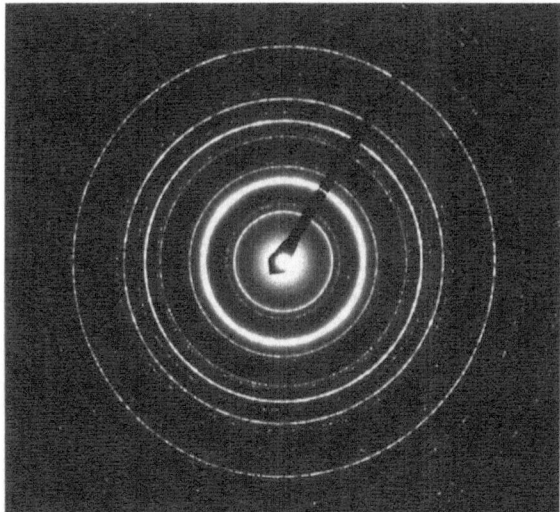

Fig. 4.26. Electron diffraction pattern of a polycrystalline Au thin film

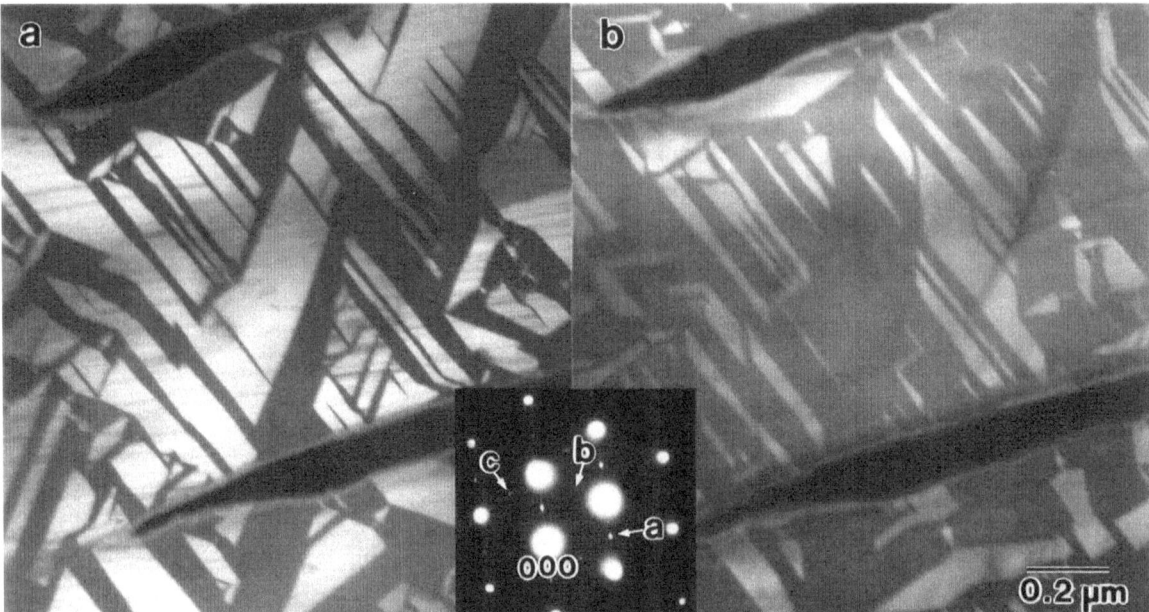

Fig. 4.27. Dark-field images **a** and **b** were obtained with the superlattice reflections a and b in the inset, respectively. Notice that bright regions are different in **a** and **b**

this crystal has an orthorhombic lattice without six-fold rotational symmetry, although the fundamental structure of the crystal has a hexagonal lattice. This difference is very important for interpreting the structure of crystals.

Effect of Multiple Scattering. The multiple scattering of electrons in materials produces extra reflections at special positions where reflections do not appear in single scattering (kinematical approximation). This phenomenon is called "*double diffraction*," and it should be taken into account when the *extinction rule* is being determined. This is the rule about the indices of reflections disappearing owing to the symmetry of the crystal. The *double-diffraction spots* (*double reflections*) are produced by the multiple-scattering process, where scattered beams can act as new sources of further scattering in the crystal. The indices of all new spots introduced by the multiple-scattering process can be given by the sum of the indices of the two or more primary diffraction spots. For example, all spots of the form $h_1 + h_2$, $k_1 + k_2$, $l_1 + l_2$ are possible double-diffraction spots, where $h_1 k_1 l_1$ and $h_2 k_2 l_2$ are any two allowed primary dif-

fraction spots. In particular, it should be noted that positions on lines through the origin have a high possibility of producing double reflections, and these reflections have a relatively strong intensity.

Figure 4.28 is diffraction patterns of a $Nd_2Fe_{14}B$ compound taken with the incident beam parallel to the main axes. In Fig. 4.28b, reflections with $h + l = 2n + 1$ disappear, so this structure has an extinction rule of $h + l = 2n + 1$. However, in Fig. 4.28a and c, reflections of $h + l = 2n + 1$ (for instance, 300, 003, etc.) appear. They are all double-diffraction spots. Therefore, when determining extinction rules from electron diffraction patterns, reflections which disappear can be believed, but reflections which appear should be doubted, and it is necessary to determine whether or not they are double reflections. The reason that double reflections with $h + l = 2n + 1$ disappear in Fig. 4.28b but appear in Fig. 4.28a and c is because in Fig. 4.28b, reflections of $h + l = 2n + 1$ cannot be made by the sum of the reciprocal lattice vectors for diffraction spots, but they can be obtained by the sum of the reciprocal lattice vectors in Fig. 4.28a and c.

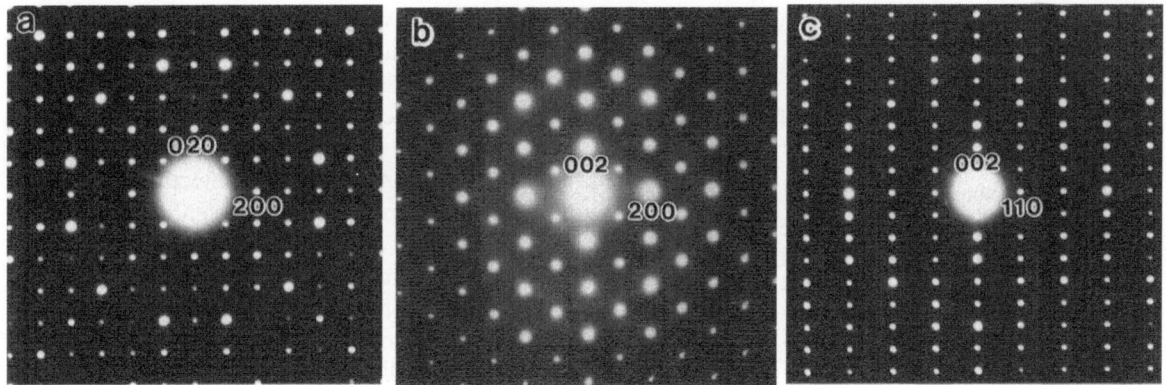

Fig. 4.28. Electron diffraction patterns of $Nd_2Fe_{14}B$ obtained with different incident beam directions. Note that the reflections satisfying the relation $h + l = 2n + 1$ (n is an integer) are observed in **a** and **c**, but are not observed in **b**. These reflections appear as a result of the multiple-scattering process

In order to determine whether diffraction spots are double reflections or not, it is important to observe diffraction patterns after removing diffraction spots contributing to the multiple-scattering process, for example, diffraction patterns with systematic reflections only.

4.3.3 Electron Diffraction Patterns of Various Structures

4.3.3.1 Electron Diffraction Patterns of Simple Structures

Appendix C shows diffraction patterns of typical structures, taken with the incident beam parallel to the main axes. Notice that in the face-centered cubic structure, only reflections where indices are all odd or all even appear, while in the body-centered cubic structure, reflections only appear when the sum of the h, k, and l indices, $h + k + l$, is even. The diamond structure is constructed of two-sublattices of a face-centered cubic lattice with a displacement of 1/4, 1/4, 1/4, so some reflections in the diffraction patterns of the face-centered cubic structure disappear. It should be noted that some double reflections appear in the diffraction patterns of the [110] incidence for the diamond structure and in the [110] incidence for the hexagonal close-packed structure.

4.3.3.2 Electron Diffraction Patterns of Ordered Alloys

As mentioned in Sect. 3.2.3, many alloys have superlattice structures with ordered arrangements of atoms in the simple structures described above. In these cases, extra reflections of relatively weak intensity (*superlattice reflections*) appear, in addition to reflections resulting from fundamental structures (*fundamental reflections*). Appendix C gives diffraction patterns of typical ordered structures. In the diffraction patterns of the Cu_3Au-type stucture, which is an ordered structure of the face-centered cubic type, and of the CsCl-type structure, which is that of the body-centered cubic type, superlattice reflections appear at positions where no reflections appear in the diffraction patterns of face-centered and body-centered cubic structures, respectively. It should also be noted that the ZnS-type structure, which is a superlattice of the diamond structure, has superlattice reflections at positions where no reflections appear in the diffraction patterns of the diamond structure.

4.3.3.3 Electron Diffraction Patterns of Incommensurate Structures

In general, superlattice reflections appear at positions divided by integers between fundamental reflections. That is, a unit cell of a superlattice has a size which is an integral multiple of a fundamental unit cell in real space. However, some ordered structures produce superlattice reflections at positions that cannot be divided by integers between fundamental reflections, that is, they appear at positions indexed with non-integers. In other words, the period of the superlattice cannot be described by an integral multiple of the period of the fundamental lattice, but has a period with a non-integral multiple. This type of structure is called an *incommensurate structure* (see Sect. 3.2.3). Figure 4.29a is an electron diffraction pattern of Cu_3Pd with a one-dimensional incom-

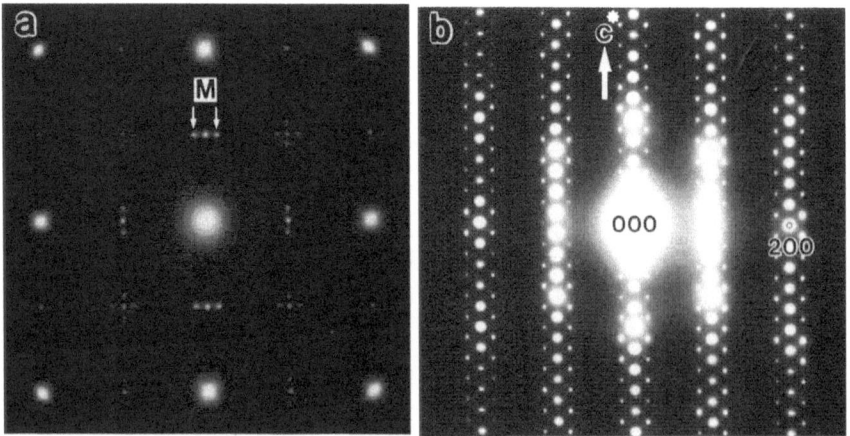

Fig. 4.29. Electron diffraction patterns of incommensurate structures. **a** Cu_3Pd with a one-dimensional long-period structure. The electron diffraction pattern consists of three variants. **b** $Tl_2Ba_2CaCu_2O_8$ with a modulated structure

mensurate structure [22]. In this case, the period M of the superlattice is estimated to be $6.8a_0$ (where a_0 is the period of the fundamental lattice) from a comparison between the intervals of the super-lattice reflections and the fundamental reflections. Figure 4.29b is also a diffraction pattern of an incommensurate structure of a Tl-based supercon-ducting oxide. From this pattern, the period of the superlattice is twice as large along the c-axis and 5.9 times as large along the a-axis as the period of the fundamental lattice. The incommensurate struc-ture of Fig. 4.29a is formed with an ordered ar-rangement of atoms, whereas that of Fig. 4.29b is produced mainly with lattice modulation.

4.3.3.4 Electron Diffraction Patterns of Twinning Structures

Diffraction patterns from some domains with a twin relation can be interpreted as an overlapping of patterns from each domain. For example, the diffraction pattern in Fig. 3.17 is formed by the overlapping of two patterns from two domains with a twinning relationship. However, it should be noted that in the special case where a twin plane is inclined against the incident electron beam, electrons scattered from the top domain are diffracted again in the different domain at the bottom, and then double diffraction between the different domains occurs and produces a complex diffraction pattern.

4.3.3.5 Diffuse Scattering

Line-shaped reflections elongated in one direction are often observed in electron diffraction patterns.

This type of reflection extended in one, two, or three dimensions is called *diffuse scattering*. Dif-fuse scattering is caused by many types of defect, and some typical examples are given below.

Crystal Shape Effect. An infinite array of a peri-odic arrangement of atoms produces sharp dif-fraction spots described as a delta function, but a periodic arrangement of finite size gives rise to broad reflections. For example, the diffraction spots for a crystal in the form of a thin plate are elongated along the direction normal to the plate, and those for a crystal with a needle shape are extended in all directions perpendicular to the needle axis and have disk-shaped intensity distri-butions. This type of effect, when the shape or size of a crystal gives rise to extended diffraction spots, is called the *shape effect*.

A crystal that becomes smaller in three-dimensional space produces diffraction spots ex-tended in three dimensions, and a needle-like crystal that becomes smaller in two dimensions perpendicular to the needle axis gives rise to disk-shaped reflections extended in two dimensions. A plate-like crystal that becomes smaller in one di-mension produces streak reflections elongated in one dimension. Keeping in mind the shape effect, the shape of small precipitates can be estimated from the shape of the diffuse scattering of those precipitates.

Defects. A crystal including stacking faults (mis-takes in the stacking of atomic planes) and micro-twins gives rise to streak reflections elongated in the direction normal to the planar defects (Fig.

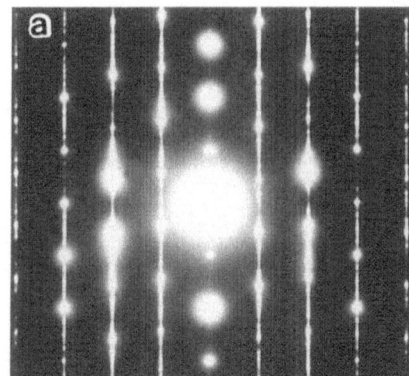 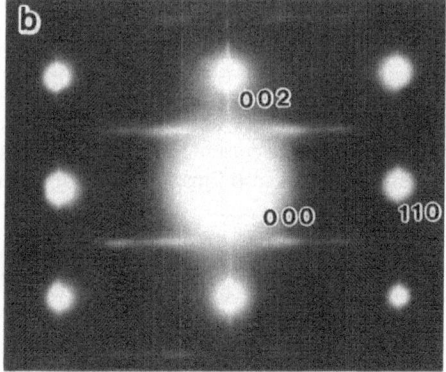

Fig. 4.30. a Diffuse scattering with a line shape caused by a stacking fault. **b** Diffuse scattering extending in the horizontal plane caused by a weak correlation of atomic columns parallel to the c-axis

Kikuchi patterns

As seen in Figs 3.64 and 4.36, in addition to diffraction spots an election diffraction pattern can show black-and-white lines and bands. These patterns were first observed by Dr. S. Kikuchi with mica in 1928 [23] and thus they are generally called *Kikuchi patterns*. The lines and bands are called *Kikuchi lines* and *Kikuchi bands*, respectively.

Kikuchi patterns result from a scattering process where inelastically scattered electrons are scattered elastically in a crystal. As schematically shown in Fig. 4.31a, inelastic scattering is basically the forward scattering, and thus the inelastically scattered wave (*S*) around the transmitted beam has a strong intensity, while the inelastically scattered wave (*W*) at a higher scattering angle has a weak intensity. If the strongly scattered wave *S* suffers Bragg scattering and results in a diffracted beam at a higher scattering angle, there will be a stronger intensity at this angle and thus a white line will be observed. On the other hand, if the inelastically scattered electrons *W* are scattered into a lower scattering angle due to the Bragg reflection, the intensity at this angle becomes smaller, resulting in a black line. When the distance between a pair of black-and-white lines is *R*, *R* satisfies the following relation (i.e. Eq. 4.23):

$$Rd = L\lambda$$

where *d* is the lattice spacing corresponding to the Bragg reflection, and *L* and λ are the camera length and the wavelength of the incident electrons, respectively. It should also be noted that if the incident electrons are parallel to a crystal axis, black-and-white bands appear instead of lines due to the dynamical diffraction effect on the inelastically scattered electrons (see Fig. 3.64).

Since Kikuchi patterns continuously change with the incident beam direction, they are quite useful for determining the crystal orientation accurately. They can also be used to determine the wavelength of the incident electrons, as will be noted in Sect. 4.5.1. However, Kikuchi patterns can be seen in relatively thick crystals, and thus the observation of high-resolution images, especially structure images (see Sect. 2.1.4), cannot appropriately be carried out in this region even if the crystal orientation is accurately determined.

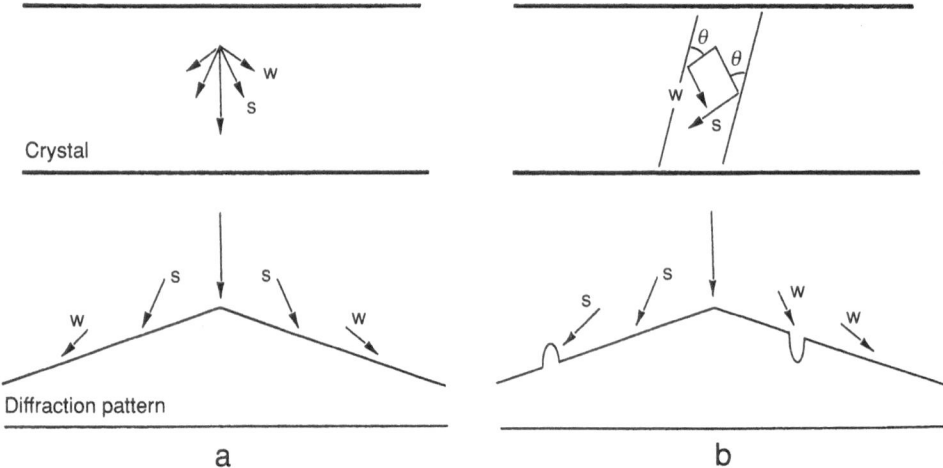

Fig. 4.31. Schematic illustration showing the formation of Kikuchi lines. **a** Background of electron diffraction due to the inelastically scattered electrons. **b** Contrast produced when the inelastically scattered electrons suffer the Bragg reflection through a scattering angle 2θ

4.30a). Stacking faults appearing in a random way produce streak reflections with a uniform intensity distribution, but faults appearing with some correlation result in reflections with a heterogeneous intensity distribution which corresponds with the correlation. If it is possible to confirm streak reflections elongated in one direction by tilting a crystal, then planar defects should be considered (including plate-like crystals).

Streak reflections elongated in one direction in a diffraction pattern are often a cross section of diffuse reflections extended in the plane. Figure 4.30b is an example of streak reflections observed as a cross section of two-dimensional diffuse scattering. This diffuse scattering results from a disordering of the two-dimensional arrangement of atom columns, in which the atomic arrangement in the one-dimensional column is fixed. Thus, the diffuse scattering is extended in the planes perpendicular to the axis of the atom column. Characteristic intensity distributions in the diffuse scattering are caused by some correlation in the atom columns, and from these intensity distributions, the interaction energy between the atom columns can be estimated [24].

It should be remembered that one-dimensional disordering of planer stacking gives rise to one-dimensional streak reflections, two-dimensional disordering for an array of atom columns to two-dimensionally extended reflections, and three-dimensional disordering of an atomic arrangement to three-dimensionally extended reflections.

4.4 Weak-Beam Method

4.4.1 Principles of the Weak-Beam Method

The dark-field method and the weak-beam method [25–27] have been used to observe lattice defects, especially dislocations. Although the resolution of these methods is not so high as that of high-resolution electron microscopy, there is no strong limitation on the incident beam direction being different from high-resolution electron microscopy, and they are quite useful for investigating the line shapes and distributions of dislocations. In order to observe dislocations by the dark-field method, a strongly excited Bragg beam, g, is used for imaging. As illustrated in Fig. 4.32, the diffraction condition where only the Bragg beam g is excited in addition to the transmitted beam is called the *two-beam condition*. In order to determine the dissociation width accurately and the Burgers vectors of partial dislocation, it is appropriate to use the weak-beam method, which has a higher resolution limit than the conventional dark-field method described above. Figure 4.33 shows typical weak-beam images of a dissociated dislocation in silicon [28]. As illustrated in Fig. 4.34a, the image was observed by selecting the weakly excited reflection g, with one of the systematic reflections $3g$ being strongly excited, where g is the 220 reflection shown in Fig. 4.33. This observation mode is described as $g/3g$ or $g(3g)$. The electron diffraction pattern in Fig. 4.35 shows an example of this diffraction condition (g = 220). Under this condition, the lattice planes of the perfect crystal region do not satisfy the Bragg condition for g, while the lattice planes of the part around the dislocation core do satisfy the Bragg condition, as illustrated in Fig. 4.34b. Thus, only the region satisfying the Bragg condition appears

bright on the dark background of the perfect crystal region in the weak-beam image.

In the weak-beam image of dislocations, if the reciprocal lattice vector of the reflection g selected is perpendicular to the Burgers vector b, i.e.,

$$b \cdot g = 0 \qquad (4.24)$$

the image contrast of the dislocation disappears. Thus, by using the relation in Eq. 4.24, the Burgers vector can be determined. In Fig. 4.33b, one of the partial dislocations seen in Fig. 4.33a disappears under the condition of Eq. 4.24. It is noted that the Burgers determination based on the relation in Eq. 4.24 is restricted to the case where the strain field around the dislocation core fits the isotropic elasticity theory. If the elasticity theory is not applicable, the dislocation contrast does not disappear even if the condition of Eq. 4.24 is satisfied [29]. Also, if one wishes to determine the magnitude of the Burgers vector, the change in the dislocation contrast with the change in the reflection selected should be investigated in detail [29].

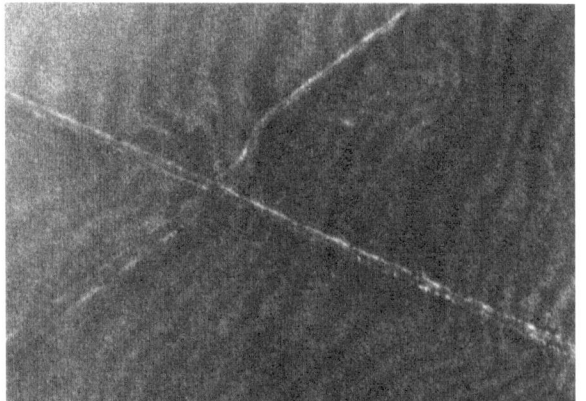

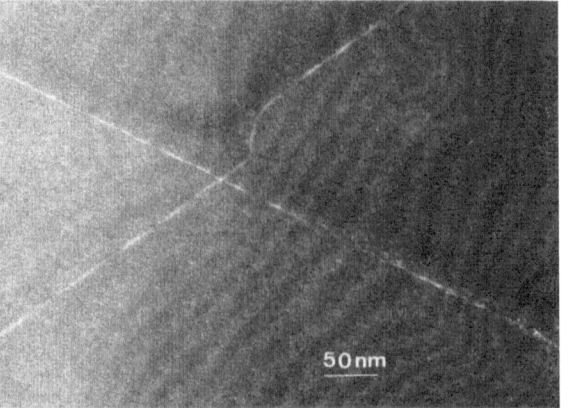

Fig. 4.33. Weak-beam images of dislocations in silicon observed with different reflections

Fig. 4.32. Two-beam condition used for dark-field imaging

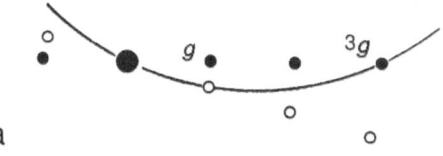

Fig. 4.34. a Diffraction condition ($g/3g$) used for observing weak-beam images. **b** Schematic illustration showing lattice planes around the dislocation core. Lattice planes near the dislocation core, indicated by *dotted lines*, form the reciprocal lattice points indicated by the *open circles* in **a**

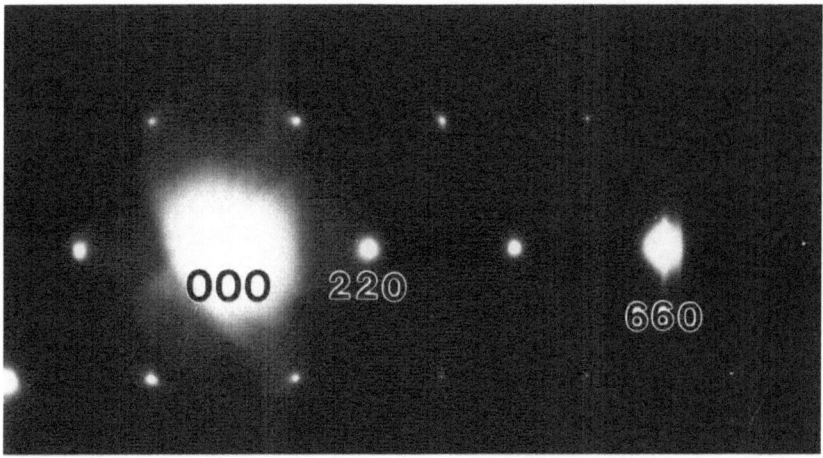

Fig. 4.35. Electron diffraction pattern showing a diffraction condition ($g/3g$, $g = 220$) used for observing weak-beam images

As demonstrated above, the weak-beam method is useful to determine the Burgers vectors of perfect and partial dislocations, but the resolution is limited to about 1.5 nm. Thus, if the dissociation width is smaller than this value, the width and its dissociation mode cannot be clarified by the weak-beam method, and high-resolution electron microscopy should be used. It should be noted that for both the conventional dark-field method and the weak-beam method, the diffraction condition can be set appropriately and easily when the Ewald sphere is relatively small. This situation is the opposite to that in high-resolution electron microscopy, where a higher accelerating voltage with a larger Ewald sphere directly results in a higher resolution.

4.4.2 Weak-Beam Method in Practice

A typical procedure for observing a weak-beam image is given below.

1. Observation of the morphology of dislocations. At a relatively low magnification, the morphology of dislocations can be observed over a wide area. It is easier to find dislocations if the focus setting is changed continuously, i.e., overfocus to underfocus and vice versa. This is because the strain field around the dislocation core is very sensitive to a change in focus, and thus the image contrast of the dislocation is easily distinguished from contaminations whose image contrast is not sensitive to the focus change.

2. Coarse setting of the diffraction condition. By inserting a selected area aperture into the area including the dislocation to be examined, a diffraction pattern of the area is obtained. By tilting the specimen, systematic reflections are excited.

3. Confirmation of the area. By changing to the imaging mode, specimen movement due to tilting can be corrected.

4. Observation of a dark-field image. By switching to the dark-field mode, which is normally supplied on the operating board, the dislocation can be observed.

5. Adjustment of the diffraction condition. By changing to the diffraction pattern with the dark-field mode, the reflection g is centered or aligned parallel to the optical axis by adjusting the so-called "dark tilt." By tilting the specimen, the reflection of $3g$ can be excited to get the condition $g/3g$.

6. Insertion of an objective aperture. By inserting an objective aperture, only the reflection of g is selected for imaging.

7. Imaging in the bright-field mode. By switching to the bright-field mode, the transmitted beam is confirmed to be in the objective aperture. If it is not, the reflection g should be centered again accurately in the dark-field mode.

8. Coarse focus setting. In the bright-field imaging mode, the dislocation can be observed by removing the selected area aperture. By magnifying the image about $50\,000$–$150\,000$ times, the focus setting should be adjusted to sharpen the dislocation image.

9. Fine focus setting. Changing to the dark-field mode, the dislocation can be observed. Normally, dislocation lines disappear under the condition of Eq. 4.24; otherwise dislocation lines appear bright. In the latter case, observe the dislocation line with a pair of binoculars and adjust the focus setting to get the sharpest image of the dislocation. If the image contrast is too weak, change the diffraction condition near to $g/2g$. Conversely, if the image contrast is too strong or broad, the condition should be nearer to $g/4g$.

10. Exposure. In order to image a weak intensity, a long exposure time is necessary; it normally takes 10–20 s to record with conventional EM film.

4.5 Evaluation of the Performance of Electron Microscopes

In order to observe high-resolution images under optimum conditions and analyze them through computer simulation, it is necessary to know the exact values of the basic parameters of the electron microscope, such as the wavelength of the electron and the spherical aberration constant. These parameters may be available through the manufacturer, but users can also evaluate them independently. We now explain how these parameters can be determined. We also show how to evaluate the resolution of the electron microscope, and discuss several important factors which limit the resolution.

4.5.1 Evaluation of Basic Parameters in Electron Microscopes

The basic parameters in an electron microscope are:

1. wavelength (or accelerating voltage);
2. spherical aberration constant of the objective lens;
3. chromatic aberration constant.

As indicated in Eq. 1.21, the wavelength of the incident electron and the spherical aberration constant of the objective lens determine the resolution, and the chromatic aberration constant affects the resolution through the chromatic aberration, as in Eq. 1.24. The principles and procedures of the measurement of these parameters are now given in order.

4.5.1.1 Wavelength

Principle of Measurement. Through the Bragg condition of Eq. 1.1, i.e.,

$$2d \sin\theta = \lambda$$

λ can be determined if one can measure the scattering angle (2θ) accurately using the electron diffraction pattern of a specimen whose lattice constants are known. If the wavelength is determined, the accelerating voltage can be given through Eq. 1.8. In practice, since the scattering angle is very small, Eq. 4.23, i.e.,

$$Rd = L\lambda$$

can be used. Here we need to know the exact value of the camera length in order to determine

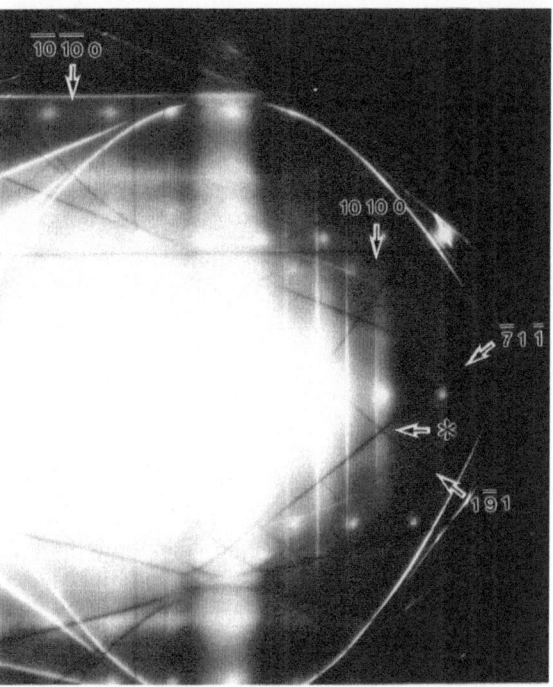

Fig. 4.36. Kikuchi lines observed in an electron diffraction pattern of silicon

the wavelength by measuring the distance between the transmitted beam and the reflection **g**. However, there are other methods by which the wavelength of the incident electron can be determined, i.e., by investigating Kikuchi lines [30–32]. We now explain the method proposed by Høier [31].

Measurement Procedure
1. Observe the Kikuchi lines on an electron diffraction pattern of a material whose lattice constants are known. Figure 4.36 shows an example of electron diffraction patterns of silicon.

2. Try to find an area where there are three Kikuchi lines corresponding to the reflections $h_i k_i l_i$ ($i = 1, 2, 3$), and these lines cross each other. In Fig. 4.36, one such area is indicated by an asterisk and is shown schematically in Fig. 4.37.

3. For the Kikuchi line $h_3 k_3 l_3$, measure the distances $2D$ and ΔD in Fig. 4.37. $2D$ is the distance between a pair of Kikuchi lines $\bar{h}_3 \bar{k}_3 \bar{l}_3$ and $h_3 k_3 l_3$, while ΔD is the distance from the Kikuchi line $h_3 k_3 l_3$ to the intersection of Kikuchi lines $h_1 k_1 l_1$ and $h_2 k_2 l_2$.

4. Using the following conditions which are satisfied for the three reflections, i.e.,

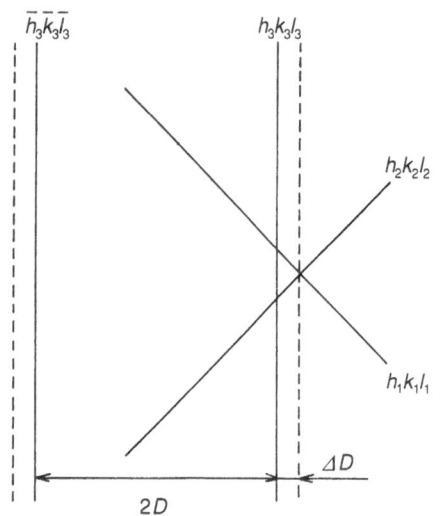

Fig. 4.37. Constitution of Kikuchi lines in the area indicated by an *asterisk* in Fig. 4.36, where the Kikuchi lines $h_3k_3l_3$ (=10$\overline{1}$00) and $\overline{h}_3\overline{k}_3\overline{l}_3$ (=$\overline{10}$1$\overline{0}$0) appear *white* and *black*, respectively

$$ph_1 + qk_1 + rl_1 = \frac{1}{2}d\lambda\left(h_1^2 + k_1^2 + l_1^2\right)^2 \quad (4.25)$$

$$ph_2 + qk_2 + rl_2 = \frac{1}{2}d\lambda\left(h_2^2 + k_2^2 + l_2^2\right)^2 \quad (4.26)$$

$$ph_3 + qk_3 + rl_3$$
$$= \frac{1}{2}d\lambda\left(1 + \Delta D/D\right)\left(h_3^2 + k_3^2 + l_3^2\right)^2 \quad (4.27)$$

$$p^2 + q^2 + r^2 = 1 \quad (4.28)$$

the wave length λ can be determined by eliminating p,q,r, which indicate the vector of the incident beam direction. In Fig. 4.36, three Kikuchi lines correspond to 1$\overline{9}$1, $\overline{7}$11, and 10$\overline{1}$00, and by measuring $2D$ and ΔD, the wavelength was determined as 8.69923×10^{-4} nm, which corresponds to an accelerating voltage of 1002.7 kV.

4.5.1.2 Spherical Aberration Coefficient of an Objective Lens

Principles of Measurement.
Principle No. 1 [33]. According to the discussion in Sect. 1.3.1, a digital diffractogram of a high-resolution image of a thin amorphous film can be given as (see also Sect. 4.5.2)

$$I(\boldsymbol{u}) \propto F(\boldsymbol{u})^2 B(\boldsymbol{u})^2. \quad (4.29)$$

In the above equation, $F(\boldsymbol{u})$ is a structure factor

and $B(\boldsymbol{u})$ is an imaginary part of a contrast transfer function given by

$$B(\boldsymbol{u}) = \sin\left(\chi(\boldsymbol{u})\right)$$
$$= \sin\left[(\pi/2)\left(-C_s\lambda^3|\boldsymbol{u}|^4 + 2\Delta f\lambda|\boldsymbol{u}|^2\right)\right].$$
$$(4.30)$$

Therefore, intensity maxima and minima of the digital diffractogram are observed under the following condition for n being either odd and even numbers:

$$\chi(\boldsymbol{u}) = n\pi/2. \quad (4.31)$$

Thus, we have the following condition for C_s and Δf at the intensity maxima and minima of the digital diffractogram:

$$-C_s\lambda^3|\boldsymbol{u}|^2 + 2\Delta f\lambda = n/|\boldsymbol{u}|^2. \quad (4.32)$$

In principle, by measuring $|\boldsymbol{u}|$ for different n at intensity maxima or minima, C_s and Δf can be evaluated.

It should be noted that for thicker films or lower accelerating voltages, the weak-phase object approximation cannot be applied, and errors in the determination of C_s and Δf become larger.

Principle No. 2. An image shift due to the defocus Δf for a reflection $\boldsymbol{u}_1$ is given by

$$\delta = C_s(\lambda/d_1)^3 - \Delta f(\lambda/d_1) \quad (4.33)$$

where d_1 is the lattice spacing for $\boldsymbol{u}_1$. Therefore by observing δ_1 and δ_2 for reflections $\boldsymbol{u}_1$ and $\boldsymbol{u}_2$, we can evaluate C_s as

$$C_s = \frac{\delta_1 d_1 - \delta_2 d_2}{\lambda^3\left[\left(1/d_1^2\right) - \left(1/d_2^2\right)\right]} \quad (4.34)$$

from Eq. 4.33.

A procedure for Principle No. 1 is now given, with an example.

Measurement Procedure
1. At a magnification of 250 000–500 000, take a few images of amorphous Ge (or amorphous C) with different underfocus conditions where the underfocus values are larger than the Scherzer focus.

2. Obtain digital diffractograms (or optical diffractograms) from the images observed. Two examples are shown in Fig. 4.38.

3. From the intensity profiles of the digital diffractograms (see the bottom of Fig. 4.38), find

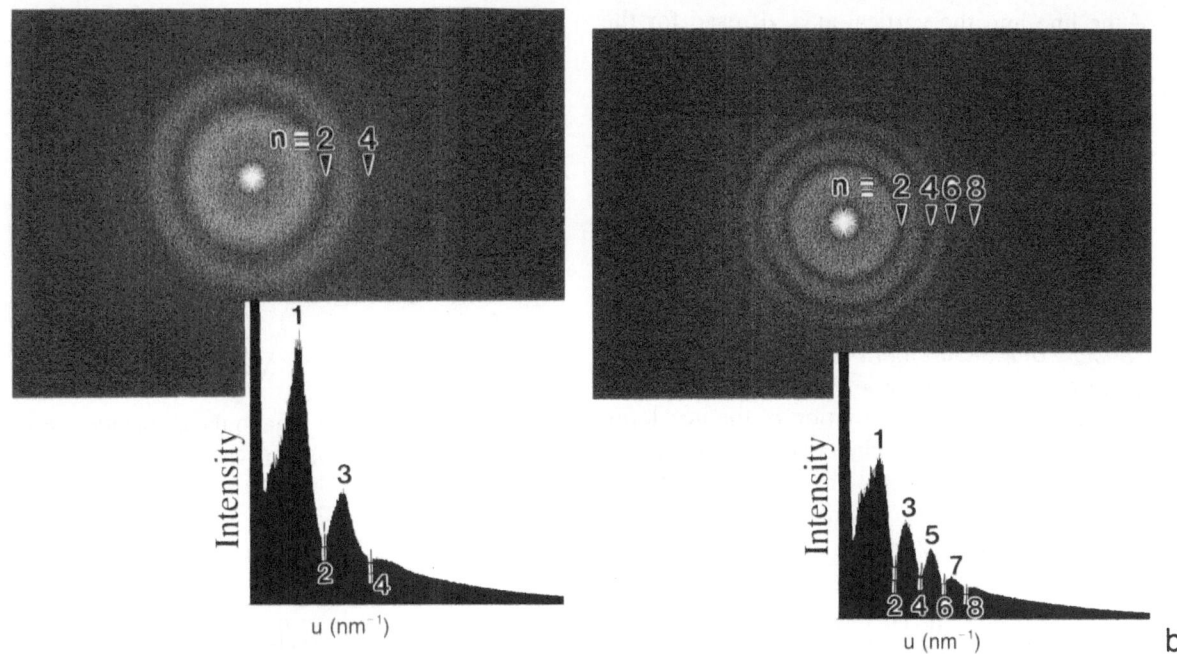

Fig. 4.38. Digital diffractograms obtained from high-resolution images of amorphous Ge observed at different defocus values. The accelerating voltage is 1.25 MV. The intensity profiles of the digital diffractograms are shown at the bottom

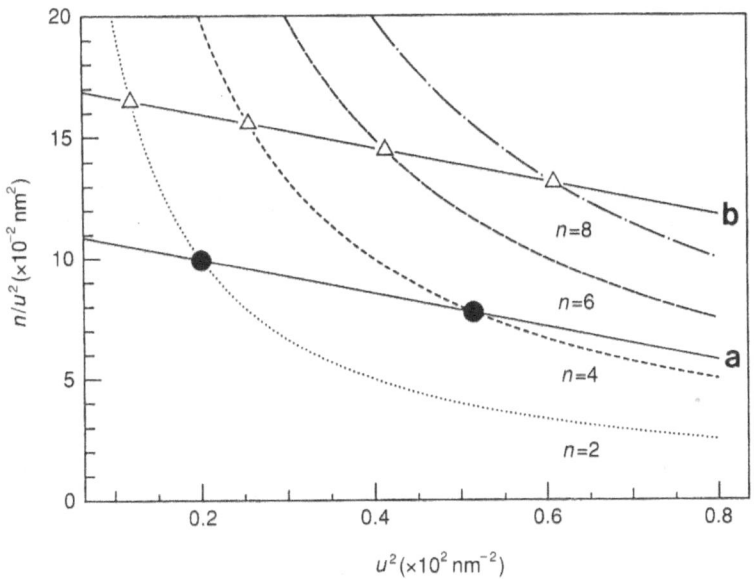

Fig. 4.39. Diagram of the relation between n/u^2 and u^2 used to evaluate a spherical aberration constant

the positions of, for example, the intensity minima for $n = 2,4,\ldots$ in Eq. 4.31 and measure their distances from the origin on a scale of nm^{-1}. Small particles, such as gold particles prepared on an amorphous film, may be useful to calibrate this distance.

4. For $n = 2,4,\ldots$ draw lines of $n/|\boldsymbol{u}|^2$ as a function of $|\boldsymbol{u}|^2$, as shown in Fig. 4.39. Now plot the data measured above.

5. Connect the data points to get the straight lines shown in Fig. 4.39. From the slope of the line, C_s can be determined, while from the intersection

of the line and the vertical axis, Δf used for the observation can be evaluated. The data shown in Figs. 4.38 and 4.39 were obtained with a high-voltage electron microscope (1250 kV). C_s was determined as 1.74 ± 0.10 mm, and the defocus values Δf were evaluated as a 77.1 nm and b 118.0 nm [34].

4.5.1.3 Chromatic Aberration Coefficient

Principle of Measurement. Chromatic aberration is caused by the fluctuation of the incident electron energy due to the variation of the accelerating voltage (V) and the objective lens current (I), and the focus shift due to the chromatic aberration is given using the chromatic aberration coefficient C_c by Eq. 1.22, i.e.,

$$\Delta = C_c \left[\left(\Delta V_r / V_r \right)^2 + \left(2\Delta I / I \right)^2 \right]^{1/2}$$

where V_c is the accelerating voltage relativistically corrected for the electron mass (Eq. 1.23). When the accelerating voltage is changed by $\Delta V_r'$ on a scale of kV, the image shift δ which corresponds to δ' at the specimen position is given by

$$\delta' = \delta / M = 2\theta C_c \left(\Delta V_r' / V_r \right). \quad (4.35)$$

The image shift δ' may correspond to the specimen shift Δ' along the optical axis, as shown in Fig. 4.40, and is approximately given by

$$\delta' = 2\theta \Delta'. \quad (4.36)$$

Thus, from Eqs. 4.35 and 4.36,

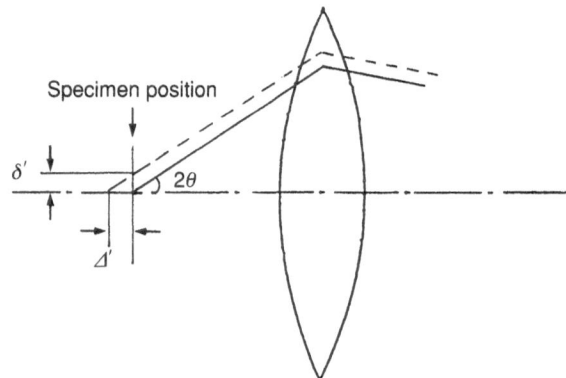

Fig. 4.40. Image shift δ' due to the chromatic aberration, and the corresponding specimen shift Δ' along the optical axis

$$C_c = \Delta' / \left(\Delta V_r' / V_r \right) \quad (4.37)$$

and C_c can be evaluated by measuring Δ'.

Measurement Procedure

1. Observe a specimen under the just-focused condition, i.e., $\Delta f = 0$.

2. Change the accelerating voltage by $\Delta V_r'$, say 1 kV. Adjust the specimen height to the just-focused condition using a so-called Z-controller (see Sect. 2.2.2.4), and measure the size of the shift Δ'. Alternatively, one can change the specimen height first and then evaluate the change in the accelerating voltage which causes the image shift.

3. Insert the measured values of $\Delta V_r'$ and Δ' in Eq. 4.37 to determine C_c.

4.5.2 Evaluation of the Resolution of Electron Microscopes

The resolution of an electron microscope is basically determined by the spherical aberration constant, as shown in Eq. 1.21. However, in practice the resolution is also affected by the fluctuation of the accelerating voltage and the convergence of the incident beam. We now consider the evaluation of the effective resolution taking into account these aberration and beam convergence, and also discuss each factor which limits the resolution.

4.5.2.1 Resolution

In principle, the resolution of an electron microscope is evaluated from a digital diffractogram of a high-resolution image of an amorphous thin film. The intensity profile of a digital diffractogram is generally given by [35]

$$I(u) \propto F(u)^2 B(u)^2 H(u)^2 D(u)^2 S(u)^2 \quad (4.38)$$

where the components are:

$F(u)$, structure factor of the amorphous thin film;
$B(u)$, imaginary part of the contrast transfer function;
$H(u)$, modulation transfer function of the recording material;
$D(u)$, damping due to chromatic aberration;
$S(u)$, damping due to beam convergence.

$B(u)$, $D(u)$, and $S(u)$ are given by Eqs. 4.30, 1.24, and 1.25, respectively. Thus, taking these factors into account, the resolution of an electron microscope is determined from the position in a digital diffractogram where the value of the effective transfer function is zero. In general, the resolution determined from Eq. 4.38 is lower than the point resolution defined by Eq. 1.21.

4.5.2.2 Factors Limiting Resolution

Chromatic Aberration. As given by Eq. 1.22, the chromatic aberration results from a fluctuation in the objective lens current and a fluctuation in the energy of incident electrons, which is attributed to the energy spread of electrons emitted from a filament and the fluctuation of the accelerating voltage. The fluctuations in the accelerating voltage and lens current consist of two components from the direct current and the alternating current. The direct current component produces a continuous focus shift (focus drift) (see Sect. 2.2.1), while the alternating current component (the so-called

ripple) causes the chromatic aberration. Usually if one needs to know the value of the ripple, the best course is to ask the manufacturer. Roughly speaking, the fluctuation in electron energy and lens current is about 5×10^{-6} and C_c is 2–3 mm. Thus the focus spread due to the chromatic aberration is 10–20 nm according to Eq. 1.22. Since the chromatic aberration directly limits the resolution, as shown in Eq. 1.24, it is always necessary for users to check the aberration by observing high-resolution images and investigating their digital diffractograms.

Beam Convergence. The intensity of the incident beam on the screen is generally affected by the beam size due to the first condenser lens (adjustable by the so-called spot-size knob), the aperture size at the condenser lens, and the beam convergence due to the second condenser lens (adjustable by the so-called brightness knob). When a large area is illuminated by adjusting the second condenser lens at low magnification, the intensity of the incident beam is mainly changed by the spot size. On the other hand, when high-resolution images are observed at high magnification, the electron beam needs to converge on a specific point in the specimen. Under this condition, the intensity of the incident beam is affected by both the *condenser aperture* size and the beam convergence due to the second condenser lens. Now, the illumination angle at the specimen or beam convergence affects the resolution, as indicated by Eq. 1.25. Thus, in order to know the effect of the beam convergence on the resolution, it is necessary to evaluate the beam convergence, which can be measured from the size of the diffraction spots. It should be noted that if the prefield of the objective lens is strong, it affects the beam convergence.

Stability of the Specimen Stage. In addition to the chromatic aberration and beam convergence, mechanical and thermal drift of the specimen stage and external vibrations also limit the resolution. External vibration results from sound and also comes through the ground. Most of the vibration through the ground can be removed by buffers. To minimize the sound, any pumping systems should be isolated from the electron microscope. It has been noted that an electron microscope tends to swing in one direction, and the resolution is thereby limited preferentially along this direction. Also, as pointed out in Sect. 2.2.1, thermal and mechanical drifts at the specimen stage tend to decrease with the passage of time.

4.6 Specimen Preparation Techniques

In order to observe fine high-resolution images, it is necessary to prepare thin films without introducing contamination or defects. For this purpose, it is important to select an appropriate specimen preparation method for each material, and to find an optimum condition for each method. In this section, we consider various specimen preparation methods for high-resolution electron microscopy. The methods are first outlined, and then their characteristics and special features are pointed out.

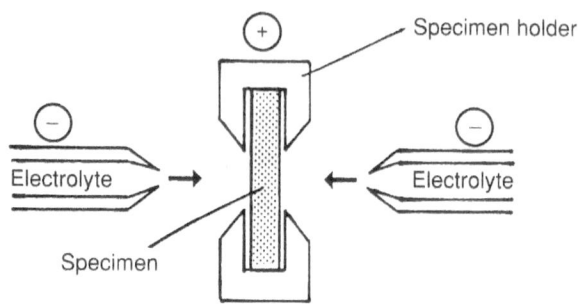

Fig. 4.41. Principle of the twin jet-polishing method

4.6.1 Crushing

Outline: A specimen is usually crushed with an agate mortar and an agate pestle. The flakes which are obtained are suspended in an organic solvent such as butyl alcohol and acetone, and dispersed with supersonic waves or simply by stirring with a glass stick. Finally, the solvent containing the specimen flakes is dripped onto a microgrid on a filter paper.

Notes: This method is limited to materials which tend to cleave. Since this is the simplest method, and it is also possible to find thin regions of a few nanometers thickness with little contamination on the surface, it is quite useful for high-resolution electron microscopy. In fact, most of the high-resolution images of high-Tc superconductors discussed in Sect. 3.2.2 were obtained from specimens prepared in this way. However, since grain boundaries are rather fragile, it is usually difficult to observe them in specimens prepared by this method.

4.6.2 Electropolishing

Outline: First, a bulk specimen is sliced into thin plates of about 0.3 mm in thickness by a fine cutter or a multi-wire saw. A thin plate is further thinned mechanically down to about 0.1 mm in thickness. Electropolishing is performed in a specific solution (electrolyte) by supplying a direct current with the positive pole at the thin plate and the negative pole at a stainless plate. In order to avoid preferential polishing at the edge of the specimen,

all the edges are covered with insulating paint. This is called the window method. The electropolishing is finished when there is a small hole in the plate with very thin regions around it. Nowadays, the so-called twin-jet polishing method is widely used. As shown in Fig. 4.41, the solution is jetted through two small nozzles onto the central part of a specimen plate on both sides. Thus the specimen plate, of 0.1–0.2 mm in thickness and 3 mm in diameter, should be prepared in advance. The plate can be obtained directly with a disc puncher if the specimen is reasonably soft, but otherwise a spark cutter should be used. In a conventional jet polishing machine, when a hole is made in the center of the disc, polishing is automatically stopped by the operation of a photo-cell. Each material needs a specific solution, and polishing is performed at the appropriate temperature and voltage. The solutions to be used for specific materials and the appropriate operation conditions are listed in the references [36].

Notes: Electropolishing is mainly used to prepare thin films of metals and alloys. After electropolishing, the specimen should be washed as soon as possible in methanol or water. When a specimen is not washed correctly, contamination such as an oxide layer forms on the surface. This surface layer produces a strong background in the high-resolution image. The existence of such an oxide layer can be checked from a high-resolution image and an electron diffraction pattern obtained from the edge of the specimen. In order to remove the oxide layer, it is sometimes useful to use ion milling (see Sect. 4.6.5).

4.6.3 Chemical Polishing

Outline: Thinning is performed chemically, i.e., by dipping the specimen in a specific solution. As for electropolishing, a thin plate of 0.1–0.2 mm in thickness should be prepared in advance. If a small dimple is made in the center of the plate with a dimple grinder, a hole can be made by etching around the center while keeping the edge of the specimen relatively thick. Appropriate solutions for specific materials are listed in the references [36].

Notes: This method is frequently used for thinning semiconductors such as silicon. As with electro-polishing, if the specimen is not washed properly after chemical etching, contamination such as an oxide layer forms on the surface. Ion milling is sometimes useful to remove this contamination.

4.6.4 Ultramicrotomy

Outline: Specimens of thin films or powders are usually fixed in resin and trimmed with a glass knife before being sliced with a diamond knife. This process is necessary so that the specimens in the resin can be sliced easily by a diamond knife. Acrylic or epoxy resin is usually used for fixing specimens. Acrylic resin is easily sliced and can be removed with chloroform after slicing. When using an acrylic resin, a gelatin capsule is used as a vessel. Epoxy resin takes less time to solidify than acrylic, and it remains strong under electron irradiation.

Notes: This method has been used for preparing thin sections of biological specimens and sometimes for thin films of inorganic materials which are not too hard to cut. In general, skill is needed to set the geometrical configurations of the dia-

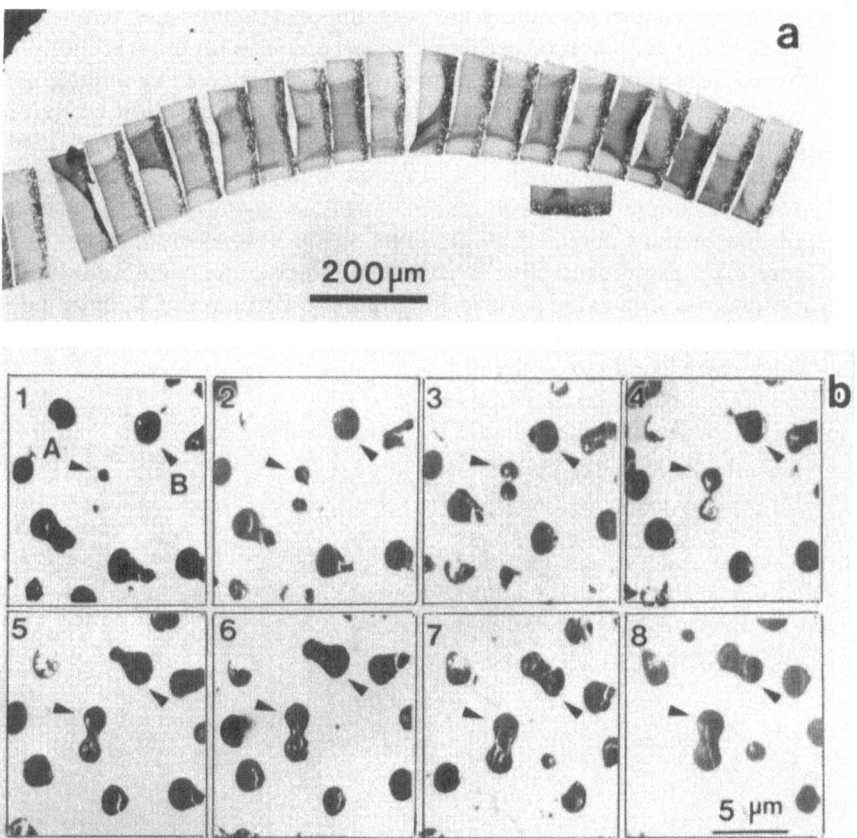

Fig. 4.42. a Successive thin sections of a resin containing small iron oxide particles obtained by ultramicrotomy. **b** Enlarged micrographs of the successive thin sections in **a** showing successive sections of peanut-shaped particles

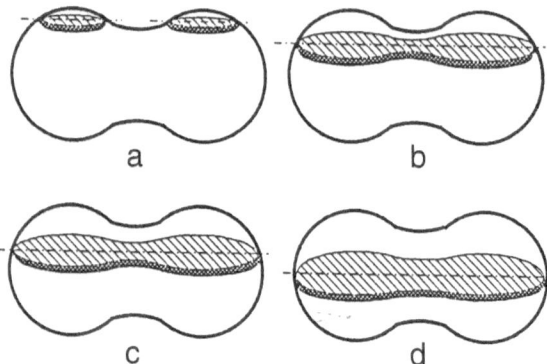

Fig. 4.43. Schematic illustration showing successive sections of a peanut-shaped particle (**a** → **d**)

4.6.5 Ion Milling

Outline: The so-called sputtering phenomenon is used, where atoms are ejected from the surface by irradiation with accelerated ions. First, a thin plate (less than 0.1 mm) is prepared from a bulk specimen by using a diamond cutter and by mechanical thinning. Then, a disk 3 mm in diameter is made from the plate using a diamond knife or a ultrasonic cutter, and a dimple is formed in the center of the surface with a dimple grinder. If it is possible to thin the disk directly to 0.03 mm in thickness by mechanical thinning without using a dimple grinder, the disk should be strengthened by covering the edge with a metal ring. Ar ions are usually used for the sputtering, and the incidence angle against the disk specimen and the accelerating voltage are set as 10°–20° and a few kilovolts, respectively. Recently, the so-called atom milling has been used, where the disk specimen is irradiated with neutral atoms instead of ions. This is especially suitable for semiconductors to reduce the irradiation damage.

Notes: This method is widely used to obtain thin regions of ceramics and semiconductors in particular, and also for cross sections of various multilayer films. When ion milling is continued for some time, a composition change sometimes occurs at the surface due to the difference of spattering efficiencies in the constituent atoms, and amorphous layers can also form on the surface due to ion irradiation damage. In order to avoid these effects, the conditions of the ion milling should be optimized, i.e., by using different ions, lowering the accelerating voltage, and adjusting

mond knife and the specimen after trimming, and to slice a specimen into homogeneous thin sections. It is also necessary to practice picking up thin sections onto a grid. If the trimming and slicing are not carried out correctly, an expensive diamond knife will be damaged. Also, during slicing, lattice strain is frequently introduced into the sections. To help to fix the sections onto the grid, a special grid covered with a collodion or carbon thin film is usually used instead of a holey carbon film, but these types of film give background images on the high-resolution images of thin sections.

Figure 4.42 shows an example of electron microscope images of thin sections obtained by ultramicrotomy. Figure 4.42a shows successive thin sections of a resin containing iron oxide particles, while Fig. 4.42b shows a series of enlarged micrographs, each of which corresponds to a part of a thin section in Fig. 4.42a. In the series of eight micrographs in Fig. 4.42b, sections such as those indicated by A and B clearly show successive sections of peanut-shaped particles, as shown schematically in Fig. 4.43. High-resolution images obtained from these thin sections are presented in Fig. 3.26. Three-dimensional structural information has been obtained from electron microscope images of these thin sections [37].

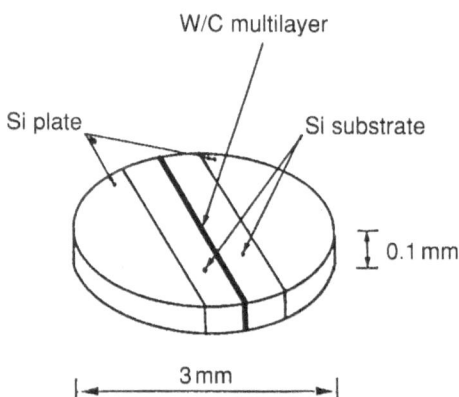

Fig. 4.44. Constituents of a disk containing multilayer films for dimple grinding and ion milling

the incident beam angle. If the incidence angle becomes too small, the metal ring used for strengthening the disk is irradiated with the ions and the specimen disk is coated with the metal. To minimize any increase in specimen temperature during sputtering, the use of a cooling stage with liquid nitrogen is effective. Ion milling is also used as the final process of thinning to remove any contamination of the thin specimen, which is prepared by electropolishing or chemical polishing. Figure 4.44 shows an example of a disk specimen before dimple grinding and ion milling. The specimen is a multilayer film on a Si substrate [38]. In general, thin plates or thin disk specimens are extremely fragile, so in order to minimize any mechanical shock on these specimens one should use the so-called *vacuum tweezers* by which a thin specimen is stuck to a fine tube with low pressure inside. It is easy to pick up a specimen with vacuum tweezers without mechanical shock.

4.6.6 Focused Ion Beam (FIB)

Outline: This method was originally developed for the purpose of fixing semiconductor devices [39]. In principle, ion beams are sharply focused on a small area, and the specimen is thinned very rapidly by sputtering. Figure 4.45 illustrates the incident beam directions of these ions, and of the electrons used in the observations. Usually Ga ions are used, with an accelerating voltage of about 30 kV and a current of about $10\,\text{A/cm}^2$. The probe size is several tens of nanometers.

Notes: This method is currently attracting much attention. It may be especially useful for specimens containing a boundary between different materials, where it may be difficult to thin the boundary region homogeneously by other methods such as ion milling. By detecting the secondary electrons emitted from the specimen while irradiating it with ion beams, a secondary electron image of the surface can be displayed as in a scanning electron microscope. Thus, by observing the secondary electron image, one can accurately select the appropriate region for thinning. Special care should be taken to avoid irradiation damage due to the strong ion beams, and also the implantation of Ga ions. This system of focused ion beams is expensive compared with other thinning instruments.

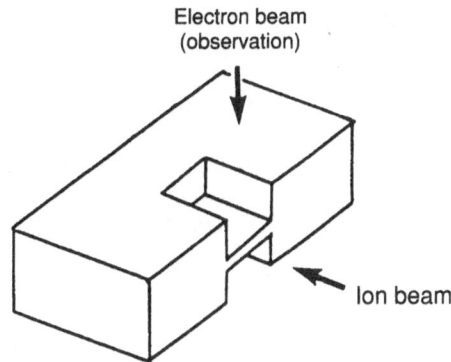

Electron beam (observation)

Ion beam

Fig. 4.45. Incident beam directions of ions for thinning and electrons for observations in the focused ion beam (FIB) method

4.6.7 Vacuum Evaporation

Outline: The specimen is set in a W-coil or basket. Resistance heating is applied by an electric current passing through the coil or basket, and the specimen is melted, then evaporated (or sublimated), and finally deposited onto a substrate. The deposition process is usually carried under a pressure of 10^{-3}–$10^{-4}\,\text{Pa}$, but in order to avoid surface contamination, a very high vacuum is necessary. A collodion film or cleaved rock salt is used as a substrate. Rock salt is especially useful in forming single crystals with a special orientation relationship between each crystal and the substrate. Salt is easily dissolved in water, and then the deposited films can be fixed on a grid. Recently, as an alternative to resistance heating, electron beam heating or an ion beam sputtering method has been used to prepare thin films of various alloys.

Notes: This method is used for preparing homogeneous thin films of metals and alloys, and is also used for coating a specimen with the metal or alloy. In the resistance heating method, assuming that the specimen is evaporated homogeneously in all directions, the specimen thickness is simply estimated by the following equation:

$$t = \frac{M}{4\pi R^2 \varrho} \qquad (4.39)$$

where R (cm) is the distance between a specimen and the substrate, and M (g) and ϱ (g/cm^2) are the specimen mass and density, respectively. The above equation gives a rough estimate of the specimen thickness, but for a more accurate thickness measurement, a quartz crystal film thickness monitor should be used.

References

1. Shindo D, Hiraga K, Iijima S, Kudoh J, Nemoto Y, Oikawa T (1993) J Electron Microsc 42:227
2. Gomyo A, Suzuki T, Iijima S (1988) Phys Rev Lett 60:2645
3. Shindo D, Hiraga K, Hirabayashi M (1984) Sci Rep RITU A32:32
4. Frank J (1980) In: Hawkes PW (ed) Computer processing of electron microscope images. Springer, Berlin, p 187
5. Saxton WO, Koch TL, (1982) J Microsc 127:69
6. Tanaka N, Maki T, Mihama K, Tsuno K (1989) J Electron Microsc 38:54
7. Saxton WO (1988) Scanning Microsc Suppl 2:213
8. Taniguchi Y, Yakai Y, Ikuta T, Shimizu R (1992) J Electron Microsc 41:21
9. Yasuami S, Koga K, Ohshima K, Sasaki S, Ando M (1992) J Appl Crystallogr 25:514
10. Sonoda M, Takano M, Miyahara J, Kato H (1983) Radiology 148:833
11. Mori N, Oikawa T, Katoh T, Miyahara J, Harada Y (1988) Ultramicroscopy 25:195
12. Ogura N, Yoshida K, Kojima Y, Saito H (1994) Proceedings of the 13th International Congress on Electron Microscopy, Les Editions de physique, Les Ulis, France vol 1, p 219
13. Mooney PE, Fan GY, Mayer CE, Truong KV, Bui DB, Krivanek OL (1990) Proceedings of the 12th International Congress for Electron Microscopy, San Francisco Press, Seattle vol 1, p 164
14. Taniyama A, Shindo D, Oikawa T (1997) J Electron Microsc 46:303
15. Zuo JM, MacCartney MR, Spence JCH (1996) Ultramicroscopy 66:35
16. Ishizuka K (1993) Ultramicroscopy 52:7
17. Zuo JM (1996) Ultramicroscopy 66:21
18. Oikawa T, Shindo D, Hiraga K (1994) J Electron Microsc 43:402
19. Hofmann D, Ernst F (1994) Ultramicroscopy 53:205
20. Shindo D, Oku T, Kudoh J, Oikawa T (1994) Ultramicroscopy 54:221
21. Shindo D, Oku T, Hiraga K, Oikawa T (1994) Proceedings of the 13th International Congress on Electron Microscopy, Les Editions de physique, Les Ulis, France vol 1, p 359
22. Shindo D, Hiraga K, Oikawa T, Mori N (1990) J Electron Microsc 39:449
23. Kikuchi S (1928) Proc Imp Acad Jpn 4:271, 275, 354, 471
24. Hiraga K, Hirabayashi M (1980) J Appl Crystallogr 13:17
25. Cockayne DJH, Ray ILF, Whelan MJ (1968) Proceedings of the 4th European Regional Conference on Electron Microscopy, Tipografia Poliglotta Vaticana, Roma, p 129
26. Howie A, Basinski ZS (1968) Philos Mag 17:1039
27. Cockayne DJH, Ray ILF, Whelan MJ (1969) Philos Mag 20:1265
28. Alexander H, Spence JCH, Shindo D, Gottschalk H, Long N (1986) Philos Mag A53:627
29. Head AK, Humble P, Clarebrough LM, Morton AJ, Forwood CT (1973) Computed electron micrographs and defect identification. North-Holland, Amsterdam
30. Uyeda R, Nonoyama M, Kojiso M (1965) J Electron Microsc 14:296
31. Høier R (1969) Acta Crystallogr A25:516
32. Pumphrey PH (1972) Proceedings of the 5th European Regional Conference on Electron Microscopy, The Institute of Physics, London, p 466
33. Krivanek OL (1976) Optik 45:97
34. Park GS, Shindo D (1996) J Electron Microsc 45:152
35. Troyon M (1978/79) Optik, 52:401
36. Hirsch PB, Howie A, Nicholson RB, Pashley DW, Whelan MJ (1965) Electron microscopy of thin crystals. Butterworths, London, p 455
37. Park GS, Shindo D, Waseda Y (1994) J Electron Microsc 43:208
38. Oshino T, Shindo D, Hirabayashi M, Aoyagi E, Nikaido H (1989) Jpn J Appl Phys 28:1909
39. Kirk ECG, Cleaver JRA, Ahmed H (1987) Inst Phys Conf Ser 84:691

Appendix

Appendix A. Physical Constants, Conversion Factors and Electron Wavelength

1. Physical Constants

		SI units	CGS units
Electron charge (e)	= 1.6022	$\times\,10^{-19}\,$C	$\times\,10^{-20}\,$emu
	= 4.8032		$\times\,10^{-10}\,$esu
Electron mass (m_e)	= 9.1094	$\times\,10^{-31}\,$kg	$\times\,10^{-28}\,$g
Proton mass (m_p)	= 1.6726	$\times\,10^{-27}\,$kg	$\times\,10^{-24}\,$g
Neutron mass (m_n)	= 1.6749	$\times\,10^{-27}\,$kg	$\times\,10^{-24}\,$g
Velocity of light (c)	= 2.9979	$\times\,10^{8}\,$m s^{-1}	$\times\,10^{10}\,$cm s^{-1}
Mass energy of an electron ($m_e c^2$)	= 8.1871	$\times\,10^{-14}\,$J	$\times\,10^{-7}\,$erg
	(= 0.51100 MeV)		
Planck's constant (h)	= 6.6261	$\times\,10^{-34}\,$Js	$\times\,10^{-27}\,$erg·s
($\hbar = h/2\pi$)	= 1.0546	$\times\,10^{-34}\,$Js	$\times\,10^{-27}\,$erg·s
Compton wavelength ($\lambda_c = h/m_e c$)	= 2.4263	$\times\,10^{-12}\,$m	$\times\,10^{-10}\,$cm
Avogadro's number (N_A)	= 6.0221	$\times\,10^{23}\,$mol^{-1}	$\times\,10^{23}\,$mol^{-1}

2. Conversion Factors

$1\,\text{eV} = 1.6022 \times 10^{-19}\,$J $1\,\text{Å} = 0.1\,$nm
$1\,\text{Torr} = 133.32\,$Pa $1\,\text{kX} = 0.10020\,$nm

3. Electron Wavelength and Interaction Constant

Accelerating voltage, V (kV)	Wavelength, λ (nm)	$\sqrt{1-\beta^2}$	Interaction constant, σ (V^{-1}nm^{-1})
80	0.00417572	0.86464	0.0100871
100	0.00370144	0.83633	0.0092440
120	0.00334922	0.80983	0.0086381
150	0.00295704	0.77307	0.0079892
180	0.00266550	0.73951	0.0075284
200	0.00250793	0.71871	0.0072884
300	0.00196875	0.63009	0.0065262
400	0.00164394	0.56092	0.0061214
500	0.00142126	0.50544	0.0058732
600	0.00125680	0.45995	0.0057072
700	0.00112928	0.42196	0.0055897
800	0.00102695	0.38978	0.0055030
900	0.00094269	0.36215	0.0054368
1000	0.00087192	0.33819	0.0053850
1250	0.00073571	0.29018	0.0052956
1300	0.00071361	0.28216	0.0052824
1500	0.00063745	0.25410	0.0052397
2000	0.00050432	0.20350	0.0051760
2500	0.00041783	0.16971	0.0051423
3000	0.00035693	0.14554	0.0051223

Appendix B. Geometry of Crystal Lattice

System		Unit cell volume, V
Cubic $a = b = c$ $\alpha = \beta = \gamma = 90°$	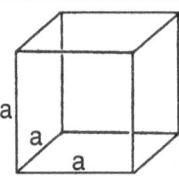	$V = a^3$
Tetragonal $a = b \neq c$ $\alpha = \beta = \gamma = 90°$	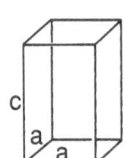	$V = a^2c$
Hexagonal $a = b \neq c$ $\alpha = \beta = 90°, \gamma = 120°$	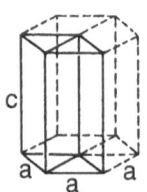	$V = \dfrac{\sqrt{3}a^2c}{2} = 0.866a^2c$
Rhombohedral $a = b = c$ $\alpha = \beta = \gamma < 120°, \neq 90°$	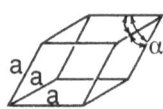	$V = a^3\sqrt{1 - 3\cos^2\alpha + 2\cos^3\alpha}$
Orthorhombic $a \neq b \neq c$ $\alpha = \beta = \gamma = 90°$	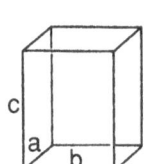	$V = abc$
Monoclinic $a \neq b \neq c$ $\alpha = \gamma = 90° \neq \beta$	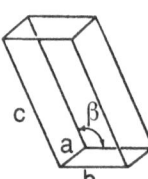	$V = abc \sin\beta$
Triclinic $a \neq b \neq c$ $\alpha \neq \beta \neq \gamma$	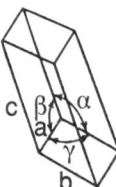	$V = abc\sqrt{1 - \cos^2\alpha - \cos^2\beta - \cos^2\gamma + 2\cos\alpha\cos\beta\cos\gamma}$

Distance of lattice planes, d	Angle of lattice planes, ϕ
$$\frac{1}{d^2} = \frac{h^2 + k^2 + l^2}{a^2}$$	$$\cos\phi = \frac{h_1 h_2 + k_1 k_2 + l_1 l_2}{\sqrt{\left(h_1^2 + k_1^2 + l_1^2\right)\left(h_2^2 + k_2^2 + l_2^2\right)}}$$
$$\frac{1}{d^2} = \frac{h^2 + k^2}{a^2} + \frac{l^2}{c^2}$$	$$\cos\phi = \frac{\dfrac{h_1 h_2 + k_1 k_2}{a^2} + \dfrac{l_1 l_2}{c^2}}{\sqrt{\left(\dfrac{h_1^2 + k_1^2}{a^2} + \dfrac{l_1^2}{c^2}\right)\left(\dfrac{h_2^2 + k_2^2}{a^2} + \dfrac{l_2^2}{c^2}\right)}}$$
$$\frac{1}{d^2} = \frac{4}{3}\left(\frac{h^2 + hk + k^2}{a^2}\right) + \frac{l^2}{c^2}$$	$$\cos\phi = \frac{h_1 h_2 + k_1 k_2 + \frac{1}{2}\left(h_1 k_2 + h_2 k_1\right) + \dfrac{3a^2}{4c^2} l_1 l_2}{\sqrt{\left(h_1^2 + k_1^2 + h_1 k_1 + \dfrac{3a^2}{4c^2} l_1^2\right)\left(h_2^2 + k_2^2 + h_2 k_2 + \dfrac{3a^2}{4c^2} l_2^2\right)}}$$
$$\frac{1}{d^2} = \frac{\left(h^2 + k^2 + l^2\right)\sin^2\alpha}{a^2\left(1 - 3\cos^2\alpha + 2\cos^3\alpha\right)} + \frac{2(hk + kl + hl)\left(\cos^2\alpha - \cos\alpha\right)}{a^2\left(1 - 3\cos^2\alpha + 2\cos^3\alpha\right)}$$	$$\cos\phi = \frac{a^4 d_1 d_2}{V^2}\Big[\sin^2\alpha\left(h_1 h_2 + k_1 k_2 + l_1 l_2\right) \\ + \left(\cos^2\alpha - \cos\alpha\right)\left(k_1 l_2 + k_2 l_1 + l_1 h_2 + l_2 h_1 + h_1 k_2 + h_2 k_1\right)\Big]$$
$$\frac{1}{d^2} = \frac{h^2}{a^2} + \frac{k^2}{b^2} + \frac{l^2}{c^2}$$	$$\cos\phi = \frac{\dfrac{h_1 h_2}{a^2} + \dfrac{k_1 k_2}{b^2} + \dfrac{l_1 l_2}{c^2}}{\sqrt{\left(\dfrac{h_1^2}{a^2} + \dfrac{k_1^2}{b^2} + \dfrac{l_1^2}{c^2}\right)\left(\dfrac{h_2^2}{a^2} + \dfrac{k_2^2}{b^2} + \dfrac{l_2^2}{c^2}\right)}}$$
$$\frac{1}{d^2} = \frac{1}{\sin^2\beta}\left(\frac{h^2}{a^2} + \frac{k^2\sin^2\beta}{b^2} + \frac{l^2}{c^2} - \frac{2hl\cos\beta}{ac}\right)$$	$$\cos\phi = \frac{d_1 d_2}{\sin^2\beta}\left[\frac{h_1 h_2}{a^2} + \frac{k_1 k_2 \sin^2\beta}{b^2} + \frac{l_1 l_2}{c^2} - \frac{\left(l_1 h_2 + l_2 h_1\right)\cos\beta}{ac}\right]$$
$$\frac{1}{d^2} = \frac{1}{V^2}\big(S_{11}h^2 + S_{22}k^2 + S_{33}l^2 + 2S_{12}hk + 2S_{23}kl + 2S_{13}hl\big)$$	$$\cos\phi = \frac{d_1 d_2}{V^2}\big[S_{11}h_1 h_2 + S_{22}k_1 k_2 + S_{33}l_1 l_2 \\ + S_{23}\left(k_1 l_2 + k_2 l_1\right) + S_{13}\left(l_1 h_2 + l_2 h_1\right) + S_{12}\left(h_1 k_2 + h_2 k_1\right)\big]$$

$$S_{11} = b^2 c^2 \sin^2\alpha, \qquad S_{12} = abc^2\left(\cos\alpha \cos\beta - \cos\gamma\right),$$
$$S_{22} = a^2 c^2 \sin^2\beta, \qquad S_{23} = a^2 bc\left(\cos\beta \cos\gamma - \cos\alpha\right),$$
$$S_{33} = a^2 b^2 \sin^2\gamma, \qquad S_{13} = ab^2 c\left(\cos\gamma \cos\alpha - \cos\beta\right).$$

Appendix C. Typical Structures in Materials and Their Electron Diffraction Patterns

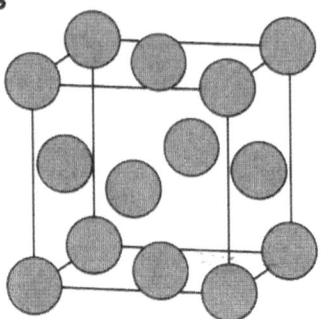

Face-centered cubic structure
(Al, Cu, Au, Pb, Ni etc.)

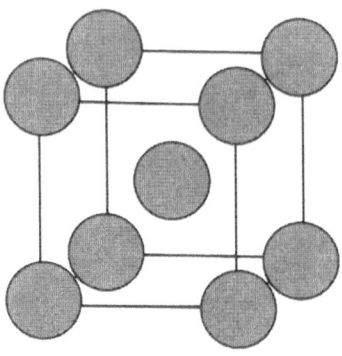

Body-centered cubic structure
(Fe, Mo, Nb, Na, V etc.)

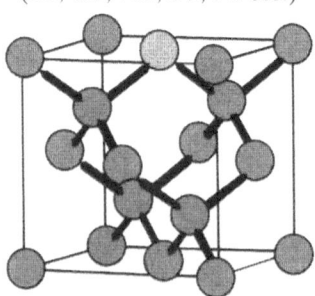

Diamond structure
(Si, Ge etc.)

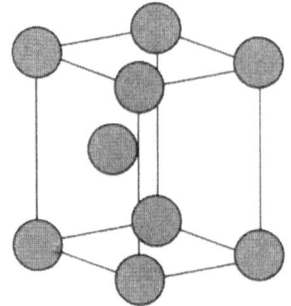

Hexagonal close-packed structure
(Cd, Mg, Ti, Zn, Zr etc.)

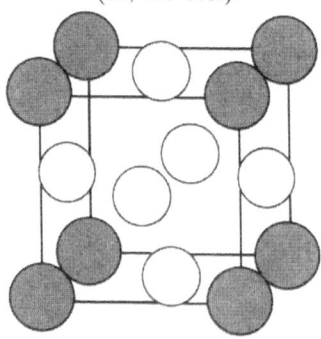

Cu_3Au-type structure
(L1$_2$-type: Cu_3Au, Ni_3Al, Ni_3Fe,
Ni_3Mn, Ni_3Pt etc.)

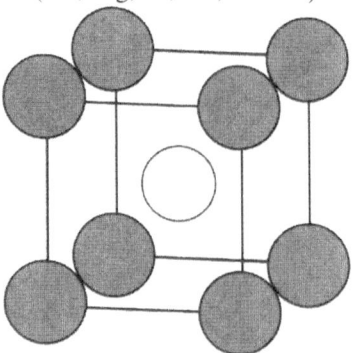

CsCl-type structure (L2$_0$-type or B2-type: CuZn,
AlFe, FeCo, CuPd, AuMn etc.)

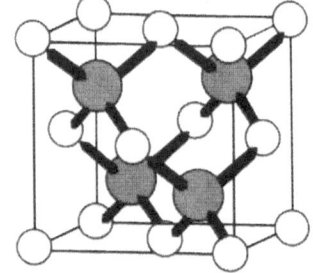

ZnS-type structure
(B3-type: ZnS, β-SiC, GaAs etc.)

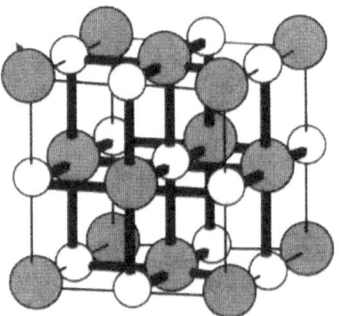

NaCl-type structure
(B1-type: TiC, NbC, VC etc.)

Fig. C.1. Typical basic structures and superstructures

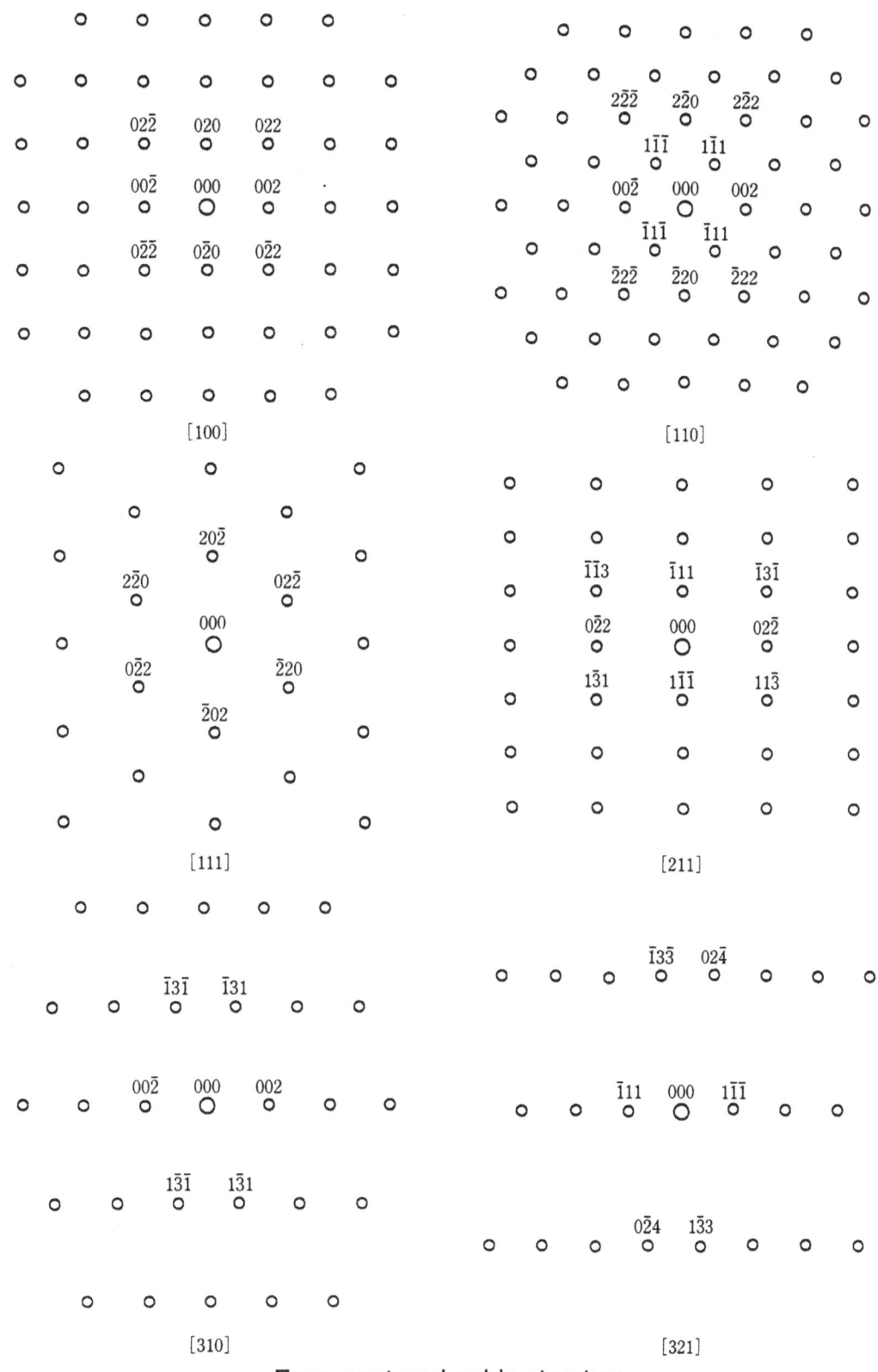

Face-centered cubic structure

Fig. C.2. Electron diffraction patterns of some basic structures. Small dark dots indicate double-diffraction spots. Incident beam directions are indicated in reciprocal space

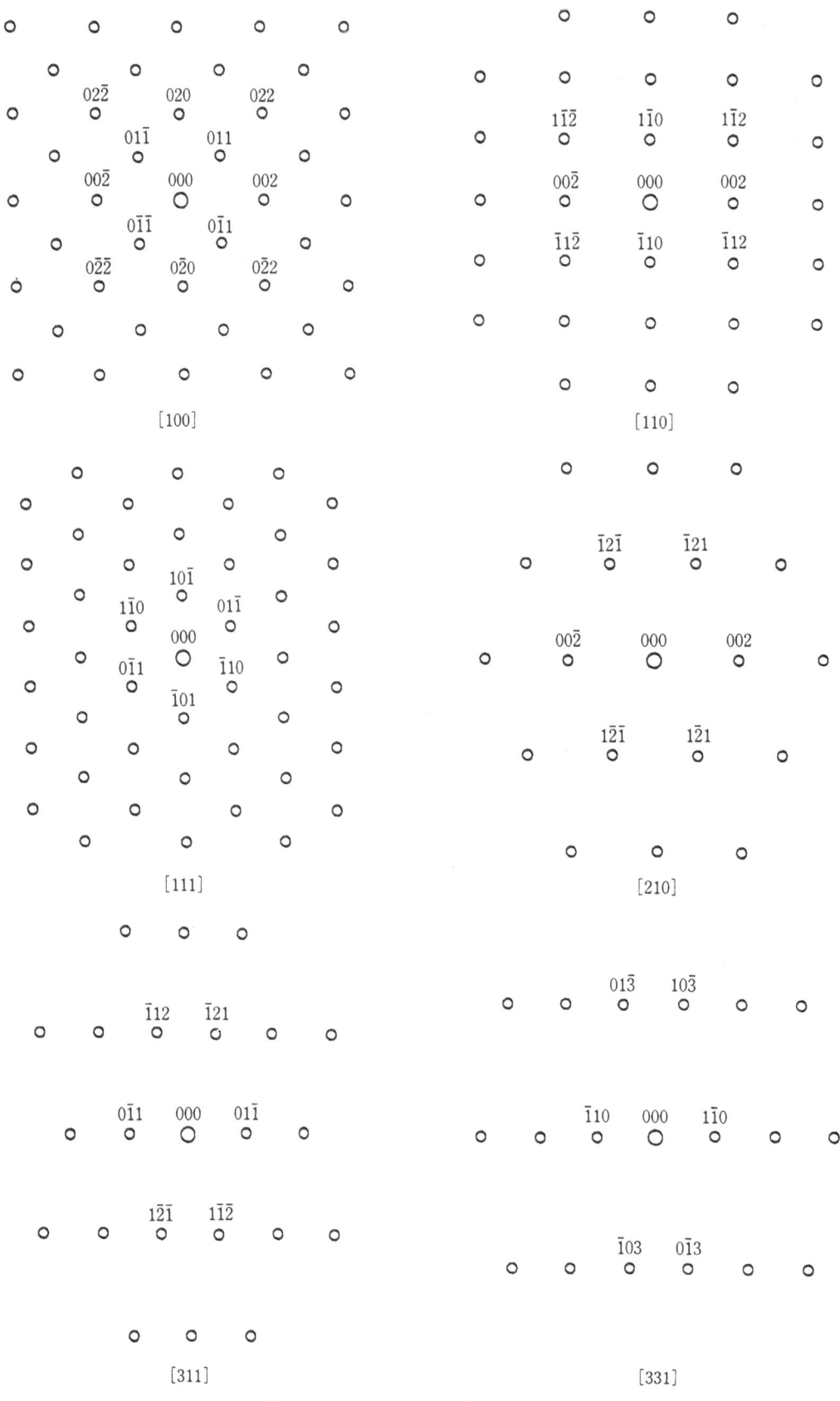

Body-centered cubic structure

Fig. C.2. *Continued*

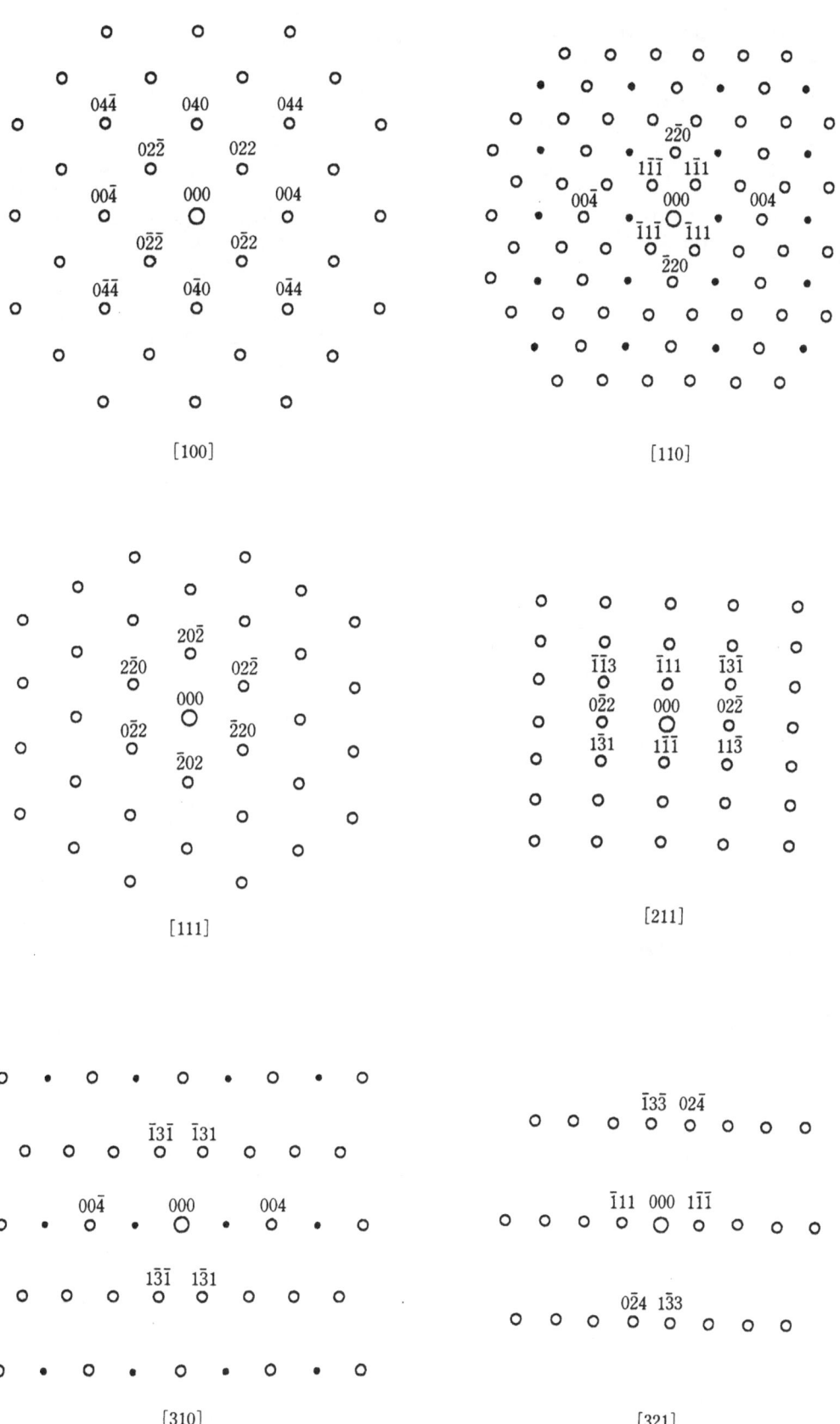

Diamond structure

Fig. C.2. *Continued*

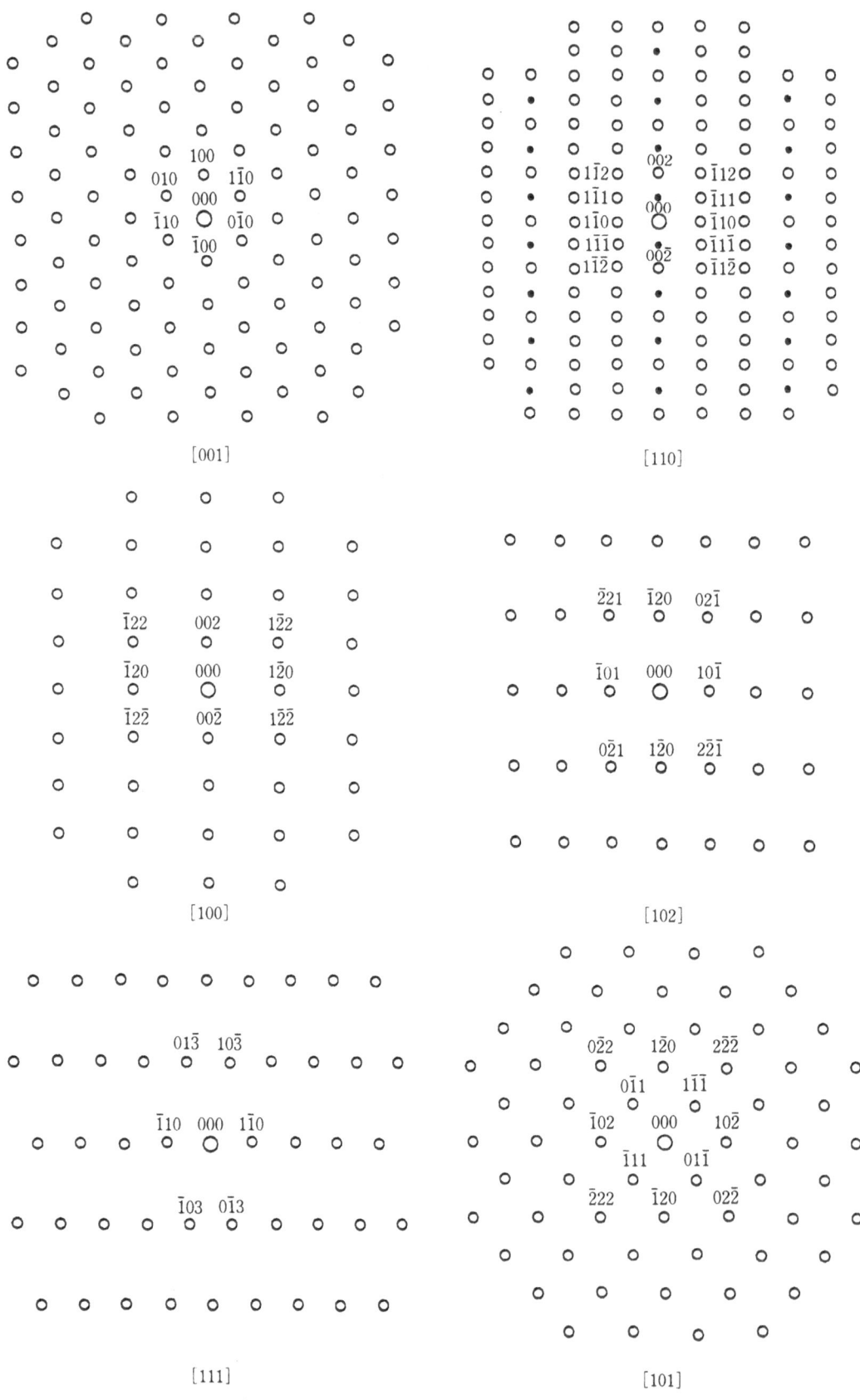

Hexagonal close-packed structure

Fig. C.2. *Continued*

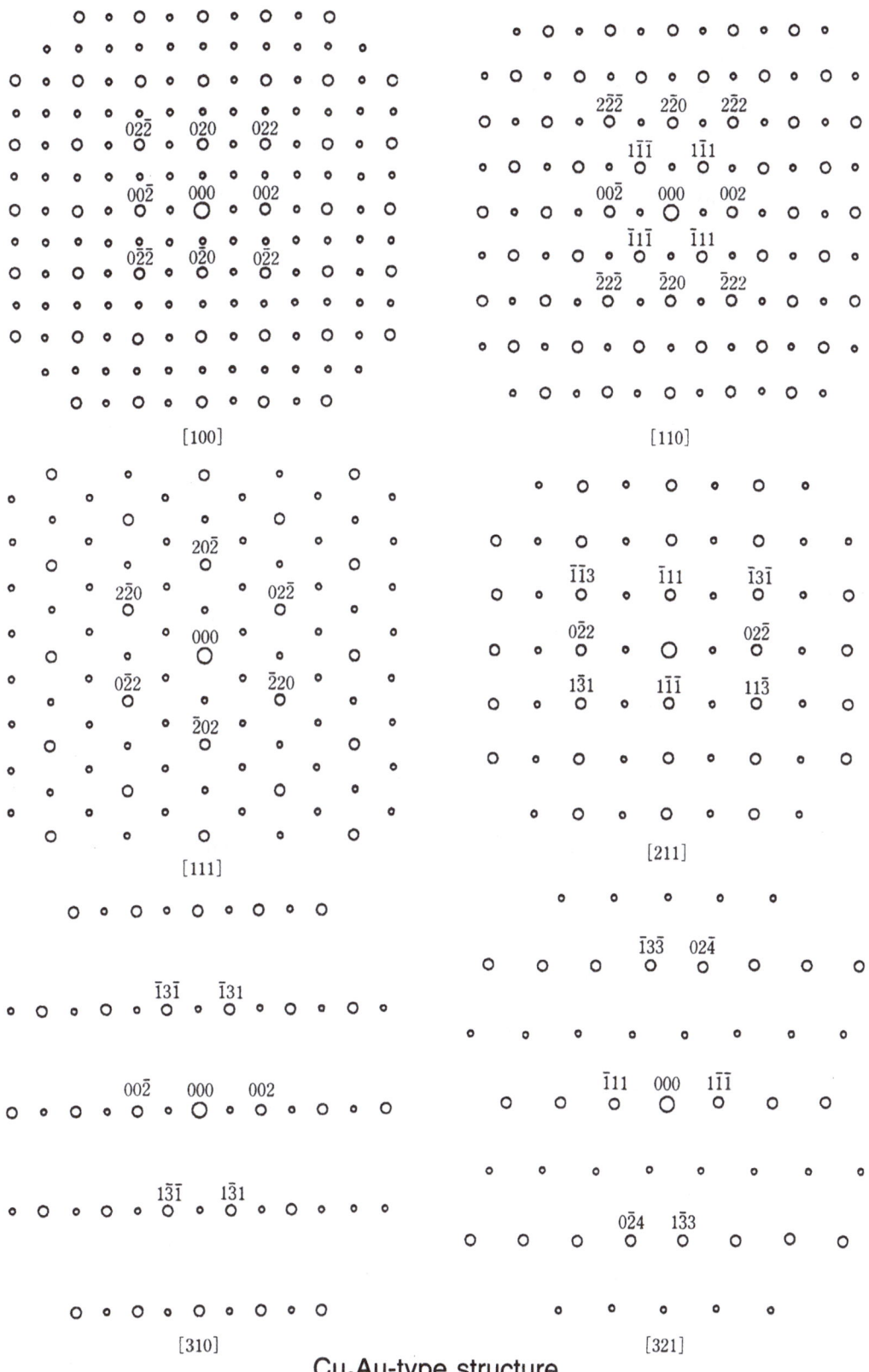

Cu₃Au-type structure

Fig. C.3. Electron diffraction patterns of various superstructures. Large and small circles indicate fundamental and superlattice reflections, respectively

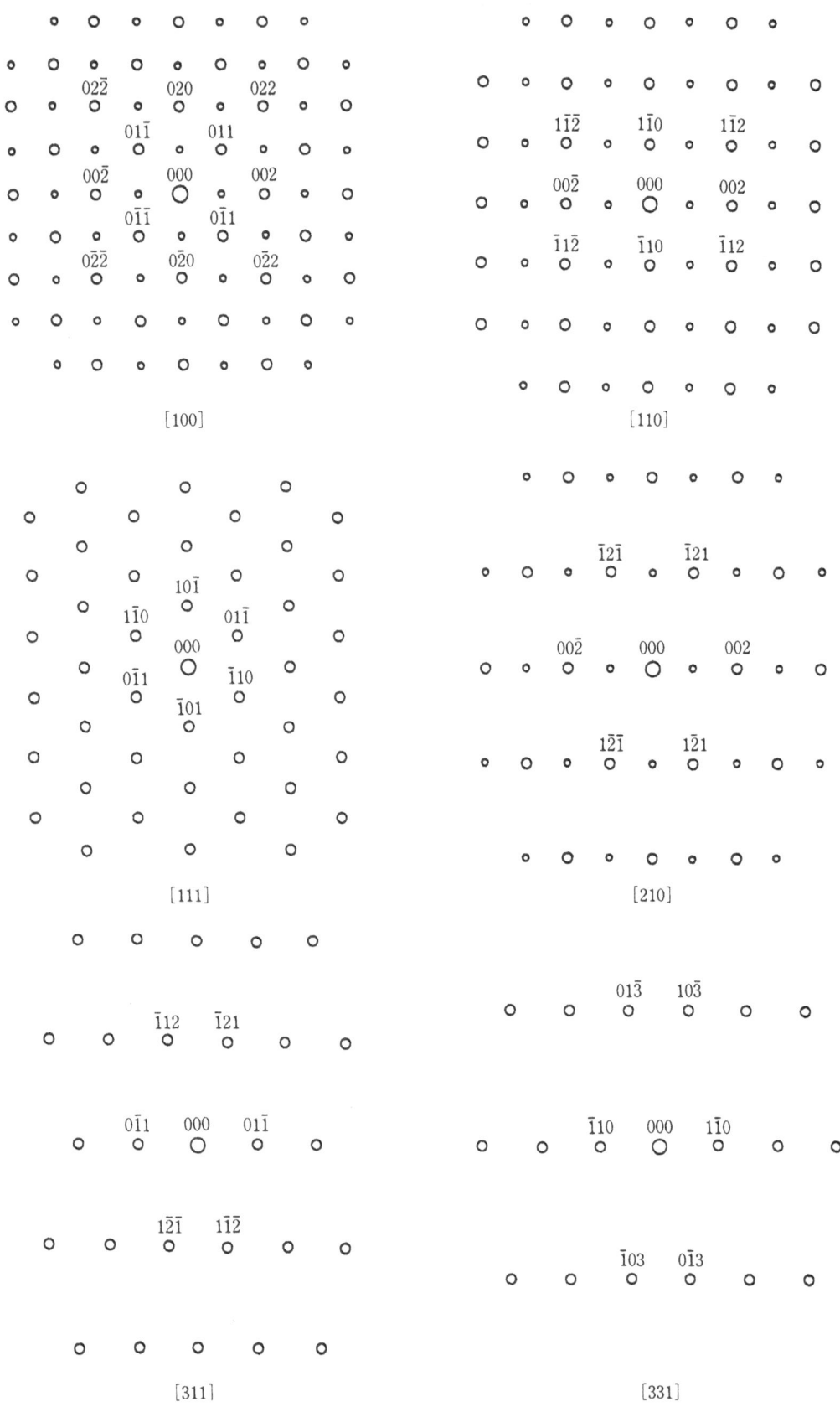

CsCl-type structure

Fig. C.3. *Continued*

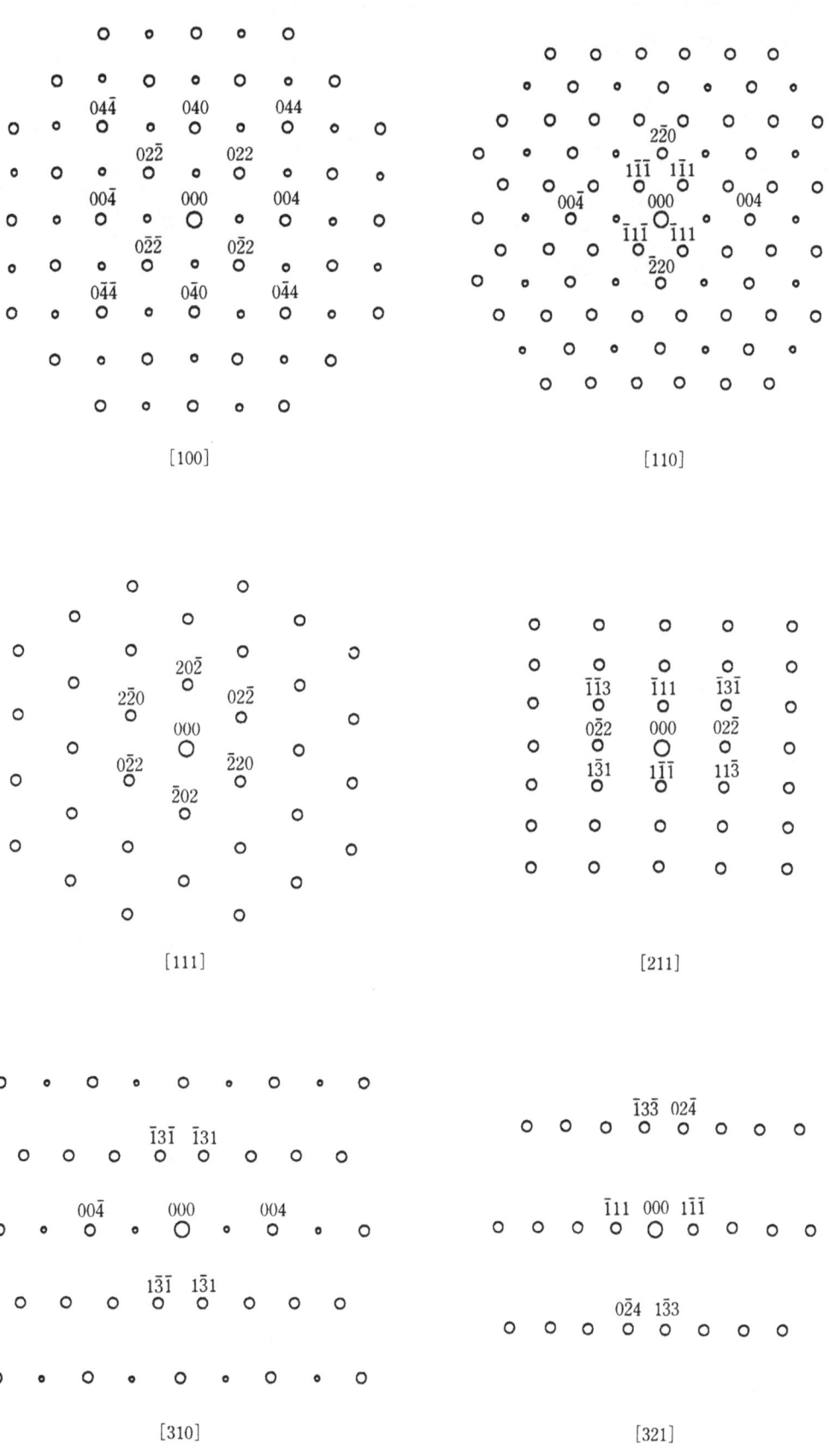

ZnS-type structure

Fig. C.3. *Continued*

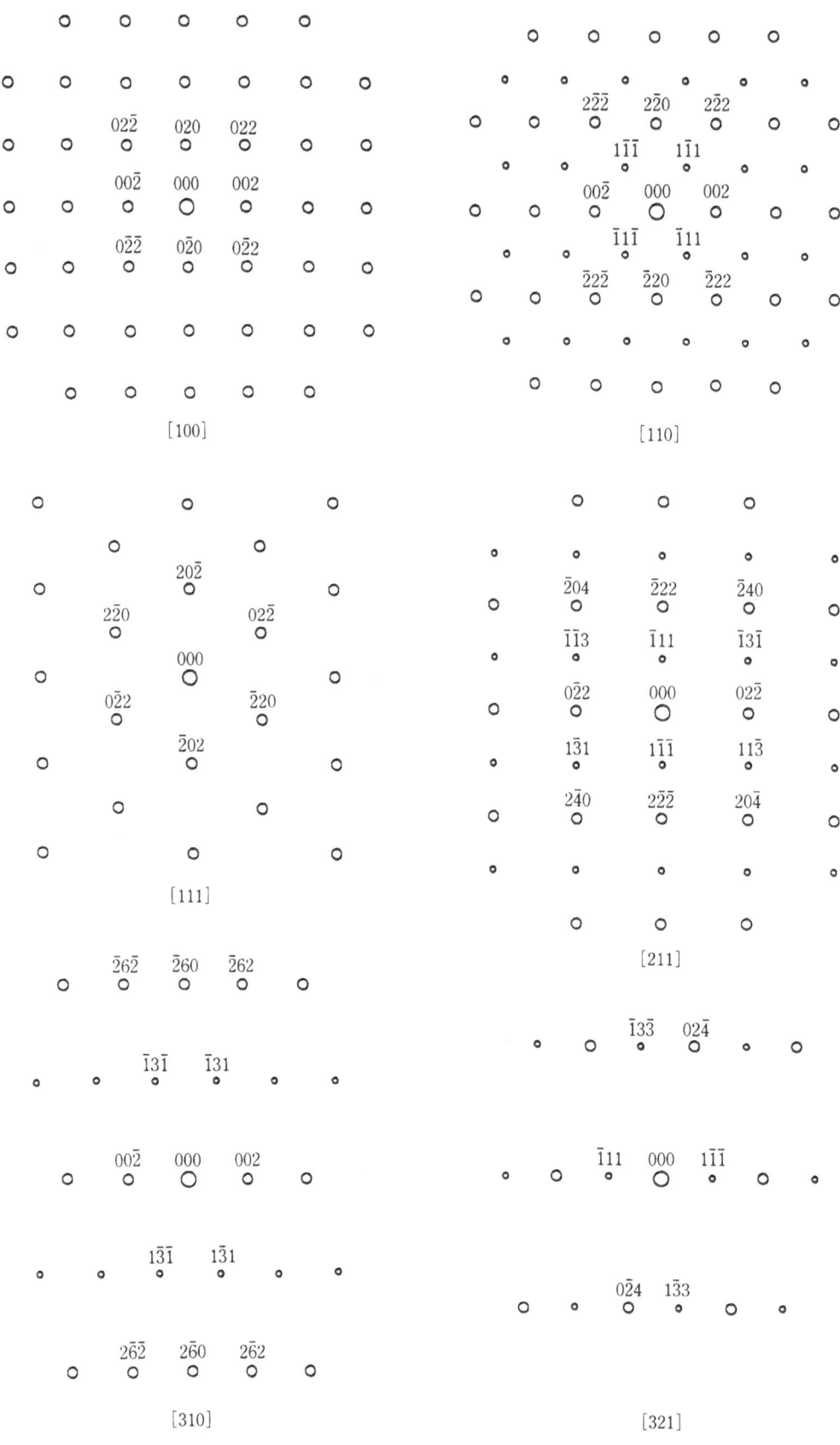

NaCl-type structure

Fig. C.3. *Continued*

Appendix D. Properties of Fourier Transform

In order to illustrate the properties of Fourier transform, some two-dimensional objects $q(x, y)$ and the amplitudes of their Fourier transform $\left|Q(u, v)\right| = \left|\mathcal{F}\left[q(x, y)\right]\right|$ are presented below. It should be noted that according to a Babinet's principle, the amplitude of a Fourier transform of a circular hole ($q_1(x, y)$) is equal to that of a circular object ($1 - q_1(x, y)$) except at the origin ($u, v = 0$).

Two-dimensional object; $q(x, y)$

$$q_1(x, y) = \begin{array}{l} 1, \ \sqrt{x^2 + y^2} \leq R \\ 0, \ \sqrt{x^2 + y^2} > R \end{array}$$

Amplitude of the Fourier transform of the object; $\left|Q(u, v)\right| = \left|\mathcal{F}\left[q(x, y)\right]\right|$

$$\left|Q_1(u, v)\right| = \pi R^2 \frac{\left|J_1\left(2\pi R\sqrt{u^2 + v^2}\right)\right|}{2\pi R\sqrt{u^2 + v^2}}$$

$\left(J_1\text{: Bessel function of the first order and kind}\right)$

(1) Circular hole

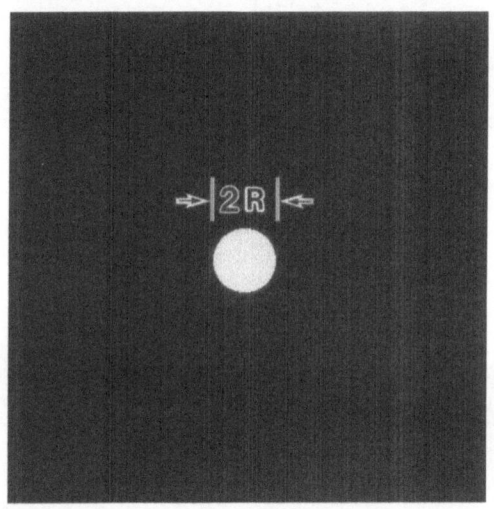

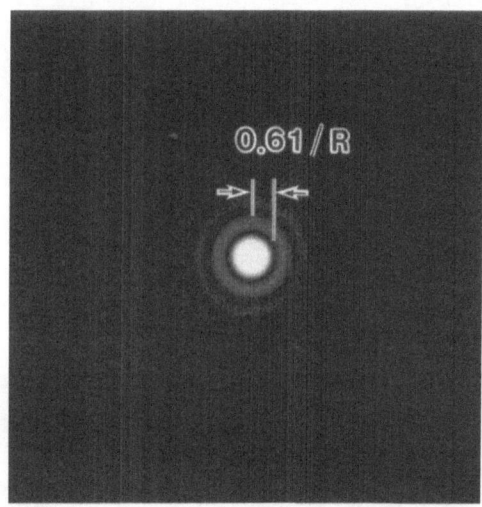

When R is large

(2) Rectangle hole

$$q_2(x, y) = \begin{array}{l} 1, \; |x| \le a/2 \text{ and } |y| \le b/2 \\ 0, \; |x| > a/2 \text{ or } |y| > b/2 \end{array} \qquad \left| Q_2(u, v) \right| = \left| ab \frac{\sin \pi au}{\pi au} \cdot \frac{\sin \pi bv}{\pi bv} \right|$$

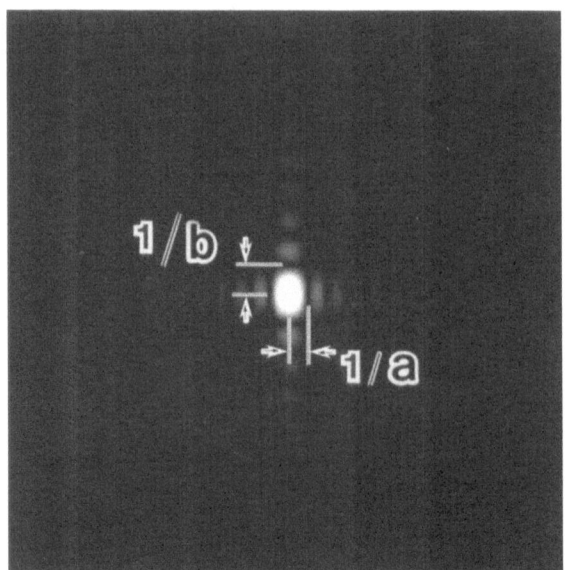

When a is much larger than b

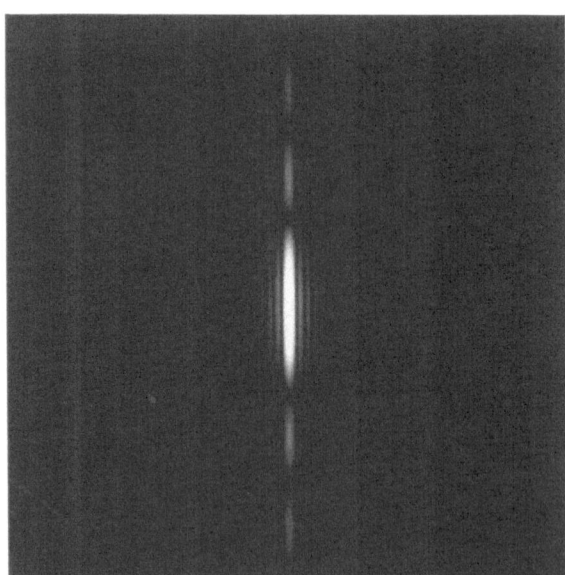

(3) Periodic array of delta functions

$$q_3(x, y) = \sum_{m,n=-\infty}^{\infty} \delta(x - md, y - nd)$$

$$|Q_3(u, v)| = \frac{1}{d^2} \sum_{m,n=-\infty}^{\infty} \delta\left(u - \frac{m}{d}, v - \frac{n}{d}\right)$$

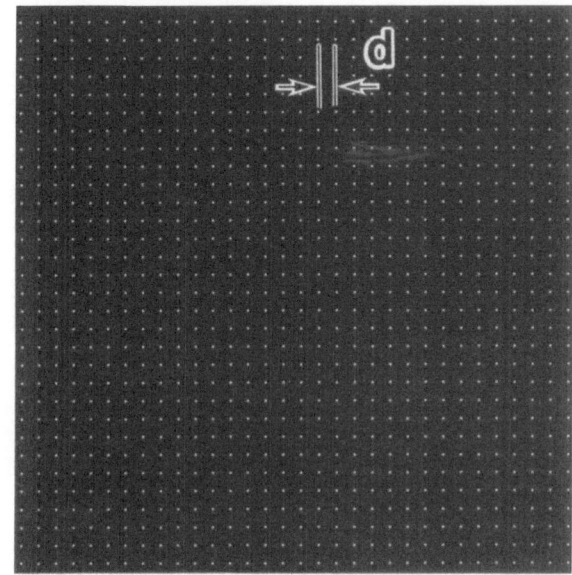
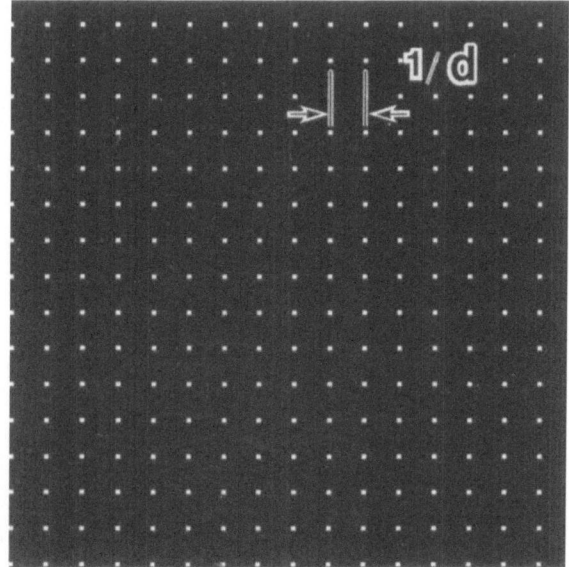

(4) Gaussian function

$$q_4(x, y) = \exp\left[-\left(x^2 + y^2\right)/s^2\right]$$

$$|Q_4(u, v)| = \pi s^2 \exp\left[-\pi^2 s^2\left(u^2 + v^2\right)\right]$$

(5) Sine wave

$$q_5(x, y) = 1 + \sin(2\pi x/p)$$

$$\left|Q_5(u, v)\right| = \delta(u, v)$$
$$+ \frac{1}{2}\left[\delta\left(u + \frac{1}{p}, v\right) + \delta\left(u - \frac{1}{p}, v\right)\right]$$

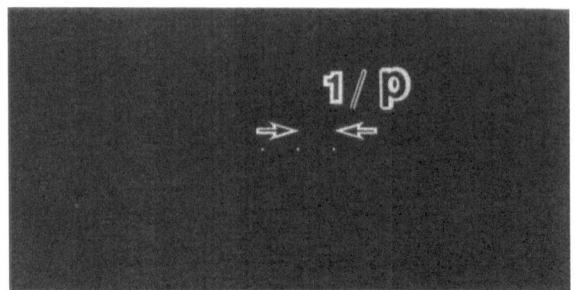

(6) Periodic array of marrow slits

$$q_6(x, y) = q_2(x, y) * q_3(x, y)$$
$$\text{for } q_2, b \rightarrow \infty$$

$$\left|Q_6(u, v)\right| = \left|Q_2(u, v) \cdot Q_3(u, v)\right|$$

(7) Periodic array of small rectangle holes in an aperture

$$q_7(x, y) = \left(q_3(x, y) * q_2(x, y)\right) \cdot q_1(x, y)$$

$$\left|Q_7(u, v)\right| = \left|\left(Q_3(u, v) \cdot Q_2(u, v)\right) * Q_1(u, v)\right|$$

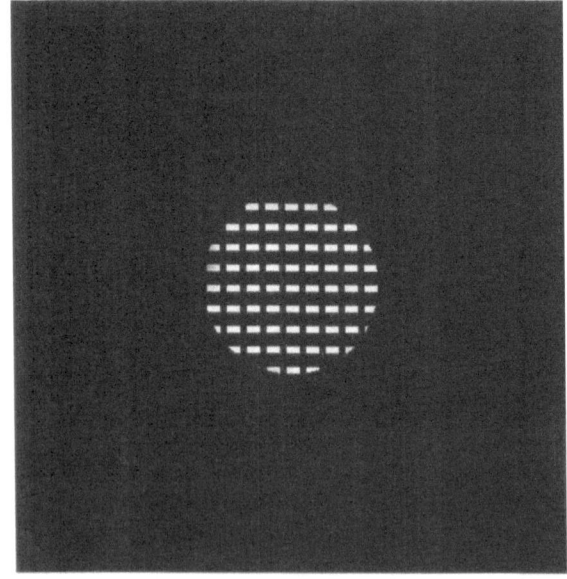
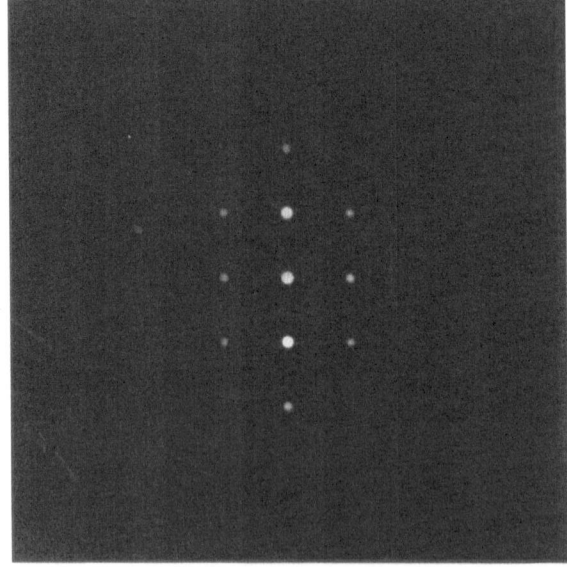

Appendix E. Sign Conventions

In this book, as noted in Sect. 1.2, the wave function of the incident beam is set to be $\exp[i(\boldsymbol{kr} - \omega t)]$. However, in some literature, it is set to $\exp[-i(\boldsymbol{kr} - \omega t)]$. A choice of sign in the phase of the wave function must be made because the sign conventions fix the signs of the transmission function (Eq. 1.6), the propagation function (Eq. 1.38), and the contrast transfer function, which are necessary for computer simulation. The two choices [1] are listed in Table E1. In a similar manner, two choices are used in the Fourier transform and the calculation of structure factors, as shown in Table E2. The first line (quantum mechanical notation) is adopted in this book, and satisfies the Schrödinger wave equation

$$ i\hbar \frac{\partial \psi}{\partial t} = -\frac{\hbar^2}{2m} \frac{\partial^2 \psi}{\partial \boldsymbol{r}^2} \qquad (E1) $$

On the other hand, the second line does not satisfy the above equation, but it has been used for a long time, especially in optics. For example, the International Tables for X-Ray Crystallography [2] adopts the second line. The second line is also consistent with quantum mechanical formulation if it does not include time. For computer simulation, it is important to confirm which notation is adopted.

If one uses the first line in Table E1 and the second line in Table E2 (or the second line in Table E1 and the first line in Table E2), a diffraction pattern rotated through 180° is obtained. Also, the defocus Δf is positive for underfocus in this book, while in some literature it is negative for underfocus.

References

1. Saxton WO, O'Keefe MA, Cockayne DJH, Wilkens M (1984) Ultramicroscopy 13:349
2. International Tables for X-Ray Crystallography (1974) Kynoch Press, Birmingham, England vol. IV

Table E1. Wave conventions

	Quantum-mechanical convention	Alternative convention
Incident wave	$\exp[i(\boldsymbol{k}\cdot\boldsymbol{r} - \omega t)]$	$\exp[-i(\boldsymbol{k}\cdot\boldsymbol{r} - \omega t)]$
Transmission function	$\exp[i\sigma\varphi(\boldsymbol{r})\Delta z]$	$\exp[-i\sigma\varphi(\boldsymbol{r})\Delta z]$
Propagation function	$\exp[-i\pi\lambda\Delta z|\boldsymbol{u}|^2]$	$\exp[i\pi\lambda\Delta z|\boldsymbol{u}|^2]$
Contrast transfer function	$\exp[i\chi(\boldsymbol{u})]$	$\exp[-i\chi(\boldsymbol{u})]$

$$ \chi(\boldsymbol{u}) = \pi\left\{ \Delta f \lambda |\boldsymbol{u}|^2 - 0.5 C_s \lambda^3 |\boldsymbol{u}|^4 \right\} $$

Table E2. Fourier transform conventions

	Quantum-mechanical convention	Alternative convention
Real space to reciprocal space	$\int \psi(\boldsymbol{r})\exp(-2\pi i \boldsymbol{u}\cdot\boldsymbol{r})d\boldsymbol{r}$	$\int \psi(\boldsymbol{r})\exp(2\pi i \boldsymbol{u}\cdot\boldsymbol{r})d\boldsymbol{r}$
Reciprocal space to real space	$\int \Psi(\boldsymbol{u})\exp(2\pi i \boldsymbol{u}\cdot\boldsymbol{r})d\boldsymbol{u}$	$\int \Psi(\boldsymbol{u})\exp(-2\pi i \boldsymbol{u}\cdot\boldsymbol{r})d\boldsymbol{u}$
Structure factor	$\sum_j f_j\exp(-2\pi i \boldsymbol{u}\cdot\boldsymbol{r}_j)$	$\sum_j f_j\exp(2\pi i \boldsymbol{u}\cdot\boldsymbol{r}_j)$

Subject Index

Main entries shown in bold

absorption function   13
acrylic resin   165
α-Si_3N_4 25
amplitude contrast   1
anti-phase boundary   43, **105**
anti-phase domain   105
astigmatism   33
autocorrelation function   133
axial illumination method   2

back focal plane   1
barium titanate   91
beam convergence   8, **163**
Bednorz and Müller   91
β-Si_3N_4 25
β-SiC   24
Bloch wave approach   9
body-centered cubic structure   152
boundary dislocation   58
Bragg condition   1
bright-field image   1
bright-field method   1
Burgers vector   41

camera length   149
CCD camera   129
chemical polishing   165
chemical vapor deposition (CVD) method   54
chromatic aberration   8
chromatic aberration coefficient   **8**, 162
cleaved surface   69
close-packed boundary   58
coincidence boundary   58
coincidence site lattice   58
computer simulation   11
condenser aperture   163
contour map   **130**, 143
contrast transfer function   **5**, 160
convolution   **10**, 147
corrected accelerating voltage   8
correction of astigmatism   33
cross-correlation function   135
cross-slip   43
crushing   164
crystalline approximant   122

CS plane   75
CsCl-type structure   152
CuAu I   103
CuAu II   103

$D0_{22}$-type structure   100
$D0_{23}$-type structure   101
$D0_3$-type structure   52
$D1_a$-type   109
dark current   138
dark-field image   **1**, 149
dark-field method   1
Debye-Scherrer ring   17
decagonal quasicrystal   113
defocus value   5
delta function   147
diamond cutter   166
diamond structure   152
diffraction contrast   **41**, 43
diffuse scattering   108, **153**
digital data   138
digital diffractogram   8
dimple grinder   166
disc puncher   164
dislocation   38, **41**
dislocation core   41
dislocation line   42
dislocation loop   43
dissociated dislocation   **42**, 156
dissociation width   41
double diffraction   150
double-diffraction spot   150
dpi   129
DQE   140
dynamic range   138
dynamical diffraction effect   4, **9**
dynamical factor   100

edge dislocation   **43**, 49
edge-on view   69
effective transfer function   8
electron beam heating   167
electron diffraction pattern   1
electron microscope image   1
electropolishing   164

end-on view   42
envelope function   8
epoxy resin   165
Ewald sphere   **148**, 157
extinction rule   150
extra half-plane   **42**, 43

face-centered cubic structure   152
fading phenomenon   141
fast Fourier transform   12
Fibonacci sequence   115
filtering   42, 46, **132**
FINEMET   17
focus drift   31
focused ion beam (FIB)   167
four-axis notation   73
Fourier transform   1, **3**
fracture toughness   83
Fraunhofer diffraction   3
Fresnel diffraction   3
Fresnel fringes   35
fundamental reflection   **29**, 152
fundamental translation vector   147

gas-atomization method   64
golden ratio   113
grain boundary   38, **54**

halo ring   19
hexagonal close-packed structure   152
high-resolution electron microscope image   1
high-resolution electron microscopy   1
high-resolution image   1
higher-order Born approximation   9
histogram   129
hot isostatic pressing (HIP)   54

icosahedral quasicrystal   113
icosahedral symmetry   113
image plane   1
image processing   129
imaging plate   138
inclination of incident electron beam   13
incommensurate structure   **105**, 152
inelastic scattering   154
inelastically scattered electron   154
information-resolution limit   8
inner potential   4
interaction constant   4
intermetallic compound   48
interphase boundary   61
ion beam sputtering method   167

ion milling   166
ionicity   13

jog   52

Kikuchi band   154
Kikuchi line   154
Kikuchi pattern   154
Kikuchi, S.   154
kinematical approximation   148
kink   49

$L1_0$-type structure   103
$L1_2$-type structure   43
lattice constant   12, **147**
lattice defect   13
lattice distortion   41
lattice fringes   17
lattice image   17
LEED   69
linear phason strain   116

magnetic coercivity   61
mass of electron   4
mean inner potential   **4**, 15
metal organic chemical vapor deposition
   (MOCVD)   133
micro-twinning structure   61
microdiffraction   1
microgrid   164
moiré fringes   67
monoclinic structure   83
multi-slice method   9
multiple scattering   150
multiple-twinning particle   116

noise   131
non-integral value   105
normalized Euclidean distance   143
nucleation type   61

objective aperture   1
off-axis illumination method   2
one-dimensional long-period superstructure   103
optical bench   129
optical diffractogram   8
optical ray diagram   1
order-disorder phase transformation   99
ordered alloy   99

over-printing   129
oxygen vacancy   **91**, 94

partial dislocation   42
perfect dislocation   42
Perovskite structure   75
phase contrast   2
phase contrast transfer function   5
phase filter   129
phase object approximation   4
pinning type   61
pixel number   129
pixel size   129
plan view   69
Planck's constant   4
plastic deformation   83
point resolution   8
precipitation hardened magnet   61
prefield of the objective lens   163
profile view   69
propagation function   10, **12**
pseudo-color image   130

quantitative analysis   138
quantum noise   131
quartz crystal film thickness monitor   167
quasicrystal   113
quasiperiodicity   113

R-factor 145
random grain boundary   54
real space   1
reciprocal space   1
relativistic effect on the electron mass   8
REM   69
ReO₃ structure   75
residual index   143
resistance heating   167
resolution limit   8
RHEED   69
rotation fault (RF)   78

sampling   129
sampling point   **15**, 129
Sato and Toth   112
saturation magnetization   61
scattering amplitude   3
scattering factor   12, **148**
Scherzer focus   5
Schrödinger wave equation   185
screw dislocation   43
selected area aperture   1

selected area diffraction   1
semi-coincidence boundary   64
shape effect   153
short-range ordered state   108
Σ value   58
sign convention   3, **185**
silver halide particle   131
slip plane   43
slow-scan CCD camera   138
smoothing   131
soft magnetic material   17
spatial filter   131
specimen drift   31
spherical aberration coefficient   5, **160**
sputtering   166
sputtering efficiency   166
stacking fault   24, **42**
stacking fault energy   43
stigmator   **33**, 39
STM   69
stress-induced martensitic transformation   88
structural material   54
structure image   17
subgrain boundary   88
superlattice dislocation   43
superlattice partial dislocation   43
superlattice reflection   **29**, 152
superstructure image   29, **101**
symmetric boundary   58
systematic reflection   151

temperature factor   12
tetragonal structure   83
three-axis notation   73
tilt boundary   54, **58**
translation process   131
translation vector   131
transmission cross coefficient   10
transmission function   4
triacontahedral atom cluster   122
triple junction   54
TV camera   129
twin   153
twin boundary   38, **58**
twin-jet polishing method   164
two-beam condition   156

ultrahigh vacuum electron microscope   69
ultramicrotomy   69, **165**

vacuum evaporation   167
vacuum tweezers   167

Vickers indentor 83
vidicon camera 129

wave function **3**, 185
wavelength **4**, 159
weak-beam method 41, **156**
weak-phase object approximation **4**, 136
Wiener filter 136
window method 164

X-ray diffraction 91

YAG 138

Z-controller 34
Z-type faulted dipole 43
zeolite 73
zinc-blende (ZnS)-type structure **136**, 152